Phytomycology and Molecular Biology of Plant–Pathogen Interactions

This book provides information on various aspects of practical fungal molecular biology. It aims to unravel the plant–pathogen interface, applications of molecular tools, genetic variability and identification of plant fungi, and mechanisms of plant–fungal interactions. It provides data on the role of fungal pathogens in plant disease development; covers aspects of classical and operational fungal taxonomy, biology study, and the transition of this study into modern molecular techniques; and discusses beneficial fungi, effects of fungal interactions with host plants on plant physiological functions, and production and industrial applications of fungi.

Phytomycology and Molecular Biology of Plant–Pathogen Interactions

Edited by
Dr. Imran Ul Haq, Dr. Siddra Ijaz,
and Dr. Iqrar Ahmad Khan

CRC Press
Taylor & Francis Group
Boca Raton   London

CRC Press is an imprint of the
Taylor & Francis Group, an **informa** business

First edition published 2023
by CRC Press
6000 Broken Sound Parkway NW, Suite 300, Boca Raton, FL 33487–2742

and by CRC Press
4 Park Square, Milton Park, Abingdon, Oxon, OX14 4RN

CRC Press is an imprint of Taylor & Francis Group, LLC

© 2023 Taylor & Francis Group, LLC

First edition published by CRC Press 2023

Library of Congress Cataloging-in-Publication Data
Names: Haq, Imran Ul, editor. | Ijaz, Siddra, editor. | Khan, Iqrar A. (Iqrar Ahmad),
    editor.
Title: Phytomycology and molecular biology of plant pathogen interactions / edited
    by Dr. Imran Ul Haq, Dr. Siddra Ijaz, and Professor Dr. Iqrar Ahmad Khan.
Identifiers: LCCN 2022002599 (print) | LCCN 2022002600 (ebook) |
    ISBN 9780367755065 (hardback) | ISBN 9780367755072 (paperback) |
    ISBN 9781003162742 (ebook)
Subjects: LCSH: Plant-pathogen relationships—Molecular aspects. | Phytopathogenic
    fungi—Research. | Phytopathogenic microorganisms.
Classification: LCC SB732.7 P49 2022 (print) | LCC SB732.7 (ebook) |
    DDC 632—dc23/eng/20220126
LC record available at https://lccn.loc.gov/2022002599
LC ebook record available at https://lccn.loc.gov/2022002600

ISBN: 978-0-367-75506-5 (hbk)
ISBN: 978-0-367-75507-2 (pbk)
ISBN: 978-1-003-16274-2 (ebk)

DOI: 10.1201/9781003162742

Typeset in Times
by Apex CoVantage, LLC

Contents

Preface

Nature has given fungi the ability to produce organic acids, enzymes, and secondary metabolites as well as bring about protein synthesis and thus have wide applications in food and pharmaceutical industries through biotechnological interventions. Metabolic engineering has emerged as a multidisciplinary approach and a powerful tool for optimizing and introducing new cellular processes. Historically, studies on industrial microorganisms—their primary metabolism processes, biochemical pathways, and reactions. Plant–microbe interaction generated a vast amount of information on the subject, which helped scientists advance the research in metabolic engineering using recent data obtained from modern technologies like proteomics, metabolomics, and genomics. In recent years, genetic engineering has provided more efficient and effective alternatives over traditional strain improvement methods such as mutagenesis and genetic recombination. This has resulted in great success in the commercial production of various fungal proteins. Advancement in molecular biology has made it possible to unravel the favorable traits in the strains of fungi encoded by heterologous genes. Numerous strategies have been developed to date, improving the functionality of genes and reducing the expression constraints. Strain improvement is related to the isolation of many genes of desired traits and the availability of wild isolates. Recently, many genes of fungal strains have been identified that are involved in disease development, and disease symptoms become partially or completely disappeared upon their disruption. Tagged mutagenesis techniques have been used for the identification of pathogenicity genes. Many genes involved in the degradation of the cell wall, infection structure formation, responses to host plant defenses, toxins production, and signal cascades have been identified. In addition to these genes, novel functional genes have also been identified. More pathogenicity-related genes can be analyzed by using molecular techniques. This confuses the outcome of host–microbe association. There is a great need to unravel a series of cellular events such as the highly complex network of genetic microbes, interactions (microbial and metabolic), and the signaling events involved in host–microbe interactions. In mutualistic relationships, the roots of plants form a mycorrhizal symbiosis with soil-inhibiting fungi. Roots of such plants must have the capability to detect the fungal partners to establish this interaction. On the basis of this overview of fungal molecular biology, we have tried to design this book with chapters that give in-depth information on various aspects of fungal molecular biology to unravel the plant–pathogen interface. This book comprises 13 chapters, which cover the following topics: journey from Koch's postulates to molecular system biology; insight into fungal biology; fungal genome and genomics; pathogenicity genes; molecular mechanisms–relayed defense responses in plants against fungal pathogens; fungal gene expression and interaction with the host plant; biotechnological interventions in fungal strains improvement; molecular identification and detection of phytofungi; underpinning the phylogeny and taxonomy of phytofungi through computational biology; application of system biology in plant–fungus interaction; multiomics approach and plant–fungal interaction; the molecular mechanism of mycorrhizal symbiosis; and metabolic engineering of plant fungi for industrial applications. This book will provide comprehensive insights to students and researchers involved directly or indirectly in the study of the molecular biology of plants and their associated fungi.

Imran Ul Haq
Siddra Ijaz
Iqrar Ahmad Khan

About the Editors

Dr. Imran Ul Haq holds a postdoctorate in plant pathology from the University of California, Davis, United States. He is currently an associate professor in the Department of Plant Pathology, University of Agriculture Faisalabad, Pakistan. He has supervised over 40 graduate students for M.phil and PhD and was responsible for the establishment of the Fungal Molecular Biology Laboratory Culture Collection, an affiliated member of the World Federation for Culture Collections in the Department of Plant Pathology, University of Agriculture, Faisalabad, Pakistan. He has published over 50 research articles, five books, four laboratory manuals, and several book chapters. He has made colossal contributions to fungal taxonomy by reporting novel fungal pathogen species in plants. His research interests are fungal molecular taxonomy and nanotechnology integration with other control strategies for sustainable plant disease management.

Dr. Siddra Ijaz holds a postdoctorate on CRISPR/Cas technology for genome editing from the Plant Reproductive Biology Laboratory, University of California, Davis, United States. She is currently an associate professor at the Center of Agricultural Biochemistry and Biotechnology, University of Agriculture Faisalabad, Pakistan. She also serves as the deputy managing editor of an international impact factor journal, *Pakistan Journal of Agriculture Sciences*. She has supervised over 50 MPhil and PhD students and has published more than 50 research articles, six books, and several book chapters. Her research focus includes plant genome engineering using transgenic technologies, genome editing through CRISPR/Cas systems, nanobiotechnology, and exploration of genetic pathways in plant–fungus interactions.

Dr. Iqrar Ahmad Khan has a distinguished career in horticulture/botany spanning more than four decades. He is a graduate of the University of Agriculture, Faisalabad, Pakistan, and the University of California, Riverside, United States. He has supervised over 70 graduate students, including 14 PhD candidates. He has undertaken 44 research projects and published 375 articles and books. He served as the vice-chancellor of two universities for more than nine years (2008–2017) and founded five new universities and district campuses between 2008 and 2017). He has also established several national and international research centers and institutes, including Chinese Confucius Institute and Bio-Energy Institute. He was Chief of Party/director of the US–Pak Center of Advanced Studies (2017–2019) and Ela Bhatt Professor at the University of Kassel, Germany (2020). His research on wheat, potato, orange, mango, and dates has benefitted industry and significantly improved the life and livelihood of farmers and rural communities. He is a Distinguished National Professor and a fellow of the Pakistan Academy of Sciences. He was awarded Sitara-e-Imtiaz for his services by Pakistan's president and Ordre des Palmes Academiques by the French government for his international contribution to education.

Contributors

Ahmet Çağlar ÖZKETEN
Department of Chemistry, Middle
East Technical University, Ankara,
Turkey

Ebtihal Alsadig Ahmed Mohamed
Agricultural Research Corporation,
Khartoum North, Sudan

Anita Payum
Rani Lakshmi Bai Central Agricultural
University, Jhansi, India

Aşkım Hediye Sekmen ÇETİNEL
Ege University, Faculty of Science,
Department of Biology, Izmir,
Turkey

Ayşe ANDAÇ ÖZKETEN
Department of Biology, Middle East
Technical University, Ankara,
Turkey

Azime GOKCE
Ege University, Faculty of Science,
Department of Biology, Izmir,
Turkey

Barbaros ÇETINEL
Bornova Plant Protection Research
Institute, Izmir, Turkey

Bukhtawer Nasir
Centre of Agricultural Biochemistry
and Biotechnology (CABB),
University of Agriculture Faisalabad,
Pakistan

Cemre TATLI
Ege University, Faculty of Science,
Department of Biology, Izmir/Turkey

Erhan ERDIK
Bornova Plant Protection Research
Institute, Izmir/Turkey

Farwa Batool
Department of Biochemistry,
University of Agriculture Faisalabad,
Pakistan

Ghedir Issam
Agricultural Sciences Department,
University of Abbes Laghrour,
Khenchela, Algeria

Guoqing Li
State Key Laboratory of Agricultural
Microbiology and Hubei Key
Laboratory of Plant Pathology,
Huazhong Agricultural University,
Wuhan, China

Ifrah Rashid
Department of Plant Pathology, Islamia
University of Bahawalpur, Pakistan

Imran Ul Haq
Department of Plant Pathology,
University of Agriculture Faisalabad,
Pakistan

Iqrar Ahmad Khan
Institute of Horticultural Sciences,
University of Agriculture Faisalabad,
Pakistan

Jing Zhang
State Key Laboratory of Agricultural
Microbiology and Hubei Key
Laboratory of Plant Pathology,
Huazhong Agricultural University,
Wuhan, China

Laith Khalil Tawfeeq Al-Ani
School of Biology Science, Universiti
   Sains Malaysia, Pulau Pinang,
   Malaysia

Liliana Aguilar-Marcelino
Tecnologico de Monterrey, School of
   Engineering and Sciences, Centre of
   Bioengineering, Campus Queretaro,
   Mexico

Maria Babar
Centre of Agricultural Biochemistry
   and Biotechnology (CABB),
   University of Agriculture Faisalabad,
   Pakistan

Mingde Wu
State Key Laboratory of Agricultural
   Microbiology and Hubei Key
   Laboratory of Plant Pathology,
   Huazhong Agricultural University,
   Wuhan, China

Mubashar Raza
State Key Laboratory of Mycology,
   Institute of Microbiology, Chinese
   Academy of Sciences, Beijing, China

Muhammad Kaleem Sarwar
Department of Plant Pathology,
   University of Agriculture Faisalabad,
   Pakistan

Nabeeha Aslam Khan
Department of Plant Pathology,
   University of Agriculture Faisalabad,
   Pakistan

Nishat Zafar
Institute of Microbiology, University of
   Agriculture Faisalabad, Pakistan

Ömür BAYSAL
Department of Molecular Biology
   and Genetics, Muğla Sıtkı Koçman
   University, Muğla, Turkey

Parisa Razaghi
State Key Laboratory of Mycology,
   Institute of Microbiology,
   Chinese Academy of Sciences,
   Beijing, China

Qaiser Shakeel
Department of Plant Pathology,
   Islamia University of Bahawalpur,
   Pakistan

Rabia Tahir Bajwa
Department of Plant Pathology,
   Islamia University of Bahawalpur,
   Pakistan

Shehla Riaz
Department of Plant Pathology,
   University of Agriculture Faisalabad,
   Pakistan

Siddra Ijaz
Centre of Agricultural Biochemistry
   and Biotechnology (CABB),
   University of Agriculture Faisalabad,
   Pakistan

Sidra Anam
UK Public Health England, National
   Institute of Health, Islamabad

Yang Long
State Key Laboratory of Agricultural
   Microbiology and Hubei Key
   Laboratory of Plant Pathology,
   Huazhong Agricultural University,
   Wuhan, China

Younus Raza
Department of Plant Pathology,
University of Agriculture Faisalabad,
Pakistan

Zaianb Malik
Department of Plant Pathology,
University of Agriculture Faisalabad,
Pakistan

Zakria Faizi,
CAB International, Afghanistan

Zemran MUSTAFA
Department of Agricultural Sciences
and Technology, Sivas University
of Science and Technology, Sivas,
Turkey

1 A Journey From Koch's Postulates to Molecular System Biology

*Imran Ul Haq, Siddra Ijaz
and Iqrar Ahmad Khan*

CONTENTS

DOI: 10.1201/9781003162742-1

1

A BRIEF CHRONOLOGICAL JOURNEY TO THE DEVELOPMENT OF KOCH'S POSTULATES

Ancient Romans considered plant diseases as divine punishment and hence advisedly sacrificed red dogs to the Wheat God, Robigus, in anticipation of quality grains without "red dust." Aristotle and Theophrastus's historical records from the 3rd century BC mention and speculate about devastating diseases of cereals, legumes, and trees. Evident plant diseases and resulting human suffering caused a constant fear of hunger and a consequent belief in supernatural causation. Early religious writings also mention various plant diseases, i.e., blast, blight, and mildews. Thenceforth, this belief was encouraged by kings and religious leaders either due to their own ignorance/same belief or to hide their poor performance, favoritism, corruption, and inability.

Consequently, evil spirits, the wrath of gods, or unfavorable star or planetary positions were considered disease-causing agents. In the Dark Ages, frequent ergot epidemics caused the death of thousands of Europeans, who were dependent on rye as the primary food source, and people were getting mad and suffering painful deaths.[1] Unfortunately, victims were considered possessed and sometimes accused of witchcraft. However, it was poisonous rye bread, made from ergot-contaminated grains, that was responsible for disease development and epidemics (Lucas et al., 1992; Iriti and Faoro, 2008).

Since Theophrastus, plant diseases were shrouded in speculations and superstitions and nothing new was written about them until the trade between Europe and its eastern lands started around the 10th century. This trade led to an educated and wealthy middle class that paved the way for curious naturalists. This, presumably, first generation of life scientists was also hampered by dogmatic religious leaders. Apart from this, until the Middle Ages, most people were illiterate, books were few, and communication was tedious and cost an arm and a leg. Due to these factors, new written observations about plants diseases traveled at a slow pace and to a few people. This continued until the remarkable invention of the printing press in the mid-15th century, which has a significance equal to that of harnessing of fire and invention of the wheel in human history. This marvelous machine started educating the masses. After two centuries, a Dutch cloth merchant, Anton von Leeuwenhoek, built himself the first microscope to observe the fine weaves of fabrics. He started examining other things and was delighted and astonished to observe "Animalcules." The microscope opened a whole new world to observation and oppugned the belief in supernatural causation. It raised a serious question in people's minds: From where do these microscopic organisms come? The notion of spontaneous generation, posited by Greek philosophers about the mysterious appearance of flies and frogs, was also applied to microbes. Hence, the microscopic organisms in infected tissues were considered the result of the disease and not the cause. This seemingly strange statement was accepted as a fact from the 16th century to the mid-18th century. During this period, various experiments, both in favor of and against "spontaneous generation," were conducted. Although Anton DeBary, founder of mycology and plant pathology, proved through scientific experiments that a fungus was responsible for the Irish Famine, his work was accepted only by a few members of the scientific community. Finally, spontaneous generation was demolished once and for all by Louis Pasteur in 1861. He devised various ingenious experiments to prove that microbes are produced by other identical microbes. Impressively, his experiments also demonstrated no known conditions in which germs can be produced without similar parents, hence deflating spontaneous generation (Aneja, 2007; Chaudhari et al., 2015).

KOCH'S POSTULATES AND RADICAL OVERHAUL IN DISEASE EPIDEMIOLOGY

Although Louis Pasture was convinced that microbes are disease-causing agents, he failed to link any microbe with any particular disease. However, Robert Koch, in the 19th century, conclusively confirmed the etiology of anthrax, tuberculosis, and cholera, establishing the germ theory of disease. He also confronted various microscopy issues and became the first to use oil immersion lenses, condensers, solid culture media, and stain and fix bacteria on a slide. These unprecedented achievements of the German physician were capped with the Noble Prize in 1905 (Blevins and Bronze, 2010).

Koch's postulates were derived from his ingenious work to confirm the causation in contagious diseases. These standards were both for skeptics to prove that microbes are infectious and for microbiologists to be meticulous before confirming the etiology and epidemiology of a disease. Koch's postulates are summarized from his presentation in 1890, which stipulate the following guidelines (Grimes, 2006; Neville et al., 2018):

1. In its characteristic form, the putative pathogen should invariably be present in all infected plants/tissues in such abundance and distribution that disease symptoms could be justified.
2. The pathogen under consideration must be isolated from all infected host samples and grown on pure culture media.
3. Inoculation of pure samples in a healthy susceptible host must manifest the characteristic symptoms and develop the disease.
4. After disease development in the inoculated host, the pathogen must be re-isolated and indistinguishable in its characteristic forms and arrangements from the initially isolated pathogen.

Significant advancements in microbiology have examined, reevaluated, and modified Koch's postulates repeatedly. It is done to include newly discovered microorganisms with unique characteristics: some are restricted to specific host cells; some establish commensalism with the host; a few do not propagate on artificial media; a few develop the disease after co-infection with other pathogens; some produce toxins that affect the host far from the site of production; and others require specific environmental conditions to develop disease. Host and pathogen disease-developing mechanisms, infectious proteins, and genetic discoveries also require a reappraisal of Koch's postulates. Despite all modern discoveries and knowledge, these meticulous postulates remain the touchstone while investigating the etiological and epidemiological aspects of infectious diseases.

Since Robert Koch, these postulates have played a crucial role in understanding the etiological relationship between host and pathogen. All four of the guidelines—the earlier postulates—are successfully implemented to confirm the etiology of most pathogenic fungi, bacteria, nematodes, viruses, and spiroplasmas. However, isolation, culturing, purification, and inoculation of plant pathogen in the host are not possible for parasitic obligate fungi, some viruses, protozoa, and phytoplasmas, so for. Therefore, Koch's postulates are not fully implemented in these scenarios, and the possible association of these microbes with diseases is tentative. Despite all these complications, Koch's postulates have and will continue to play a critical role, occasionally fine-tuned, in understanding all plant diseases (Agrios, 2005).

EXCEPTIONS TO KOCH'S POSTULATES

Koch's stringent postulates provided a framework for contemporary scientists to prove the etiology of possible microbial diseases. These were quite successful for various human and plant diseases, as Koch was able to isolate and grow the bacterium responsible for tuberculosis on artificial media, inoculate the pathogen into healthy hosts, and reproduce the characteristic disease symptoms. Healthy individuals lacked the pathogen in their tissues. Koch himself was, however, concerned about his postulates' shortcomings, as demonstrated by his experiments on cholera for identifying disease causation. He was sure that the disease was caused by a particular microbe but could not culture it on artificial media. In addition, the suspected microbe was also isolated from healthy hosts; thus, it defied the second and third postulates. Thus, in his succeeding papers, he stated that if the consistent and exclusive presence of a microbe is associated with a disease, it validates the causation relationship between the microbe and the disease (Tabrah, 2011).

Later research in disease causation revealed that these postulates are invalid for a disease involving multiple pathogens. Multiple microbial infections contradict the third postulate, re-inoculation should develop the disease, which is not the case if synergism is absent. Fulfillment of the second postulate—a pathogen should be associated with organisms having a disease rather than with healthy ones—is also very challenging when the pathogen is also isolated from healthy hosts, e.g., *Xylella fastidiosa*, which can cause various diseases in more than 500 plant species can also be isolated from the xylem tissues of apparently healthy plants. Pathogens responsible for powdery mildew of wheat and downy mildew of *Arabidopsis* are obligate parasites belonging to ascomycetes and oomycetes, respectively; hence, they defy the second condition of pathogenicity. The fourth postulate is impossible to achieve without using the latest molecular techniques, as identification at sub-specie and strain levels requires electron microscopy and genomics (McDowell, 2011).

Koch devised his famous disease pathogenicity rules on the basis of the tools and technology available at the time to prove the pathogenicity of demonstrably alive microbes that metabolize, grow, and reproduce independently. Modern research has revealed that these rules could not be successfully exercised when the pathogen was impossible to grow on artificial culture media or when, unknowingly, innocuous microorganisms turned to be pathogenic. Inability to prove these postulates for a disease misled the researcher to erroneous conclusions. For years, due to lack of evidence, viruses were considered a pathogen of aster yellow and Pierce's disease of grapes caused by phytoplasmas and vascular bacteria. As mentioned earlier, a predominant association of a putative pathogen with a particular disease is strong evidence for an etiological relationship but not confirmation (Walker et al., 2006).

SYNERGISM AND KOCH'S POSTULATES

Koch's first two postulates propose that particular symptoms can only be produced by particular pathogens, and vice versa, i.e., the moncausa model of disease development. Later research shows that this strict classification of causal specificity does not apply to diseases involving multi-microbial interactions or other environmental or abiotic causes (Ross and Woodward, 2016).

Plant disease complexes involve a network of various microbial interactions. Usually, monoculture inoculations are performed to estimate the pathogenic potential of a given microbe. However, possible synergistic relations among pathogens led to more severe disease development and consequent symptoms, which are not justifiable with monoculture inoculations. Synergism studies are crucial to fully understanding pathogenesis, disease development, and control strategies (Begon et al., 1986). Tomato pith necrosis is caused by the synergism of eight different bacterial species, which in turn cause different levels of disease severity. A similar phenomenon is observed in broccoli soft rot, where inoculation of one bacterial species does not produce characteristic symptoms and severity (Kůdela et al., 2011).

Similarly, fungal pathogens co-exist to produce economically significant diseases such as severe "grapevine decline," *Septoria* leaf spot, *Fusarium* head blight, and foot and crown rot of wheat. Other significant diseases of the synergistic complex are black spot of pea and leaf spot of eucalyptus. Likewise, viral pathogens such as cucumber mosaic virus, in association with cowpea mosaic virus, increase the extent and severity of cowpea disease in the field. Mixed co-infection of fungal (*Fusarium, Alternaria,* and *Colletotrichum*) and bacterial (*Xanthomonas*) pathogens have also been observed, as in the case of walnut necrosis (Ma et al., 2013; Lamichhane and Venturi, 2015).

Conclusively, it can be stated that Koch's postulates are challenging to fulfill where pathogen presence precedes symptom development; mycotoxins produce symptoms far from their site of production; asymptomatic carriers are involved; a pathogen cannot grow on artificial culture media; or synergism or multi-risk factors are involved.

KOCH'S POSTULATES AND FLAGELLATES

The major constraint to satisfy Koch's postulates during present times is their inability to incorporate obligate parasites. Notably, plant viruses are a considerable threat to sustainable agricultural goals. Intercontinental trade, climatic changes, rapid mutation, and subsequent evolution of viruses cause frequent viral epidemics. Hence, fast and accurate identification of the virus responsible, to the strain level, is indispensable for timely disease management. Since viruses assume control of host cells and usurp their machinery for proliferation, their propagation in artificial media is impossible, which violates Koch's postulates. Thus, blind adherence to the dogmatic belief that viruses should meet these criteria of pathogenesis impeded the early development of virology instead of aiding it.

Flagellates of Kingdom Protozoa, which were associated with plant diseases since the 1900s, also left scientists in the same quandary. Researchers of these parasites suggested that their association with the host is pathogenic. These are obligate parasites and could not be grown artificially; hence, their version was rejected despite compelling shreds of evidence. Until now, flagellates have not been fully recognized as plant parasites. Although phytoplasmas and vascular-inhabiting bacteria are associated with flagellates because of unfulfilled Koch's postulates, contrarily, their pathogenicity is universally accepted. Ironically, the pathogenicity of flagellates is as evident as that of phytoplasmas and vascular-inhabiting bacteria. It will be appropriate to deduce that a few, if not all, protozoa are capable of causing economically debilitating plant diseases and deserve attention as prospective plant parasites. The fastidious prokaryote *Candidus liberobacter* was associated with citrus-greening diseases. Earlier, this was deduced by using microscopy, whereas, currently, polymerase chain reaction (PCR) has proved to be more effective. However, Koch's postulates are still to be completed for this fastidious pathogen. Economically debilitating diseases since early times—rusts, smuts, and mildews—are also caused by obligate parasites (Davis and Brlansky, 2007; Gillies, 2016).

ALTERNATIVE APPROACHES INFERRING PATHOGENICITY

In 1936, Thomas Rivers stated that for viral diseases, Koch's postulates are a hindrance rather than aid, and to confirm the pathogenicity of a virus against a particular disease, he proposed a new set of guidelines for researchers: the putative virus pathogen must be constantly associated with a particular disease, and its association with the disease should be causative not incidental. He also proposed various guidelines to satisfy the second postulate. In addition, he proposed methods to distinguish among the simple causative association of a virus; co-infection of a pathogenic virus with non-pathogenic viruses and persisting of a virus as a carrier in a particular host. Although these postulates/guidelines were not widely accepted, River challenged classical Koch's postulates and provided an innovative approach for future studies of disease causation in virology (Fredricks and Relman, 1996).

Conventional techniques to test the pathogenicity of a suspected pathogen for a particular host disease, including isolation, purification, re-inoculation, and microscopy, are comparatively tedious and require expertise. Hence, plant–pathogen diagnosis has moved from traditional approaches to novel and practical molecular approaches. These molecular assays can recognize and quantify pathogens more efficiently and accurately and overcome the limitations of conventional methods. Typical molecular identification approaches include PCR, restriction fragment length polymorphism (RFLP), DNA sequence-based amplifications, loop-mediated isothermal amplification (LAMP), and DNA micro/macro arrays. Bioinformatics databases are platforms to document, store, and retrieve nucleotide sequences of plant pathogens, which help delimitate existing and evolving pathogen species (Pryce et al., 2003; Tan et al., 2008; Johnston-Monje and Lopez Mejia, 2020).

SIR BRADFORD HILL'S VIEWPOINTS ON INFERRING PATHOGENICITY

In 1965, Bradford Hill, an English epidemiologist and statistician, published nine principles/postulates for inferring the pathogenicity of a putative pathogen and provided epidemiological evidence to prove its relationship with a specific disease/host (Höfler, 2005; Fedak et al., 2015). These nine criteria/viewpoints, in light of plant pathology, are listed here:

1. A strong association between a microbe and a host indicates a causal relationship. At the same time, a microbe's weak association does not exclude the possibility of a causal relationship with the host.
2. Reproducibility of a scientific experiment/observation by different researchers at different places strengthens the causation relation.
3. The probability of an etiological relationship is stronger if the host and the putative pathogen populations are isolated and there is no alternative explanation of disease development.
4. Symptoms must appear after the pathogen attack/cause; however, this could be temporarily delayed in some cases.
5. Tremendous inoculum pressure should lead to severe symptoms; however, in a few cases, greater exposure could also lower the effect/disease incidence.
6. Seemingly plausible mechanisms (of etiology and disease development) in pathogen and host should be observed, which, he said, was lacking due to the then-prevalent technology/methods.
7. The lack of laboratory observations does not nullify the causation relationship; however, coherent observations strengthen it.
8. Experimental observations are quite helpful in deciding the possibility of pathogenicity.
9. Use of analogies for complex diseases; pathogenicity confirmation of a specific pathogen lowers the evidence standards for any other possible causal agents/putative pathogens.

Although Hill's postulates are more lenient, these viewpoints also have limitations for final assessment and methodology to assign checkmarks. The magnitude of pathogen association is affected by the basic methodology used during observation/experiment and other crucial factors. The third specificity postulate deals with monocausal association and does not apply to multi-microbial infections. Various symptoms could result from a single pathogen, or various pathogens can produce indistinguishable symptoms, as discussed in "limitations of Koch's postulates." The fifth viewpoint is not satisfied for most plant diseases as host susceptibility in a large population has significant variability. Biological plausibility is challenging to explain due to the involvement and interplay of various contributory factors (van Reekum et al., 2001; Phillips and Goodman, 2004).

Koch's and Hill's postulates provide guidance on inferring the pathogenicity of plant diseases, but these guidelines should not be practiced as heuristic approaches, because no postulate/viewpoint can encompass all causations, and each criterion must be applied according to each distinct situation. Symptomology is crucial in both of these guidelines of inferring pathogenicity for a particular pathogen. Hence, disease symptoms must be distinctly articulated along with clear microscopic images of the putative pathogen(s) (Bhunjun et al., 2021).

PATHOGENICITY IN THE GENOMIC ERA

During the genomic era (1990–2003), the pace of traditional research work in plant pathology has significantly accelerated due to observational and experimental data utilization. Genomic studies of host–pathogen interactions are crucial for emerging pathogens as they explain the pathogen–host biology and virulence mechanisms by exposing the underlying potential effectors and toxins. Genomics has revealed the entirely transferrable chromosome, critical for pathogenicity, of *Fusarium oxysporum* and hence shifted conventional paradigms (Ma et al., 2010). Thus, molecular technologies help in mining information and better understanding pathogenicity, host–pathogen biology, and potential genome evolution mechanisms.

During disease development, surface proteins play an important role and are characterized first when studying pathogen–host interaction. These sequences predict virulence proteins more effectively than traditional spectrometry or enzyme assay techniques used to identify the corn smut fungus, *Ustilago maydis* (Kämper et al., 2006). These in silico–identified effectors, their phenotypes, and previously unidentified proteins could be evaluated for their potential role in pathogenicity. The role of cell wall degrading enzymes in pathogenicity, although well-established in the late 19th century, was confirmed through pathogen genome sequences, especially of pectinolytic enzymes. A comparison of cell wall degrading enzymes of different strains reveals the degree of disease severity in the host and resistance to fungicides. As described, genome sequencing also helps to devise new and efficient genetic tools, i.e., genetic markers that are used to unambiguously identify the pathogen (Liu et al., 2015; Aylward et al., 2017).

MOLECULAR APPROACHES AND PATHOGENICITY: IMPORTANT DISCOVERIES IN THE 20TH CENTURY

In 1935, Stanley Falkow isolated the tobacco mosaic virus in crystalline form and demonstrated that viruses are proteins and have the ability to propagate in living cells. His discovery revolutionized virology and earned him the Noble Prize in 1946. In 1952, D. Hershey and Martha Chase proved experimentally that DNA is the carrier of genetic information rather than proteins. The double-helix structure of chromosomes and how each codon is read to generate a specific amino acid were discovered in 1953 and 1961, respectively. In 1968, Marshal Nirenberg and his team sequenced the first tRNA molecule, which paved the way for DNA-sequencing techniques. In 1977, Frederick Sanger developed ingenious techniques of DNA sequencing and successfully sequenced the first whole genome—that of a bacteriophage virus (Sanger et al., 1977). In 1983, Karry Mullis, in California, developed the well-known DNA-amplifying technique PCR. The first bacterium (*Haemophilus influenza*), fungal (*Saccharomyces cerevisiae*), and nematode (*Caenorhabditis elegans*) genomes were sequenced in 1995, 1996, and 1998, respectively. Discoveries in physiology, biotechnology, genomics, data processing, artificial intelligence, and other associated sciences have a profound impact on plant pathology, helping feed the exploding world population (Baldi and Hatfield, 2003; Weissenbach, 2016).

STANLEY FALKOW'S MOLECULAR POSTULATES

Koch's postulates have been updated time to time, as previously described in the chapter, to encompass pathogens that failed to meet their criteria of etiological relationship with the host. in In 1988, Stanley Falkow, an American microbiologist, invoked his guidelines to identify putative pathogens on a genetic and molecular basis. These proposed molecular postulates (Falkow, 1988), with special reference to plant diseases, are described here:

1. The symptomatic host should always be associated with putative pathogenic species or strains. Furthermore, all pathogenic strains or species must have the gene concerned, whereas non-pathogenic strains should be devoid of the gene.
2. Inactivating the gene suspected to be associated with a particular virulence trait should result in considerable loss of pathogenicity in the clone (loss of function). Additionally, clones with inactivated genes must have measurably less virulence compared with that of the unaltered putative pathogen.
3. Regression of the mutated allele/gene should restore the pathogenicity in the clone. Simply put, reintroduction of the original gene in the clone/mutated microbe must restore the virulence (gain of function).

APPLICATION OF MOLECULAR KOCH'S POSTULATES TO ENHANCE PLANT DISEASE RESISTANCE

Genome studies and bioinformatics analysis alone do not confirm the pathogenicity of a particular pathogen against a plant host. The discovery of pathogenicity genes in a pathogen requires a blend of the latest genomic information and sophisticated molecular tools. It is worth mentioning here that most of the genes in a pathogen are a consequence of competition among microorganisms over the span of billions of years, rather than having emerged to deal with specific host factors. Fortuitously for pathogenic microbes, these evolved genes might increase their survivability in a new environment (potential host). These genes are marvelous artifacts, as they confer an evolutionary advantage on the pathogen upon its first encounter with a potential host (Falkow, 2004).

If complex pathogen genes are involved in virulence, then it is implied that the host's genetic factors are even more complex and numerous, ultimately deciding the outcome of the host–pathogen interaction. In addition, the production of host mutants through positional cloning, gene targeting, and mapping of quantitative trait loci involved in host–pathogen interaction helps researchers dissect the complex molecular pathways of disease development into individual components (Lengeling et al., 2001). Mapping quantitative trait locus (QTL), an interesting genetic technique that starts with a heterogeneous system, helps identify the genes involved in complex host–pathogen interactions. In an experiment, previously unknown resistance and susceptibility loci were identified in inbred strains of the host (mouse) against the *S. typhimurium* infection (Sebastiani et al., 2000).

The aforementioned approaches evaluate and measure the particular host defects upon infection with a specific mutant pathogen. When we find out, through sophisticated molecular tools, that a specific virulent gene impedes a particular gene product or single pathway in the host, arguably, molecular Koch's postulates are satisfied in every sense. Notably, the virulence gene must not be defined only in terms of the pathogen; instead, its pathogenicity must be observed because a pathogen might be a transitory or an opportunistic pathogen in another host. This is required also because the same pathogen with a particular virulence gene might produce a different set of symptoms in another host.

This detailed study of molecular pathways, compared with other methods, could produce stable, resistant crops; decrease pesticide usage; and lead to sustainable crop production. For instance, resistance achieved through R-genes in potato plants against late blight of potato is easily jeopardized as the R-genes are race-specific and can be quickly compromised by pathogen evolution. Alternatively, loss of susceptibility genes, important for pathogenesis, confers resistance against the pathogen, as observed in wheat, rice, and citrus. In the latest experiment by Kieu et al. (2021), seven susceptibility genes and two potato homologs were screened for late blight. During this study, the CRISPR-Cas9 system was manipulated to generate potato mutants with tetra-allelic deletion, which were assayed to be resistant against late blight pathogen. Functional knockouts of the susceptibility genes StDMR6–1 and StCHL1 produced highly resistant and stable mutants, which paved the way for their successful usage in other potential crops. StDMR6–2 was found to be exhibiting a growth phenotype without increased resistance, indicating that it has a function different from that of StDMR6–1 (Kieu et al., 2021).

Resistance to Brassica black leg disease, caused by *Leptosphaeria manculans*, a hemibiotrophic fungus, is determined by R genes, QTL, and adult-plant resistance. It is relatively more difficult to capture QTL during a breeding program compared with the introgression of R genes. As adult plant resistance confers resistance to the plant adult stage, it is crucial to identify well-defined QTL with significant influence on host plant phenotype, even in changing environments, to effectively deploy adult-plant resistance. The double-haploid population of *Brassica napus* was evaluated, and three QTL were identified in each population. Additionally, detailed QTL analysis identified four QTL suitable for marker-assisted disease resistance selection, along with the identification of genes with potential defense responses against pathogenic fungus (Larkan et al., 2016).

In another study, nine sheath blight–resistant QTL were identified using plant height and heading date as covariates during mapping. It was concluded that four QTL were directly related to sheath blight resistance; two of these four QTL were found to be previously unidentified (Goad et al., 2020).

Although it is not workable to knock out/manipulate every unwanted host gene, they might have a crucial role in the host's biology. Complete genomic sequences of pathogen and host offer sophisticated strategies that can effectively evaluate the effect of the pathogen's virulence gene on the host's response to infection. One of these tools is DNA microarrays, which by exploiting the sequence data calculate each gene's transcript levels both in host and pathogen, simultaneously. If the putative virulence gene overexpresses during host infection, the use of DNA microarrays of the pathogen genome sequence is considerably effective by comparing the transcriptional response of wildtype and mutant pathogen under artificial conditions that mimic the host (Cummings and Relman, 2000). DNA microarrays accelerate our understanding of events during host–pathogen encounters through the manipulation of transcriptional induction and expulsion.

Precise identification of the pathogen species responsible is crucial for appropriate handling and developing management approaches. Conventional approaches of pathogen isolation are laborious and time-intensive due to bizarrely specific growing media and conditions. PCR-based molecular strategies for complex diseases are technically challenging; furthermore, primer interactions restrict the identification of multiple targets. Oligonucleotide probes, a microarrays technology, have significantly improved detection of various plant pathogens, i.e., bacteria, fungi, oomycetes, and viruses. The high discriminative ability of this technique even differentiates single nucleotide polymorphisms in DNA fragments, making it a befitting approach to discriminate between closely related pathogenic species. Sugar beet roots are exposed to several pathogen microorganisms, including fungal, oomycete, and even bacterial species, causing up to 60% yield losses. A microarray technology based on three marker genes (ITS, TEF1alph, and 16SrDNA) identified field, storage, and spoilage fungi. Identification of these pathogens was also confirmed through traditional isolation and detection methodologies. DNA microarray proved to be an innovative and effective technique, especially when multiple suspected pathogens were involved in disease development. For plant viruses, RNA from infected plants is isolated and converted into complementary DNA, by using pathogen-specific primers amplified through PCR. Suitable and labeled molecules are used for identification after DNA arrays are hybridized. Developed DNA arrays, after washing, are used to produce results with the help of software (Lievens et al., 2006; Wei et al., 2008; Chen et al., 2013; Liebe et al., 2016).

MOLECULAR KOCH'S POSTULATES TO SYSTEM BIOLOGY

The advent of genomics equipped us with complex and dynamic knowledge that constantly shifts the significance of virulence factors in the virulence system. Although satisfying conventional or molecular Koch's postulates has played and will continue to play a significant role in our understanding of pathogenesis, grasping the factor's role in the system of pathogen–host interaction is comparatively more pragmatic. It is important to understand the factor as a system component, i.e., a virulence "factor," while investigating a plant disease, as various effectors collaborate in pathogen propagation. The aforementioned brief historical overview explains that technical limitations function as filters to discovering virulence factors, e.g., the attribution of plant diseases to gods and demons in ancient Rome. Understanding of system biology is indispensable to formulate effective disease management strategies, as plant pathologists, breeders, molecular biologists, and genomicists individually grapple with and seek to answer questions during host–pathogen interactions. For convenience, molecular plant pathology could be divided into three eras: the Disease Physiology Era (grind and find), existing from the early 20th century to its third quarter; the Molecular Genetics Era (screen for the gene), in the last quarter of the 20th century; and the Genomics Era (patterns that matter), from the start of the 21st century (Schneider and Collmer, 2010).

DISCOVERY OF VIRULENCE FACTORS

Cell wall–degrading enzymes, phytoalexins–detoxifying enzymes, toxins, and cutinases are abundantly produced and extracted from necrotrophs (bacteria, fungi, and water molds) in cell-free culture media. During the early days of scientific studies, the destructive role of these virulence factors in plant physiology and disease development was elucidated, but their deployment and extant were not known. Host-specific toxins (HST), victorin, and helminthosporium carbonum (HC) toxins were known to have been produced by fungi and have a critical role in disease development, but how HST exploited the vulnerability of host plant defenses was still to be understood. However, at the end of the 20th century, to fulfill molecular Koch's postulates, genes responsible for these virulence factors were deleted to produce clones with the inability to develop the disease, but, in fact, clones retained the ability to develop the disease. Surprisingly, some pathogens produce new toxins in plant cells, different from when growing in culture media. Almost all of these virulence factors failed to satisfy molecular Koch's postulates, which urged scientists to anticipate that a more extensive pathogen–host system is involved in disease development. The discovery of cytoplasmic effectors from biotrophic pathogens was a new class of virulence factors delivered by various bacteria, fungi, nematodes, and oomycetes in host cells. The screening of *Pseudomonas* mutants yielded a hypersensitive response and pathogenicity that lacked in type 3 secretion system but delivered effectors in host cells; however, the genes responsible were not detected. These systems were characterized by secretion and deployment of non-redundant protein secretion in host plant cells; they made a minor contribution to pathogenicity and were contemplated to be common in necrotrophs, biotrophs, and proteo-bacterial pathogens (Purdy and Kolattukudy,

1975; Matthews and Van Etten, 1983; Lindgren et al., 1986; Ried and Collmer, 1988; Rogers et al., 1994; Tonukari et al., 2000; Lorang et al., 2007).

Due to its fastidious nature, the causal organism of citrus variegated chlorosis, *Xylella fastidiosa*, was not amenable to the tools of the physiological and molecular eras. The mystery of this black box was revealed in the genomic era. Due to the known homology of virulence genes in culturable pathogens, scientists tested various hypotheses on it. Due to genomics and the resulting sequences of various pathogens, various putative virulence factors based on linkage regions, homolog, and paralog-specific amplifications have been constantly found and researched. Various suspected bacterial cytoplasmic effector (CE) genes have been found through promoter patterns. Hence, previously unknown factors are continuously being discovered by systematically applying the same criteria. For example, possible CE genes of Oomycetes have been identified using protein sequence patterns of pathogen secretions and translocations and confirmed by translocation tests. Hence, the weak phenotype is no more a hurdle in identifying and evaluating the part a candidate CE gene plays in the pathogen–plant interaction system, expectedly expanding CE inventory to thousands (Simpson et al., 2000; Schechter et al., 2004; Rehmany et al., 2005; Lindeberg et al., 2006).

THE JOURNEY FROM COMPONENTS TO THE SYSTEM

Plant–pathogen interactions emerged by combining molecular genetics and biochemical studies of specific molecules.

CYTOPLASMIC EFFECTORS

It could be assumed that plant pathogenic bacteria evolved to evade recognition by plant defense systems and that CEs help in nutrient release. Conversely, CEs wage a war against the plant, resulting in various counterattacks and plant defense suppression. First, bacteria thriving on plant surfaces enter the apoplast through stomas and disturb plants with specific molecular patterns, i.e., flagellin and peptidoglycan, which is detected by plant recognition receptors and immunity is triggered (pathogen-associated molecular patterns [PAMP]-triggered immunity). As a counterattack, bacterial pathogens translocate suppressive CEs and overpower PAMP-triggered immunity (PTI). However, plants recognize this CE attack through one of the largest proteins in plants cells, R proteins, on a susceptibility decoy target that induces effector-triggered immunity, resulting in a hypersensitivity response. In response to effector-triggered immunity (ETI), the plant pathogen mutates genes, encoding detectable effectors, or deploys a new set of effectors that suppress triggered immunity. The mutation is a successful strategy because CEs, due to their high redundancy, are dispensable. To tackle this new attack, plants acquire new proteins that recognize the new ETI suppressive or more abundant CEs in pathogen through evolution or recombination. This battle turns to war, as these steps could be repeated indefinitely. Most biotrophic bacteria deploy up to 50 cytoplasmic effectors, and astonishingly, *Phytophthora infestans* deliver around 700. Both R and CE genes are the most polymorphic genes in plants and pathogens, respectively. A few biotrophs, such as *Cladosporium fulvum* (leaf mold pathogen), are exceptions to this model as they rely on apoplastic effectors to prohibit the PTI system. It is noteworthy here that PTI was the final piece of the puzzle in the CE/PTI/ETI model in the case of biotrophic pathogen–plant interactions. It was crystallized with the discovery of flagellin (PAMP), which triggers plant receptors. *Pseudomonas syringae* produces the CE AvrPto, which interacts with a tomato-resistance gene, Pto, and a co-receptor complex to inhibit resistance gene's activity and suppress PTI, respectively (Abramovitch et al., 2006; Dodds et al., 2006; Göhre and Robatzek, 2008; Boller and Felix, 2009; Haas et al., 2009; Poueymiro and Genin, 2009).

CELL WALL DEGRADATION ENZYMES

Cell wall degradation enzymes (CWDEs) assist plant pathogenic fungi by decomposing host cell wall and acquisition of nutrinets. Enterobacteria produce pectolytic enzymes that break the glycosidic linkage between the middle lamella and the primary cell wall of the host plant, which in response activate their defense. Pathogens release defense suppressors when an enzymatic attack is successful. These pectolytic enzyme-producing necrotrophs, although similar to biotrophic pathogens in amplification, differ with regard to pathogen–plant interaction systems, as these are broad ranges without any resistant R genes. Thus, environmental factors play an important role in disease development, and when a bacterial quorum is achieved, plants detect damage-associated molecular patterns (DAMP) and initiate DAMP-triggered immunity (Darvill and Albersheim, 1984; Jones et al., 1993; Palva et al., 1993; Toth and Birch, 2005; Toth et al., 2006; Liu et al., 2008).

TOXINS

Toxin-producing fungi, such as the *Cochliobolus* genus, produce host-specific toxins (HSTs) due to inadvertent breeding in cereal crops, consequently making them susceptible to various pathogenic races. *Cochliobolus carbonum* produces HC-toxin that inhibits defense gene (histone deacetylase) expression; in response, plants encode the enzyme HC-toxin reductase that detoxifies the toxin. The mutant maize hm1 is, however, highly sensitive and is aggressively affected by HC-toxin. Hence, to prevent an aggressive attack of the pathogen, specific mutants of hm1 are removed from the field. Another species of this genus, *C. victoriae*, produces HST victorin in oats and causes the Victoria blight of oats. Oats carrying the rust-resistant gene Pc-2 are susceptible to victorin, as it elicits programmed cell death in Pc-2-carrying plants. Removal of Pc-2 gene carrying oats extinguishes the disease outbreak and consequent epidemic (Johal and Briggs, 1992; Meeley et al., 1992; Ransom and Walton, 1997; Sindhu et al., 2008).

In various crop–pathogen systems, the internal redundancy of pathogen CEs allows them to lose any CE with minimal impact on virulence, which renders newly developed R-gene resistance unstable. This characteristic feature of CEs explains why most of molecular Koch's postulates fail for these virulence factors. A genomic analysis of CE repertoire of different strains of *P. syringae* reveals that these effectors are significantly different from each other, and surprisingly, even for the same host pathogens. Internal redundancy in these effectors might be structured around redundant effector groups (REGs), which impact a specific step/process in PTI (Saijo et al., 2009; Tsuda et al., 2009).

HOW TO DISTINGUISH BETWEEN VIRULENCE FACTORS AND EFFECTORS

Plant pathologists' primary goal in the 20th century was to distinguish between pathogenicity factors—qualitatively essential/responsible for disease development/pathogenesis—and virulence factors—quantitatively essential for pathogen propagation and symptom development. With the proposed molecular Koch's postulates, it was possible to distinguish between virulence and pathogenicity factors on the basis of the virulence defects in the clone. On the basis of function, three classes of virulence were proposed: virulence genes encode host-interaction factors that contribute significantly to pathological damage; virulence and associated genes deploy true virulence genes products; virulence genes code for factors that assist pathogen in host colonization/tolerance to host defenses, i.e., reactive oxygen. According to a recent definition, effectors are pathogen molecules that promote disease or trigger host defenses upon direct interaction with host cells. Virulence related is that set of genes that performs critical cellular functions, i.e., defeat host defenses, host-tissue penetration, trigger defense response in host, are crucial during pathogenesis. As described earlier, exceptional pathosystems are important in studying the impact of individual virulence factors as components of the virulence system, of which effector is also a component. *E. amylovora*, fire blight pathogen, secrete pathogenicity factor, dspE, that is homologous to *P. syringae* effector, AvrE. AvrE can restore virulence to *E. amylovora* dspE mutants, proving that the difference between these two effectors is the system they are part of. It can be concluded that a particular factor promoting pathogenesis or quantitatively promoting virulence is a property of the system rather than the factor itself. System biology is defined as behavior and relationship investigations of all the individual elements in a functioning biological system and consequent data integration and graphical display to form a computational model. To attain this, all related components are defined and monitored; the observed response is reconciled with the predicted model, and the designing of new perturbation experiments, aiming to differentiate between competing or model hypotheses (Bocsanczy et al., 2008; Hogenhout et al., 2009; Schneider and Collmer, 2010).

RESOURCES TO EXPLORE SYSTEM BIOLOGY

Plant–pathogen interaction properties at system level are anticipated by two phases: first phase centered on individual plant–pathogen system will provide tests, such as signal transduction, and PAMP perceptions, for standardized assays; followed by second phase of data integration, from various patho-systems involving standardization and formal structure. The virulence system's emerging properties are integrated with molecular models, resulting in data-enhancing hypothesis formation in components biology. Web-based resourced and computational tools are indispensable for grasping the complexity and dynamics of plant–pathogen interaction models. Computational modeling, being robust, can distinguish and predict emerging properties, indicate expected vulnerabilities, and provide a more detailed analysis of these biological systems.

PRINCIPLES OF SYSTEM BIOLOGY

In system biology, properties that arise from the interaction of various system components are analyzed. Complex mathematical models are used to integrate the observations about different components at the cell, organ, and population levels. Data collection tools are employed to collect both quantitative and qualitative data from various system components at the genome and cell scale levels, i.e., from genes, proteins, and RNAs. After data collection, mathematical models are constructed describing the system components and their interaction. Lastly, computer algorithms are designed, which, based on experimental data, calculate and predict the outcomes of complex pathogen–host interaction (Sauer et al., 2007).

GENOMICS, PROTEOMICS, AND METABOLOMICS

Characterization of cell nucleic acids and mass spectrometry of proteins are techniques that helped scientists study system biology. Due to innumerable repetitive regions, genome assembly was challenging until fast, accurate, and reliable techniques were discovered. These techniques have made it possible to reach non-sequenced microbes, and mass spectrometry has made it possible to achieve proteomics data even from complex samples. Phosphorylation and other related methods are used to quantify and characterize the modified proteins. In metabolomics, small molecules (lipids and sugars) with diverse chemical properties are separated, identified, and quantified from complex samples using liquid and gas chromatography and nuclear magnetic resonance. Fluxes of atoms across the cell are quantified with isotope-enriched molecules, and their mapping through "fluxomics" analysis is done to reveal the physiology of functional cell (Eid et al., 2009; Heuberger et al., 2014).

MODEL RECONSTRUCTION AND ITS PREDICTIVE CAPACITY

It should be noted that the phenotype of a particular gene could be masked during reverse genetic approaches due to redundancies. These observations are frequent because individual biological systems and plant–pathogen interaction systems are complex and interconnected through many molecules with different properties. Hence, two independent pathways, without any common component, could yield the same results due to environmental or genetic factors. So, systemic characterization alone cannot explain complex biological systems; instead requires an understanding of the interaction between same (protein–protein) and different types (DNA–protein) of components. This integration of cellular components and their understanding is termed "model reconstruction" (Pritchard and Birch, 2011).

MODELS IN SYSTEM BIOLOGY

In system biology, mechanistic models are used as these involve molecular components in biologically active cells, and based on the interaction between systems components, these models can be divided into three main classes (constrain-based models, logical models, and kinetic models):

Constrain-based models focus on metabolic reaction and are developed to increase the productivity of a metabolic pathway. Hence, increase range and procutions of desirable microbial industry products. Accordingly, these are perfect for predicting phenotypic characters based on environmental conditions and studying the impact of any variation in any system component (Schatschneider et al., 2013).

Logical modeling is suitable for large system networks as it employs logical interaction rules between components. It describes the state of a particular component concerning change in one or multiple other components. This model is particularly used to study regulatory events, such as activation of signaling. Kinetic modeling is used to study the dynamics of a biological system, and it describes the concentrations of cell components on the basis of chemical equilibrium laws (Poolman et al., 2009).

Regulatory proteins are continuously recycled in various signaling pathways across organisms; hence, their respective interactions in genomics are challenging to predict and require the incorporation of biochemical data. Statistical algorithms are employed to determine the interaction/dependency between a gene and a transcription factor. Statistical algorithm was successfully used to predict the role of the *Arabidopsis thaliana* transcription factor TGA3 in colonization by *Botrytis cinerea* (Windram et al., 2012).

System biology approaches, by considering universal physical laws of interaction and specific biochemical components in both pathogen and plant, utilize laws of thermodynamics to predict possible outcomes with high accuracy and reliability. The growth of an organism is fully constrained by the matter and energy available in its local environment. Conservation laws of energy and matter suggest that the maximum growth capacity of an organism cannot be exceeded, and on the flip side, there is a limit to the metabolic capacity of an organism, due to transporter repertoires, to acquire nutrients from specific niches. Hence, during plant–pathogen interactions, there is a limit to both partners' growth, as matter and energy are limited in their local environment (Barbacci et al., 2015).

CHALLENGES IN MODELING OF PLANT–PATHOGEN SYSTEMS

The plant immune system, producing PAMP-triggered and effector-triggered immunity, is a tightly regulated system that responds according to the invader's potential while maintaining the viability of its cells. Various host cell immune system components, and their relationship with one another, are still to be discovered. In these frameworks, the discovery of the signaling network of *Arabidopsis* response upon exposure to the *P. syringae* effector AvrRp2 was crucial for system studies. This response is regulated by 22 components in *Arabidopsis*, as evident by mRNA profiling. While responding to invaders, plants only use a fragment of a signaling network to balance robustness in immune response and its adverse impacts on plant health. Researchers, by using the logical-modern approach, formed a mathematical model of *A. thaliana* immune response. For this, they reconstructed and merged the interaction of the hormonal and immune networks. Logical rules were converted into ordinary differential equations and resultantly predicted and validated that cytokinin does not play any significant role in early PTI during pathogen infection. The addition of system kinetics and the integration of pathogen components in the system are ways forward for the success of the model (Peyraud et al., 2017).

PLANT CELL WALL AND ITS DEGRADATION ENZYMES

In the true sense of system biology, plant and pathogen as interacting systems are not studied together, as plant cell wall and CWDE studies are conducted separately. The pathogen genome having plant cell wall-degrading enzymes does not feed only on dead plants; instead, it can use these enzymes to overcome plant barriers to colonize the internal tissues. These pathogen and plant components should be studied together as these evolve to function in response to each other; therefore, they are crucial for understanding the dynamics of plant–pathogen interactions (Kraepiel and Barny, 2016).

MODEL INTEGRATION ACROSS TEMPORAL AND SPATIAL SCALES

SINGLE CELL–BASED MODELS

Plant–pathogen integration is characterized from infected cell to organism level, over the temporal and spatial scales challenge multiscale data and disordered modeling approaches, which are significantly affected by environmental factors. In earlier stages, infection is at the molecular level within the first host cell, which gradually leads to systemic infection or host response. Therefore, prediction based on single-cell penetration by pathogen and response of that single host cell could not predict authentic disease development model. The stochasticity and rapid fluctuations in the micro cell environment lead to heterogeneity, producing a high variability in interaction output. Current models are based at cell levels and conform to the first localized plant–pathogen interactions. However, kinetic computing of such a system is challenging, especially when considering stochasticity, because omics data is scarce and investigations are mostly done microscopically (Cunniffe et al., 2015).

MULTICELLULAR-BASED MODELS

Contrarily, at the multicellular level, there are significant challenges related to modeling rather than data collection. Although metabolomic analysis is performed for intercellular and organ investigation, its ability to distinguish between inter- and intracellular metabolites is limited. Spatial dimensions are crucial during these analyses, such as for communication and cooperation among cells, but are hardly incorporated in mechanistic models.

ECOSYSTEM-BASED MODELS

The ecosystem plays an important role in plant–pathogen interactions and consequent disease epidemics development as it carries beneficial microbiota and influences crucial factors for pathogen growth, i.e., nutrient availability, temperature, and humidity, as seen during an experiment when bacterial diversity in the rhizosphere was manipulated to prove that competing bacterial species prevented disease development in tomato by *Ralstonia solanacearum* (Müller et al., 2016).

System biology approaches will enrich the plant–pathogen interaction field if models for plants and pathogens are freely available. Although, microbes have smaller genome size and consequent, easier metabolic reconstruction as compared to plants. But, till date, due to unavailability of efficient software tools and research models, only a few plant pathogens' models are available, i.e., X. *compasteris*, R. *solanacearum*. For various plants—Arabidopsis, barley, rice, millet, maize, and tomato—models are reconstructed on the basis of a complete genome-scale and specific metabolic pathway.

SYSTEM BIOLOGY AND PLANT PATHOLOGY

Microbial pathogens possess an arsenal of effectors that can compromise host immunity after translocation in cells. Generally, the inactivation of single or multiple effector genes does not impede the pathogen from infecting the host, exhibiting robustness in effector networks. This robustness and redundancy in effector groups could be part of the evolutionary process. In addition, frequent gain and loss of function help pathogens from being detected by plant defense systems. Correspondingly, the plant defense system produces invariant immune responses and detects invasion in case of pathogen escape recognition. This is achieved by deploying redundant receptors at the site of entry/contact of the pathogen, for example, rice has various receptors at the cell surface to detect chitin in the fungal cell wall, working as a center point in chitin signaling. Another notable immune system redundancy is evident from decoy proteins in plants which have the ability to detect pathogen effectors. Unsurprisingly, the output of plant–pathogen interaction cannot be predicted with a limited knowledge of plant or pathogen components.

Plants pay a substantial cost of resistance in the form of compromised growth. Hence, manipulation of R-genes or other associated genes for defense in plants could trade-off growth for immunity. Plant growth hormones, such as gibberellins, auxins, ethylene, and cytokinin, which have been thoroughly described to control plant growth function, are associated with plant immunity as well (Naseem and Dandekar, 2012). Although mechanistic-level studies have been profound as compared to pre-gemonic and metabolomis era, there is a need to combine various aspects of plant–pathogen interactions to fully understand the effect of growth and immunity trade-offs on plant growth and production. System biology quantifies the cost and constraints in various plant–pathogen interaction models and could significantly evaluate the trade-offs among agricultural productivity–related traits.

CONCLUSION

Traditional and molecular Koch's postulates were formulated according to the technology, tools, and knowledge of the time. They have served and will continue to serve their purpose under appropriate conditions/when needed. Latest advances have discovered various new virulence factors associated with several disease-development steps. The redundancy and robustness of these effectors and plant defenses need to be studied extensively. Advances in molecular biology have revealed that pathogens frequently turn their genes off and on to acquire specific phenotypes. Evolution has equipped them with ingenious mechanisms to detect varying environments and make adaptations to survive. Pathogens sense the local environment, process information, and shift phenotype to adjust, resulting in pathogenic and non-pathogenic behavior. Infectious pathogenesis results from various crosstalk, attacks, and defenses, as it is a continuous war between pathogen and plant with still unpredictable possibilities.

International trade and climate changes are associated with increased disease outbreaks. In these circumstances, expert knowledge of pathogen and host adaptability in a new environment is a prerequisite to fully understand the possible outcomes of host–pathogen interactions and formulate management strategies to tackle the issue of food security. There is a need to develop sophisticated models, which incorporate climate predictions, for significant crop plants against all of their possible pathogens. However, we are still struggling to find the impact of changing environment on plant and pathogen adaptability. System biology offers a workable plan to execute this mission by predicting the fitness and possible evolution of plants and pathogens.

NOTE

1 Ergot poisoning could lead to confusion, vision problem, and hallucinations. A recent ergotism account, of 1951 France, described the villagers running madly in streets. Whereas, later research proved that manifestation of this bizarre behavior was due to disturbance in serotonin receptors.

REFERENCES

Abramovitch, R.B., J.C. Anderson and G.B. Martin. 2006. Bacterial elicitation and evasion of plant innate immunity. *Nature Reviews Molecular Cell Biology* 7:601–611

Agrios, G. 2005. *Plant Pathology*, 922. Elsevier Academic Press, Burlington, MA.

Aneja, K. 2007. *Experiments in Microbiology, Plant Pathology and Biotechnology*. New Age International, New Delhi.

Aylward, J., E.T. Steenkamp, L.L. Dreyer, F. Roets, B.D. Wingfield and M.J. Wingfield. 2017. A plant pathology perspective of fungal genome sequencing. *IMA Fungus* 8:1–15

Baldi, P. and G.W. Hatfield. 2003. *DNA Microarrays and Gene Expression from Experiments to Data Analysis and Modeling*. Cambridge University Press, Cambridge, United Kingdom.

Barbacci, A., V. Magnenet and M. Lahaye. 2015. Thermodynamical journey in plant biology. *Frontiers in Plant Science* 6:481

Begon, M., J.L. Harper and C.R. Townsend. 1986. *Ecology. Individuals, Populations and Communities*. Blackwell Scientific Publications, Oxford.

Bhunjun, C.S., A.J. Phillips, R.S. Jayawardena, I. Promputtha and KD. Hyde. 2021. Importance of molecular data to identify fungal plant pathogens and guidelines for pathogenicity testing based on Koch's postulates. *Pathogens* 10:1096

Blevins, S.M. and M.S. Bronze. 2010. Robert Koch and the 'golden age' of bacteriology. *International Journal of Infectious Diseases* 14:e744-e751

Bocsanczy, A.M., R.M. Nissinen, C.S. OH and S.V. Beer. 2008. HrpN of Erwinia amylovora functions in the translocation of DspA/E into plant cells. *Molecular Plant Pathology* 9:425–434

Boller, T. and G. Felix. 2009. A renaissance of elicitors: Perception of microbe-associated molecular patterns and danger signals by pattern-recognition receptors. *Annual Review of Plant Biology* 60:379–406

Chaudhari, P., A. Shetty and R. Soman. 2015. The concepts that revolutionized the field of infectious diseases. *The Journal of the Association of Physicians of India* 63:90–92

Chen, W., Z.R. Djama, MD. Coffey, F.N. Martin, G.J. Bilodeau, L. Radmer, G. Denton and C.A. Lévesque. 2013. Membrane-based oligonucleotide array developed from multiple markers for the detection of many Phytophthora species. *Phytopathology* 103:43–54

Cummings, C.A. and D.A. Relman. 2000. Using DNA microarrays to study host-microbe interactions. *Emerging Infectious Diseases* 6:513

Cunniffe, N.J., B. Koskella, C.J.E. Metcalf, S. Parnell, T.R. Gottwald and C.A. Gilligan. 2015. Thirteen challenges in modelling plant diseases. *Epidemics* 10:6–10

Darvill, A.G. and P. Albersheim. 1984. Phytoalexins and their elicitors-a defense against microbial infection in plants. *Annual Review of Plant Physiology* 35:243–275

Davis, M.J. and R. Brlansky. 2007. Culturing fastidious prokaryotes-points to consider when working with citrus Huanglongbing or greening. In: *Proceedings of the Florida State Horticultural Society*, 136–137. Citrus Research and Education Center, Lake Alferd, FL.

Dodds, P.N., G.J. Lawrence, A.-M. Catanzariti, T. Teh, C.-I. Wang, M.A. Ayliffe, B. Kobe and J.G. Ellis. 2006. Direct protein interaction underlies gene-for-gene specificity and coevolution of the flax resistance genes and flax rust avirulence genes. *Proceedings of the National Academy of Sciences* 103:8888–8893

Eid, J., A. Fehr, J. Gray, K. Luong, J. Lyle, G. Otto, P. Peluso, D. Rank, P. Baybayan and B. Bettman. 2009. Real-time DNA sequencing from single polymerase molecules. *Science* 323:133–138

Falkow, S. 1988. Molecular Koch's postulates applied to microbial pathogenicity. *Reviews of Infectious Diseases*:S274–S276

Falkow, S. 2004. Molecular Koch's postulates applied to bacterial pathogenicity—a personal recollection 15 years later. *Nature Reviews Microbiology* 2:67–72

Fedak, K.M., A. Bernal, Z.A. Capshaw and S. Gross. 2015. Applying the Bradford Hill criteria in the 21st century: How data integration has changed causal inference in molecular epidemiology. *Emerging Themes in Epidemiology* 12:1–9

Fredricks, D.N. and D.A. Relman. 1996. Sequence-based identification of microbial pathogens: A reconsideration of Koch's postulates. *Clinical Microbiology Reviews* 9:18–33

Gillies, D.A. 2016. Establishing causality in medicine and Koch's postulates. *International Journal of History Philosophy of Medicine* 6:10603

Goad, D.M., Y. Jia, A. Gibbons, Y. Liu, D. Gealy, A.L. Caicedo and K.M. Olsen. 2020. Identification of novel QTL conferring sheath blight resistance in two weedy rice mapping populations. *Rice* 13:1–10

Göhre, V. and S. Robatzek. 2008. Breaking the barriers: Microbial effector molecules subvert plant immunity. *Annual Review of Phytopathology* 46:189–215

Grimes, D.J. 2006. Koch's postulates-then and now. *Microbe-American Society for Microbiology* 1:223

Haas, B.J., S. Kamoun, M.C. Zody, R.H. Jiang, R.E. Handsaker, L.M. Cano, M. Grabherr, C.D. Kodira, S. Raffaele and T. Torto-Alalibo. 2009. Genome sequence and analysis of the Irish potato famine pathogen Phytophthora infestans. *Nature* 461:393–398

Heuberger, A.L., F.M. Robison, S.M.A. Lyons, C.D. Broeckling and J.E. Prenni. 2014. Evaluating plant immunity using mass spectrometry-based metabolomics workflows. *Frontiers in Plant Science* 5:291

Höfler, M. 2005. The Bradford Hill considerations on causality: A counterfactual perspective. *Emerging Themes in Epidemiology* 2:1–9

Hogenhout, S.A., R.A. Van der Hoorn, R. Terauchi and S. Kamoun. 2009. Emerging concepts in effector biology of plant-associated organisms. *Molecular Plant-microbe Interactions* 22:115–122

Iriti, M. and F. Faoro. 2008. Ancient plant diseases in Roman Age. *Acta Phytopathologica et Entomologica Hungarica* 43:15–21

Johal, G.S. and S.P. Briggs. 1992. Reductase activity encoded by the HM1 disease resistance gene in maize. *Science* 258:985–987

Johnston-Monje, D. and J. Lopez Mejia. 2020. Botanical microbiomes on the cheap: Inexpensive molecular fingerprinting methods to study plant-associated communities of bacteria and fungi. *Applications in Plant Sciences* 8:e11334

Jones, S., B. Yu, N.a. Bainton, M. Birdsall, B. Bycroft, S. Chhabra, A. Cox, P. Golby, P. Reeves and S. Stephens. 1993. The lux autoinducer regulates the production of exoenzyme virulence determinants in *Erwinia carotovora* and *Pseudomonas aeruginosa*. *The EMBO Journal* 12:2477–2482

Kämper, J., R. Kahmann, M. Bölker, L.-J. Ma, T. Brefort, B.J. Saville, F. Banuett, J.W. Kronstad, S.E. Gold and O. Müller. 2006. Insights from the genome of the biotrophic fungal plant pathogen *Ustilago maydis*. *Nature* 444:97–101

Kieu, N.P., M. Lenman, E.S. Wang, B.L. Petersen and E. Andreasson. 2021. Mutations introduced in susceptibility genes through CRISPR/Cas9 genome editing confer increased late blight resistance in potatoes. *Scientific Reports* 11:1–12

Kraepiel, Y. and M.A. Barny. 2016. Gram-negative phytopathogenic bacteria, all hemibiotrophs after all? *Molecular Plant Pathology* 17:313

Kůdela, V., V. Krejzar and I. PáNKoVá. 2011. *Pseudomonas corrugata* and *Pseudomonas marginalis* associated with the collapse of tomato plants in rockwool slab hydroponic culture. *Plant Protection Science* 46:1–11

Lamichhane, J.R. and V. Venturi. 2015. Synergisms between microbial pathogens in plant disease complexes: A growing trend. *Frontiers in Plant Science* 6:385

Larkan, N.J., H. Raman, D.J. Lydiate, S.J. Robinson, F. Yu, D.M. Barbulescu, R. Raman, D.J. Luckett, W. Burton and N. Wratten. 2016. Multi-environment QTL studies suggest a role for cysteine-rich protein kinase genes in quantitative resistance to blackleg disease in Brassica napus. *BMC Plant Biology* 16:1–16

Lengeling, A., K. Pfeffer and R. Balling. 2001. The battle of two genomes: Genetics of bacterial host/pathogen interactions in mice. *Mammalian Genome* 12:261–271

Liebe, S., DS. Christ, R. Ehricht and M. Varrelmann. 2016. Development of a DNA microarray-based assay for the detection of sugar beet root rot pathogens. *Phytopathology* 106:76–86

Lievens, B., L. Claes, A.C. Vanachter, B.P. Cammue and B.P. Thomma. 2006. Detecting single nucleotide polymorphisms using DNA arrays for plant pathogen diagnosis. *FEMS Microbiology Letters* 255:129–139

Lindeberg, M., S. Cartinhour, C.R. Myers, L.M. Schechter, D.J. Schneider and A. Collmer. 2006. Closing the circle on the discovery of genes encoding Hrp regulon members and type III secretion system effectors in the genomes of three model Pseudomonas syringae strains. *Molecular Plant-Microbe Interactions* 19:1151–1158

Lindgren, P.B., R.C. Peet and N.J. Panopoulos. 1986. Gene cluster of *Pseudomonas syringae* pv. "phaseolicola" controls pathogenicity of bean plants and hypersensitivity of nonhost plants. *Journal of Bacteriology* 168:512–522

Liu, H., S.J. Coulthurst, L. Pritchard, P.E. Hedley, M. Ravensdale, S. Humphris, T. Burr, G. Takle, M.-B. Brurberg and P.R. Birch. 2008. Quorum sensing coordinates brute force and stealth modes of infection in the plant pathogen *Pectobacterium atrosepticum*. *PLoS Pathogens* 4:e1000093

Liu, J., Y. Yuan, Z. Wu, N. Li, Y. Chen, T. Qin, H. Geng, L. Xiong and D. Liu. 2015. A novel sterol regulatory element-binding protein gene (sreA) identified in *Penicillium digitatum* is required for prochloraz resistance, full virulence and erg11 (cyp51) regulation. *PLoS One* 10:e0117115

Lorang, J.M., T.A. Sweat and T.J. Wolpert. 2007. Plant disease susceptibility conferred by a "resistance" gene. *Proceedings of the National Academy of Sciences* 104:14861–14866

Lucas, G.B., C.L. Campbell and L.T. Lucas. 1992. History of plant pathology. In: *Introduction to Plant Diseases*, 15–19. Springer, Boston, MA.

Ma, L., Y.H. Cao, M.H. Cheng, Y. Huang, M.H. Mo, Y. Wang, J.Z. Yang and F.X. Yang. 2013. Phylogenetic diversity of bacterial endophytes of *Panax notoginseng* with antagonistic characteristics towards pathogens of root-rot disease complex. *Antonie Van Leeuwenhoek* 103:299–312

Ma, L.-J., H.C. Van Der Does, KA Borkovich, J.J. Coleman, M.-J. Daboussi, A. Di Pietro, M. Dufresne, M. Freitag, M. Grabherr and B. Henrissat. 2010. Comparative genomics reveals mobile pathogenicity chromosomes in Fusarium. *Nature* 464:367–373

Matthews, D.E. and H.D. Van Etten. 1983. Detoxification of the phytoalexin pisatin by a fungal cytochrome P-450. *Archives of Biochemistry Biophysics* 224:494–505

McDowell, J.M. 2011. Genomes of obligate plant pathogens reveal adaptations for obligate parasitism. *Proceedings of the National Academy of Sciences* 108:8921–8922

Meeley, R.B., G.S. Johal, S.P. Briggs and J.D. Walton. 1992. A biochemical phenotype for a disease resistance gene of maize. *The Plant Cell* 4:71–77

Müller, D.B., C. Vogel, Y. Bai and J.A. Vorholt. 2016. The plant microbiota: Systems-level insights and perspectives. *Annual Review of Genetics* 50:211–234

Naseem, M. and T. Dandekar. 2012. The role of auxin-cytokinin antagonism in plant-pathogen interactions. *PLoS Pathogens* 8:e1003026

Neville, B.A., S.C. Forster and T.D. Lawley. 2018. Commensal Koch's postulates: Establishing causation in human microbiota research. *Current Opinion in Microbiology* 42:47–52

Palva, T., K.-O. Holmström, P. Heino and E.T. Palva. 1993. Induction of plant defense response by exoenzymes of *Erwinia carotovora* subsp. carotovora. *Molecular Plant-microbe Interactions* 6:190–196

Peyraud, R., U. Dubiella, A. Barbacci, S. Genin, S. Raffaele and D. Roby. 2017. Advances on plant—pathogen interactions from molecular toward systems biology perspectives. *The Plant Journal* 90:720–737

Phillips, C.V. and K.J. Goodman. 2004. The missed lessons of sir Austin Bradford Hill. *Epidemiologic Perspectives Innovations* 1:1–5

Poolman, M.G., L. Miguet, L.J. Sweetlove and DA Fell. 2009. A genome-scale metabolic model of Arabidopsis and some of its properties. *Plant physiology* 151:1570–1581

Poueymiro, M. and S. Genin. 2009. Secreted proteins from Ralstonia solanacearum: A hundred tricks to kill a plant. *Current Opinion in Microbiology* 12:44–52

Pritchard, L. and P. Birch. 2011. A systems biology perspective on plant—microbe interactions: Biochemical and structural targets of pathogen effectors. *Plant Science* 180:584–603

Pryce, T., S. Palladino, I. Kay and G. Coombs. 2003. Rapid identification of fungi by sequencing the ITS1 and ITS2 regions using an automated capillary electrophoresis system. *Medical Mycology* 41:369–381

Purdy, R. and P. Kolattukudy. 1975. Hydrolysis of plant cuticle by plant pathogens. Purification, amino acid composition, and molecular weight of two isoenzymes of cutinase and a non-specific esterase from *Fusarium solani* f. pisi. *Biochemistry* 14:2824–2831

Ransom, R.F. and J.D. Walton. 1997. Histone hyperacetylation in maize in response to treatment with HC-toxin or infection by the filamentous fungus Cochliobolus carbonum. *Plant Physiology* 115:1021–1027

Rehmany, A.P., A. Gordon, L.E. Rose, R.L. Allen, M.R. Armstrong, S.C. Whisson, S. Kamoun, B.M. Tyler, P.R. Birch and J.L. Beynon. 2005. Differential recognition of highly divergent downy mildew avirulence gene alleles by RPP1 resistance genes from two Arabidopsis lines. *The Plant Cell* 17:1839–1850

Ried, J. and A. Collmer. 1988. Construction and characterization of an *Erwinia chrysanthemi* mutant with directed deletions in all of the pectate lyase structural genes. *Molecular Plant-microbe Interactions* 1:32–38

Rogers, L.M., M.A. Flaishman and P.E. Kolattukudy. 1994. Cutinase gene disruption in *Fusarium solani* f sp pisi decreases its virulence on pea. *The Plant Cell* 6:935–945

Ross, L.N. and J.F. Woodward. 2016. Koch's postulates: An interventionist perspective. *Studies in History Philosophy of Science Part C: Studies in History Philosophy of Biological Biomedical Sciences* 59:35–46

Saijo, Y., N. Tintor, X. Lu, P. Rauf, K. Pajerowska-Mukhtar, H. Häweker, X. Dong, S. Robatzek and P. Schulze-Lefert. 2009. Receptor quality control in the endoplasmic reticulum for plant innate immunity. *The EMBO Journal* 28:3439–3449

Sanger, F., G.M. Air, B.G. Barrell, N.L. Brown, A.R. Coulson, J.C. Fiddes, C. Hutchison, P.M. Slocombe and M. Smith. 1977. Nucleotide sequence of bacteriophage φX174 DNA. *Nature* 265:687–695

Sauer, U., M. Heinemann and N. Zamboni. 2007. Getting closer to the whole picture. *Science* 316:550–551

Schatschneider, S., M. Persicke, S.A. Watt, G. Hublik, A. Pühler, K. Niehaus and F.-J. Vorhölter. 2013. Establishment, in silico analysis, and experimental verification of a large-scale metabolic network of the xanthan producing Xanthomonas campestris pv. campestris strain B100. *Journal of Biotechnology* 167:123–134

Schechter, L.M., K.A. Roberts, Y. Jamir, J.R. Alfano and A. Collmer. 2004. Pseudomonas syringae type III secretion system targeting signals and novel effectors studied with a Cya translocation reporter. *Journal of Bacteriology* 186:543–555

Schneider, D.J. and A. Collmer. 2010. Studying plant-pathogen interactions in the genomics era: Beyond molecular Koch's postulates to systems biology. *Annual Review of Phytopathology* 48:457–479

Sebastiani, G., G. Leveque, L. Larivière, L. Laroche, E. Skamene, P. Gros and D. Malo. 2000. Cloning and characterization of the murine toll-like receptor 5 (Tlr5) gene: Sequence and mRNA expression studies in Salmonella-susceptible MOLF/Ei mice. *Genomics* 64:230–240

Simpson, A.J.G., F.d.C. Reinach, P. Arruda, F.A.d. Abreu, M. Acencio, R. Alvarenga, L.C. Alves, J.E. Araya, G.S. Baia and C. Baptista. 2000. The genome sequence of the plant pathogen *Xylella fastidiosa*. *Nature* 406:151–157

Sindhu, A., S. Chintamanani, A.S. Brandt, M. Zanis, S.R. Scofield and G.S. Johal. 2008. A guardian of grasses: Specific origin and conservation of a unique disease-resistance gene in the grass lineage. *Proceedings of the National Academy of Sciences* 105:1762–1767

Tabrah, F.L. 2011. Koch's postulates, carnivorous cows, and tuberculosis today. *Hawaii Medical Journal* 70:144

Tan, D.H., L. Sigler, C.F. Gibas and I.W. Fong. 2008. Disseminated fungal infection in a renal transplant recipient involving *Macrophomina phaseolina* and *Scytalidium dimidiatum*: Case report and review of taxonomic changes among medically important members of the Botryosphaeriaceae. *Medical Mycology* 46:285–292

Tonukari, N.J., J.S. Scott-Craig and J.D. Waltonb. 2000. The *Cochliobolus carbonum* SNF1 gene is required for cell wall—degrading enzyme expression and virulence on maize. *The Plant Cell* 12:237–247

Toth, I.K. and P.R. Birch. 2005. Rotting softly and stealthily. *Current Opinion in Plant Biology* 8:424–429

Toth, I.K., L. Pritchard and P.R. Birch. 2006. Comparative genomics reveals what makes an enterobacterial plant pathogen. *Annual Review of Phytopathology* 44:305–336

Tsuda, K., M. Sato, T. Stoddard, J. Glazebrook and F. Katagiri. 2009. Network properties of robust immunity in plants. *PLoS Genetics* 5:e1000772

van Reekum, R., D.L. Streiner and D.K. Conn. 2001. Applying Bradford Hill's criteria for causation to neuropsychiatry: Challenges and opportunities. *The Journal of Neuropsychiatry Clinical Neurosciences* 13:318–325

Walker, L., H. Levine and M. Jucker. 2006. Koch's postulates and infectious proteins. *Acta Neuropathologica* 112:1–4

Wei, T., G. Lu and G. Clover. 2008. Novel approaches to mitigate primer interaction and eliminate inhibitors in multiplex PCR, demonstrated using an assay for detection of three strawberry viruses. *Journal of Virological Methods* 151:132–139

Weissenbach, J. 2016. The rise of genomics. *Comptes Rendus Biologies* 339:231–239

Windram, O., P. Madhou, S. McHattie, C. Hill, R. Hickman, E. Cooke, D.J. Jenkins, C.A. Penfold, L. Baxter and E. Breeze. 2012. Arabidopsis defense against Botrytis cinerea: Chronology and regulation deciphered by high-resolution temporal transcriptomic analysis. *The Plant Cell* 24:3530–3557

2 An Insight Into Fungal Biology

Imran Ul Haq, Nabeeha Aslam Khan and Muhammad Kaleem Sarwar

CONTENTS

DOI: 10.1201/9781003162742-2

INTRODUCTION TO FUNGI

Fungi are the second largest group of living organisms, with 51,000 genera and more than 70,000 known species (Blackwell, 2011; Hawksworth, D.L., 2012). Fungi are eukaryotic, multicellular heterotrophs. Fungi are organisms with chitin in their cell walls as the main structural component. The other components of their cell wall are mainly glucan, mannan, and chitosan. About 22–44% of the fungal cell wall is chitin as distinguished from the cell wall of plants (Khale and Deshpande, 1992; Muzzarelli et al., 1994; Kirk et al., 2008), and fungi exhibit the absorptive mode of nutrition (as distinguished from animals). The body of fungi is filamentous, called "mycelium." The mycelium is composed of tubular cells known as "hyphae" (aseptate or septate). The body of the fungus is called "thallus." Some fungi are biotrophs (obtain nutrition from living cells), some are saprotrophs (get nutrition from the dead host), and some are necrotrophs (attack and kill the host, then obtain nutrients). They are parasites, decomposers, and also exhibit symbiosis. The true fungi are divided into four main phyla: Basidiomycota, Ascomycota, Zygomycota, and Chytridiomycota (Alexopoulos et al., 1996; Webster and Weber, 2007). The basic unit of reproduction in fungi is spore. Fungi can reproduce by both sexual and asexual methods.

Fungi are present everywhere in the environment (soil, air, and water). They can decompose organic matter into simpler elements such as oxygen, carbon, phosphorus, and nitrogen. They play an essential role in industrial processes, e.g., in the making of cheese, bread, and wine. Mushrooms (high protein source) are edible fungi. Fungi have had great medicinal relevance since 1928, when penicillin (antibacterial activity) was discovered by Alexander Fleming. Some chemicals obtained from fungi are used to produce drugs, and some are used to produce vitamins, enzymes, and organic acids.

PHYSIOLOGY AND BIOCHEMICAL GENETICS

Fungal physiology relates to nutritional requirement, growth, metabolic processes, reproduction, and death of cells. It refers to the interaction between fungi and their surroundings, including biotic and abiotic factors. The physiological processes of fungi considerably impact human health, the ecology, and industrial processes. Carbon cycling in the natural environment is possible due to fungi's decomposition of organic material. Furthermore, fungi exhibit significant roles as saprophytes, pathogens, symbiotic mutualists, biocontrol agents, nutrient mobilization, and bioremediation of heavy metals. Yeast, a fungus, is used in many industrial products, fermented beverages, food additives, probiotics, pharmaceuticals, pigments, antibiotics, enzymes, biofuels, organic fatty acids, vitamins, and sterols. On the other hand, fungi can also harm surroundings by causing catastrophic losses of crops due to several diseases, mycotoxins may be carcinogenic (Calvo et al., 2002), and spoilage of food products.

NUTRITIONAL REQUIREMENT, UPTAKE, ASSIMILATION, AND GROWTH

The nutritional needs of fungi are relatively simple. Some fungi can survive in aerobic conditions with few nutrients and growth factors. The macronutrients required include nitrogen, carbon, sulphur, oxygen, potassium, phosphorus, and magnesium, and the micronutrients iron, calcium, manganese, copper, and zinc are required in trace amounts. Fungi need a high amount of sugars (simple hexoses, such as glucose, to polysaccharides, such ascellulose and starch) for growth, and some utilize lignins. Some fungi are oligotrophic and can grow with limited nutrients; chemo-organotrophic fungi require fixed organic compounds, while nondiazotrophic fungi need nitrogenous compounds in organic or inorganic forms. Proteolytic fungi can hydrolyze the molecules of proteins to obtain amino acids and utilize them for growth. Based on their oxygen requirement, fungi are referred to as aerobic and anaerobic. Some fungi require oxygen to synthesize sterols and unsaturated fatty acids, which are the main constituents of their membrane.

Fungi obtain sulfur for growth from sulphite, sulphate, glutathione, methionine, and thiosulphate. Phosphorous is utilized for the biosynthesis of nucleic acid, glucophosphates, phospholipids, and adenosine triphosphate. Fungi can store phosphate in vacuoles as polyphosphates. Certain minerals, e.g., Li, Cs, As, Hg, Ag, Ba, Cd, and Pb, are toxic to fungi at over 100 μM concentration.

Fungal cells utilize different strategies for the acquisition of food. Fungi are chemo-organotrophic organisms, saprophytic, and parasitic. Fungi exhibit energetic interactions with nutrition. Nutrient uptake may be disturbed due to cellular barriers such as the cell wall, cell membrane, capsule, and periplasm. The cell wall is porous and can absorb small nutrient molecules, while the cell membrane is a highly selective permeable barrier. The membrane's nutrient transport mechanisms are most important in physiology, as they govern metabolization rate, growth, and cell division. Fungi exhibit both active and passive modes of nutrient uptake through the plasma membrane (diffusion channels, free diffusion, and facilitated diffusion).

The parasitism of fungi is more dominant than other modes of nutrition. Fungi are classified into two main groups on the basis of their lifestyle: necrotrophic and biotrophic.

Biotrophic fungi are dependent on a living host for nourishment and completion of their life cycle by the formation of haustoria (specialized structure between host and fungus) (Mendgen and Hahn, 2002), while specialized structures are absent in necrotrophs (Lewis, 1973). Another mode of obtaining nutrition is hemibiotrophic, in which the fungus can switch from a biotrophic to a necrotrophic lifestyle (Perfect and Green, 2001). Germination, sporulation, and proliferation of fungi all require food.

SPORE GERMINATION AND PENETRATION

Plant pathogenic fungi are usually in the starvation phase during spore germination and rely on internal nutrient storage. Glycogen, polyols, and trehalose are the major components of spores (Thevelein, 1984; Thines et al., 2000). During penetration, fatty acid metabolism occurs.

BIOTROPHIC PHASE

The pathogens establish themselves into hosts by the mechanism of mobilization in order to obtain nutrients. The haustorium is formed to establish a relationship between the host and the fungi (Voegele, 2006). Hemibiotrophs have no feeding structures; they obtain nutrients by establishing new tissues. In the hemibiotrophic mode of nutrition, hexose and invertases play a key role as sugar transporters. Rapid response to stress stimuli is the upregulation of invertase (Biemelt and Sonnewald, 2006).

The γ-aminobutyric acid (GABA) is a significant constituent of the amino acid accumulated in the apoplast during the host (tomato) fungal interaction. The GABA accumulation associated with glutamate decarboxylase and GABA transaminase of fungus metabolizes GABA into succinic semialdehyde. GABA is utilized as a nutrient by fungi without forming haustoria, an adaptation for the biotrophic mode of nutrition. The reactive oxygen species play a strong role in activating plant defense systems (Oliver and Solomon, 2004).

NECROTROPHIC PHASE

In the necrotrophic phase, fungi obtain nutrients with the help of hydrolytic enzymes, toxin production, and proteins causing the lysis of cells and death of the host plant. The lysis of cells makes high nutrients available to the fungus. Histidine and methionine are not sufficiently supplied to auxotrophic mutants by the host plant (Sweigard et al., 1998; Seong et al., 2005; Balhadere et al., 1999). Nutrient-acquisition genes expression was revealed by analysis of the transcriptome of *Fusarium graminearum*. The gene of transporters of nitrogenous compounds and sugars expressed at early stages of infection (Guldener et al., 2006). In the metabolic pathway, an important intermediate is ornithine. The uptake of ornithine is followed by the expression of arginine and proline genes. The disruption in the gene that encodes ornithine decarboxylase causes a reduction in virulence of *Stagonospora nodorum* in wheat pathogen (Bailey et al., 2000). Ornithine acts as a precursor in polyamine synthesis, which is involved in plant stress responses (Walters, 2003).

SPORULATION

Completion of the life cycle directly relates to sporulation. True biotrophic fungi maintain the biotrophic state throughout their life cycle; necrotrophic and hemibiotrophic fungi live in the necrotrophic state and also complete their life cycle in this state. The sporulation rate can be enhanced by increasing carbon and nitrogen sources (Griffin, 1994). In the interaction between wheat and the fungus *Mycosphaerella graminicola,* the fungus resides in the hemibiotrophic state of the plant. Fungal expression of genes by analyzing cDNA during plant growth resembles the transcript obtained from in vitro growth on a nutrient-rich artificial medium (Keon et al., 2005). Nutrient availability is necessary in all growth stages of the fungus; this has been revealed by disruption in the synthesis of glycerol-3-phosphate dehydrogenase exhibited by hemibiotrophic *Colletotrichum gleosporioides* (Wei et al., 2004). Mutants with low glycerol show defects in metabolism, i.e., carbon utilization, and negatively affect conidiation. The defect due to glycerol deficiency can be addressed by adding glycerol.

FUNGAL METABOLISM

NITROGEN METABOLISM

In the 1960s, nitrogen metabolism was investigated by gene action studies (D. J. Cove, George Marzluf, and Claudio Scazzocchio, among others). The assimilation of nitrogenous sources accomplishes the biosynthesis of proteins. Ammonium ions are utilized, and assimilated glutamine and glutamate act as precursors for biological synthesis. Glutamate is an essential compound in nitrogen metabolism, while the glutamine synthetase enzyme plays an important role by catalyzing the initial step leading to cellular molecule synthesis. Glutamate synthase and glutamate dehydrogenase are also necessary enzymes in nitrogen metabolism. Glutamine synthetase, coupled with glutamate synthase, assimilates ammonia into amino acids. The pathway adopted by fungi for ammonium assimilation depends on the availability of ammonium ions. Amino acids are assimilated into proteins or dissimilated by fermentation, deamination, decarboxylation, or transamination. Fungi can degrade amino acids into glutamate and ammonium. Higher alcohols are produced during fermentation. The regulatory gene in the pathway promotes the expression of only those genes that are suitable for the substrate (Marzluf, 2004). In tomato infection due to *Cladosporium fulvum,* the metabolism of GABA has been explained (Solomon and Oliver, 2001).

CARBON CATABOLISM

As fungi are chemoorganotrophs, they obtain energy by breaking down organic sources. Fungi secrete enzymes to the extracellular breakdown of polymeric compounds. Enzymes are assembled by golgi bodies, transported by vesicles, and secreted from hyphal tips. Enzymes are linked to the cell wall or excrete decay substances into the local environment to degrade polymeric enzymes such as pectin by pectinase degrades into galacturonic acid, inulin into fructose by inulinase, starch into glucose by glucoamylase, cellulose into glucose by cellulases, hemicelluloses into xylose and glucose by xylanase and hemicellulases, respectively, lipids into fatty acids by lipases, proteins into amino by proteinases, chitin into N-acetylglucosamine by chitinase, and lignin into phenolic products by ligninase. Basidiomycetes and ascomycetes can degrade lignin, e.g., white-rot fungi can secrete oxidative and peroxidative enzymes, and these enzymes with hydrogen peroxide create enzyme system. The enzyme systems of lignin and manganese peroxide release reactive oxygen, which later reacts with lignin, produces a chain of oxidation reactions, and generates phenolic products. White-rot fungi have many applications in biofuel production, paper production, bleaching, and water bioremediation. Brown-rot fungi can degrade hemicellulose and cellulose of wood products by the synergistic activity of enzymes. The catabolic reactions involve oxidative processes that excrete electrons from carbon compounds, which generates ATP. The conversion of glucose into pyruvic acid is an enzyme-catalyzed reaction called "glycolysis"; it provides energy to fungal cells. Dissimilation of the glucose molecule in aerobic conditions leads to the respiration process, which is a significant energy-generating route, including citric acid cycle, glycolysis, and phosphorylation. Yeast is an organism that can switch from the respiration process to fermentation.

CELL WALL FORMATION

The cell wall is the outer part of the cell that acts as a barrier between the individual and its environment. It plays several roles for cells, i.e., gives shape, protects cellular structures, and the cytoplasm. It maintains the turgidity of the cell, movement of nutrients, nutritional physiology, and enzymatic secretions.

COMPOSITION

Cell wall is composed of polysaccharides (homopolymers and heteropolymers), proteins, melanins, and lipids. The skeletal crystalline homopolymers of polysaccharides are not soluble in water. Cell wall also has chitin and glucan. In contrast, matrix polysaccharides are water-soluble, slightly crystalline, or amorphous. Polysaccharides consist of homopolymer, heteropolymer, and complexes. In yeast fungi, the skeletal component consists of glucan with different linkages, which cause branch formation. The predominant molecule has a larger molecular weight and is highly crystalline. The skeletal component of the cell wall of filamentous fungi, except for Oomycetes, consists of chitin linked with glucan, i.e., N-acetyl-D-glucosamine homopolymer. Oomycetes have cellulose as a significant component of the cell wall. In yeasts, the main component of the cell wall is the mannan–protein complex linked with glucans (Pastor et al., 1984). *Candida albicans* have 40 mannoproteins (Elorza et al., 1985), while *Saccharomycopsis lipolytica*, *Zygosaccharomyces roweii*, *Schizosaccharomyces pombe*, and *Hansenula wingei* have a mixture of mannoproteins (Herrero et al., 1987). Glycoproteins are also a critical component of the cell wall of filamentous fungi; they are more heterogeneous than the glycoproteins present in yeast fungi. Filamentous fungi have a well-organized carbohydrate–protein linkage. The polysaccharides are either homopolymer or heteropolymer, e.g., in *Neurospora erassa*, the glycoprotein exhibits glycosidic linkages. The application of genetic technology to the biochemistry of fungi has been studied (Beadle and Tatum, 1945). Mutant strains with changed colonial form and hyphal branching patterns have also been studied (Garnjobst and Tatum, 1967). A study has been conducted on the involvement of defective enzymes in cell wall formation (Scott, 1976). Mutants with altered cell wall polymers change enzymatic activities (Scott et al., 1973; Mishra, 1977). Changes in the composition of the cell wall can change many functions, such as cause polarized growth patterns (Valentine and Bainbridge, 1978).

REPRODUCTION

Reproduction of fungi is the biological process through which new fungal organisms are formed from parents. Reproduction is a fundamental feature of all living organisms. There are three main methods of reproduction in fungi:

- Sexual
- Vegetative
- Asexual

VEGETATIVE MODE OF REPRODUCTION

Vegetative reproduction of fungi involves the somatic portion of the thallus, in which new fungal organisms are formed, exclusive of seeds and spores, by syngamy or meiosis. Vegetative reproduction can occur by the following methods:

- Fragmentation
- Budding
- Fission
- Rhizomorphs
- Oidia
- Sclerotia
- Chlamydospores

FRAGMENTATION

The mycelium converts into small bits by accident, mechanical injury, or several hyphal segments in this method. Each segment develops into new hyphae under suitable conditions on the further division of fungal cells.

BUDDING

The budding method of reproduction occurs mainly in yeast cells. In budding, a small outgrowth, called "bud," comes from the parent cell. The bud enlarges and is separated by cross wall formation from the mother cell. This detached portion of the cell develops into new, independent yeast. Occasionally, a budding chain is produced when the first bud does not get separated from the parent cell.

FISSION

Some fungi reproduce when a single fungal cell multiplies by fission, in which the mother cell becomes elongated and divides transversely into two daughter cells of the fungus. In this method, first the nucleus divides, then the cytoplasm, and finally the cell wall is formatted. The two new cells then become independent.

RHIZORNORPHS

In higher fungi, sometimes hyphae exist in rope-like structures known as rhizo-morphs. When conditions become favorable, these rhizomorphs grow and give rise to new mycelia and fruiting bodies. Rhizomorphs remain dormant under unfavorable conditions. These dark brown rhizomorphs are rope-like structures.

OIDIA

The hyphae of filamentous fungi break into new individual cells known as "arthro-spores" or "oidia." These oidia are usually rounded or oval-shaped and look like beads; the oidium gives rise to individual new mycelium.

SCLEROTIA

Sometimes a compact mass with hard covering, called "sclerotia," is formed by interwoven hyphae, e.g., Claviceps in ergot disease forms sclerotia under unfavor-able growth conditions and remains dormant until favorable conditions prevail again, then germinating into new mycelium. Sclerotia act as perennation bodies with vari-able shapes, i.e., cylindrical, rounded, cushion like, or irregular.

CHLAMYDOSPORES

Some fungi produce chlamydospores, which act as perennating structures. They are single-celled and exist either singly or in the form of chains in the hyphae. They have to resist thick walls, accumulate food, and overwinter or oversummer during unfavorable conditions. Under favorable conditions, these spores develop into new mycelia. Chlamydospores formation is standard in Fusarium and Mucor.

ASEXUAL REPRODUCTION

The asexual reproduction method is more common in fungi than sexual reproduction, and it is usually repeated several times in a single season. The spore is a fundamental reproductive unit in the asexual reproduction of fungi. The spore formation in fungi is called "sporulation," and the spore germinates and gives rise to a new mycelium. Asexual spore is also known as "mitospore," because it is formed by mitosis in the cell. These mitospores exhibit variation in shape (oblong, globose, oval, needle, helical), size, color (hayline, orange, green, brown, yellow, black, red), and arrangement of hyphae. Mitospores are unicellular. *Curvularia* and *Alternaria* are multinuclear. The spores are of three types:

- Sporangiospores
- Conidiospores
- Zoospores

SPORANGIOSPORES

Sporangiospores usually occur inside a sporangium (sac-like structure). The sporungium-bearing hypha, called "sporangiophore," is branched. Sporangiospores are motile or non-motile (also known as "aplanospores"). Examples of these fungi are *Rhizopus* and *Mucor.*

CONIDIOSPORES

The conidium is non-motile and is formed externally on hyphae as separate cells. Conidia formed on hyphae are called "conidiophores" (conidiophores are septate or aseptate and simple or branched).

ZOOSPORES

Motile biflagellate spores are called "zoospores," and the hypha bearing zoospores is known as "zoosporangium." These spores are produced by aquatic fungi members of Oomycete, do not have cell walls, and are motile, while aplanospores exhibit cell walls.

SEXUAL REPRODUCTION

Sexual reproduction in fungi is less common than asexual reproduction. The sexual stage is called the "perfect state," whereas the asexual stage is the "imperfect state." Sexually compatible gametes or cells of opposite strains fuse in sexual reproduction. The sex organs of fungi are known as "gametangia variables" in morphological aspects. Female gametangia are called "ascogonia" in ascomycetes, while the male ones are "antheridia."

Fungi are either homothallic (opposite gametes reside on the same mycelium) or heterothallic (opposite fusing gametes come from different mycelia).

In fungi, the following three phases during sexual reproduction are involved:

- Plasmogamy
- Karyogamy
- Meiosis

These phases occur at a specific time with regular sequence during the sexual reproduction of different fungal species.

PLASMOGAMY

In the plasmogamy phase, a fusion of the cytoplasm of reproductive cells occurs, in which female and male cells come closer to make a pair of parents. However, in this phase, nuclei do not fuse. A cell in which the nucleus of the opposite sex is present without fusion is called a "dikaryotic cell." The dikaryotic condition is not common in fungi.

KARYOGAMY

Karyogamy is the fusion of the two compatible nuclei after plasmogamy. It may occur immediately after plasmogamy in lower fungi, while it is delayed for a long time in higher fungi.

MEIOSIS

Meiosis occurs after karyogamy in sexually reproducing fungi and produces genetically different spores.

PLANOGAMETIC COPULATION

Planogametic copulation involves the fusion of two compatible gametes. The sexual union may be anisogamous, isogamous, or oogamous. Oogamy and anisogamy are known as heterogamous sexual reproduction in fungi, while isogamy is the most straightforward kind of sexual reproduction, in which the morphology of fusing gametes is similar. Examples include *Catenaria* and *Olpidium*. The fusing gametes are not similar in one of the genera in the case of anisogamy. In oogamy, the antherozoid enters the oogonium and fuses with the egg, forming a zygote. For example, *Pythium* and *Albugo* exhibit oogamy.

GAMETANGIAL CONTACT

The oogonium and the antheridium make contact by forming a tube; then, one or more nuclei of the antheridium transfer into the oogonium. In this type of contact, the two gametangia remain intact and do not fuse. Gametangial contact can be seen in *Penicillium*.

GAMETANGIAL COPULATION

The oogonium and the antheridium contact each other, and the contents of both gametes fuse entirely. In this copulation, the protoplast of the male gametangium flows into the female gametangium. Gametangial copulation occurs in *Rhizopus* and *Mucor*.

SOMATOGAMY

The degeneration of sexuality can be seen in members of Basidiomycetes and Ascomycetes. This process is simplified as a fusion of two mycelia of opposite fungal strains. The fruiting body formation occurs as a result of post-fertilization changes.

SPECIALIZATION

The spermatia (minute gametes) are produced on specialized hyphae externally. Spermatia-bearing hyphae are called "spermatiophores." Sometimes, spermatia develop within the cavity, that cavity is called "spermatogonia." The female gamete bearing hyphae is called "gametangium."

MORPHOGENESIS

In kingdom fungi, the predominant modes of vegetative propagation are hyphal and yeast growth.

The members of this kingdom exhibit remarkable morphological complexity. The morphological diversity of fungi is highly noticeable during their developmental stages as the formation of fruiting bodies (diverse sizes and shapes provide the basis for classification). The vegetative growth shows remarkable variation (from multicellular hyphae to compact unicellular yeast growth). Three significant transitions have been studied by reviewing an evolutionary perspective of morphogenesis. The first transition of morphogenesis is the fungi; lineages exhibited hyphal growth forms that evolved from the Chytridiomycota. The second transition of morphogenesis was the loss in the growth of hyphae as fungi that grow as yeasts. It is highly prominent in the phylum Ascomycota, in which two subphyla consist of yeasts. The third and final transition of morphogenesis was the emergence of obligate fungal hyphal growth within a clade of the yeasts.

Hyphae emerge from germinating spores of the fungus and from reproductive structures (sclerotia). Polarized extension in hyphae occurs only when specialized structures are present in the hyphal tip (Virag and Harris, 2006; Steinberg, 2007). The vesicles apical cluster was first described in 1957 (Girbardt, 1957). Polarity establishment and polarity maintenance are the two main events involved in the emergence of hypha (Harris, 2010). The fundamental aspect of polarity establishment is symmetry breaking. The polarity axis is highly specified at the germination site, where the morpho-genetic machinery is reorganized asymmetrically. The formation of elongated hyphae depends on polarity maintenance and requires precise transfer of exocytic vesicles to the tip of hyphae, coupled with surface components (Araujo et al., 2008; Upadhyay and Shaw, 2008). Branch formation also occurs in hyphal morphogenesis, in which new hyphae appear, and two hyphae fuse, exchanging their cytoplasmic material during this hyphal fusion (Read et al., 2010). Intercalary hyphal extension has been reported instead of extension from the tip (Christensen et al., 2008).

Complex spore-bearing multicellular structures develop during reproduction (asexual and sexual) in filamentous fungi. (Taylor and Ellison, 2010). Mechanisms of reproduction in fungi are different from those in plants and animals as different sporulation structures form by modifications in the growth pattern of hyphae, e.g., conidiophores of ascomycetes exhibit a condense branching system (Kendrick, 2003), and even a simple change in the pattern of branching (even a change in the polarized pattern) could cause a degree of variation in the cell shape. The main difference between hyphal and yeast cells is the extent of polarized growth.

Unlike the sustained polarized hyphal growth, budding in cells of yeast alternate between periods of isotropic expansion and polarized tip growth followed by cytokinesis. The yeast growth mode may occur in fungi as they lose the hyphal mode of growth. Blastic pattern morphogenesis in ascomycetes is similar to budding patterns (Cole and Samson, 1979). A considerable number of fungi can propagate by both modes of growth. These fungi (dimorphic and rarely polymorphic) include basidiomycetes, ascomycetes, and zygomycetes. This transition from the yeast mode of growth to hyphal morphogenesis depends upon abiotic factors, e.g., oxygen and temperature (Hoog et al., 2000; Klein and Tebbets, 2007; Casadevall, 2008).

Mathematical modeling and experiment have been reported to show the shapes of hyphal and yeast cells (Bartnicki et al., 1989). Recent research reveals the proteins are involved in the synthesis of cell wall chitin and glucans (Verdin et al., 2009; Riquelme et al., 2007) and in the formation and functioning of microfilaments (Harris et al., 2005). Some scientists have identified complexes that have an association with the formation of hyphal tips (Jones and Sudbery, 2010; Taheri et al., 2008)

The exocyst and polarisome components are present as discrete spots or crescents on the plasma membrane of the fungus, while Spitzenkorper components occur as subapical spots.

CHITOSAN

On deacetylation of chitin, a natural polysaccharide—chitosan—is obtained. Chitosan is a cationic hydrocolloid. It is significant due to its binding capabilities to fats and minerals, its biological activities, biocompatibility, and biodegradability (Elsabee and Abdou, 2013; Xia et al., 2011). It has wide application in food items and in medicines for tissue engineering and drug delivery (Agnihotri et al., 2004; Ding et al., 2012), as well as in seed treatment, biopesticides, and plant growth enhancers (Linden et al., 2000). Fungal chitosan is the best alternative for the chitosan from animal sources used in food items (animal source is not appealing for vegetarians) (Regenstein et al., 2003; Dunham, 2012). Zygomycetes and pathogenic fungi have a potential source for the production of chitosan (White et al., 1979). The mushroom cell wall is a high source of chitosan (Wu et al., 2004).

PATHOGENESIS

Kingdom fungi contains plant pathogenic species that cause more significant losses than viruses or bacteria. In the case of animals, however, viruses and bacteria cause more significant losses than fungal pathogens. Fungal pathogenicity in animals may depend upon the immunity of the host animal. In pathogenesis, two partners are involved with favorable environmental conditions. A recent concept is the damage-response framework in animal pathogens (Casadevall and Pirofski, 2003.). Research on host degradation, regulatory genes, pathogen differentiation, and signal transduction has been published (Hamer and Holden, 1997). When one organism obtains food from another organism to secure its nourishment, this is known as parasitism. Pathogenesis involves virulence factors, secondary metabolites (toxins), enzymes (cell wall-degrading enzymes), and many other mechanisms through which the pathogen accesses the host. As interaction develops between the host and the pathogen, the host's physiology is exploited.

HOST SPECIFICITY

Host ranges of fungal pathogens vary as some fungi can attack a number of plants, e.g., *Sclerotinia sclerotiorum* and *Botrytis cinerea* cause stem rot and grey mold in many species of plants (Bolton et al., 2006; Van, 2006). The specificity phenomenon of interaction between pathogen and plant is controlled by the resistance gene of the plant and the effector (avirulence genes) of the pathogen (Flor, 1971). Some fungi can attack and infect both animals and plants, e.g., *Chaetomium globosum*, a soil-born fungus can attack roots of plants and its airborne spores can infect immuno-compromised patients, causing pneumonia (Park et al., 2005; Paterson et al., 2005). *Fusarium oxysporum* f. sp. *lycopersici* can attack both tomato plants and mice as it has mitogen-activated protein (MAP) kinase gene and zinc finger transcription factor gene for virulence in tomato and mice, respectively (Ortoneda et al., 2004). *Aspergillus flavus* can infect insects, plants (nuts, cotton, and corn), and animals by producing carcinogenic toxins (St. Leger et al., 2000).

DEFENSE SYSTEMS

Plants and animals have an inducible defense against pathogens. The defense systems of host plants or animals against fungi are effective. In animals, the immune system causes activation of the antibody-mediated defense system. The pathway in which factors and receptors can cause pore complex formation in the cell membrane of pathogen ultimately leads to lysis and opsonization occurs. This process triggers chemotaxis stimulation (Roozendaal and Carroll, 2006, Speth et al., 2004). Immunocompromised hosts are more subject to fungal diseases. The severity of fungal diseases varies from superficial (tinea caused by *Candida* spp.) to serious infection (coccidioidomycosis, histoplasmosis, and blastomycosis) (Chu et al., 2006). The pathogenic fungi of plants have no mechanisms to overcome host defenses, and plants have resistance against pathogens, which is controlled by resistance genes. Plants also exhibit inducible resistance or inducible systemic-acquired resistance, which is activated due to previous plant exposure to the pathogen; in this, signaling pathways those involve in resistance become active through ethylene, salicylic acid, and jasmonate. These volatile molecules spread throughout the plant, triggering defense responses (pathogen-related protein express) and producing glucanases or chitinases, thereby increasing resistance against the pathogen (Pieterse and Van, 2004). This result is similar to that obtained from the preexposure of animals to the pathogen. The jasmonate pathway's oxygenation of fatty acids in plants is similar to the eicosanoid pathway in animals (Shea and Poeta, 2006). Programmed cell death is a typical defense response of different hosts against fungal pathogens (Mur et el., 2006).

Fungal and bacterial pathogens of plants and animals release some molecules related to pathogenicity. Specific-pattern proteins of fungi are recognized by receptors present in plants and animals, which initiate the defense responses. These receptor proteins manifest resistance gene products (Belkhadir et al., 2004).

Molecular knowledge (whole genome sequences, tagged mutant banks) about fungi supports the study of fungal pathogenicity. The characterization of mutants created by the targeted or random insertion of markers provides information on pathogenicity. The pathogen host interactions (PHI) database catalogs the phenotypes of the mutants of pathogens that have been developed (Winnenburg et al., 2006). Genes responsible for the synthesis of toxins can be predicted in the fungal genome through the PHI database, and virulence genes can be identified by screening mutagenized fungi in model hosts.

REGULATION OF PATHOGENESIS

A microorganism is called a "pathogen" only when it can complete its disease cycle or life cycle on the host, as spores of pathogen germinate on the surface of the host plant, hyphae penetrate and colonize into the tissue, reproduce, disperse, and can alter host plant physiology, ultimately causing symptoms and disease. Environmental conditions play a vital role in completing all the aforementioned steps. The germination of spores depends upon surface recognition, host-specific molecules, moisture level, temperature, nutrients, and metabolic and physical changes (polarized growth, swelling) (Osherov and May, 2001). For example, conidial germination of *Penicillium marneffei* occurs at 25°C in the non-pathogenic phase and at 37°C in the pathogenic phase.

Dimorphic switching is a feature of animal pathogenic fungi, and this feature is not present in plant pathogenic fungi (Bassilana et al., 2003; Bassilana et al., 2005; Borges et al. 2002; Gantner et al., 2005). Genes involved in germination activate nutritional mechanisms (Doehlemann et al., 2006). In *Colletotrichum trifolii*, an alfalfa pathogen, Res protein regulates spore germination, growth, and development (Chen et al., 2006). The change in hydrophobin MHP1 affects the morphogenesis of fungal pathogens in terms of spore germination, reduction in conidial, appressorium germination, and infection (Kim et al., 2005). Melanine (cell wall molecule of a fungus) has a significant role in viability and protection of the spores (Icenhour et al., 2006). More melanine content in the cell wall of conidia of *Aspergillus fumigatus* can cause hypervirulence, a high rate of conidial germination (Maubon et al., 2006). The PHI database has at least five fungal pathogenic mutants with loss of pathogenicity (melanin biosynthesis has been disturbed) (Winnenburg et al., 2006). It has been reported that fungicides disturb melanin synthesis (Liao et al., 2000). Glucan and chitin, which are the main components of the fungal cell wall, have a prime role in the virulence and growth of many pathogens. Some of these molecules may trigger pathogenesis-related protein and defense mechanisms (Casadevall and Pirofski, 2006; Shinya et al., 2006).

INVASION OF PATHOGEN

Plant pathogenic fungi enter the host plant through stomata or wounds (natural opening), by secreting toxins, exerting mechanical forces, or forming appressoria (creating turgor pressure to gain entry into leaves) (Thines et al., 2000). Several signals and molecules, such as cyclophilins, tetraspanins, and P-type ATPases, are involved in appressorium formation (Balhadere and Talbot, 2001). Tetraspanins are involved in the virulence of the fungal pathogen and perform a significant role in penetration peg formation (Clergeot et al., 2001). Hydrolytic enzymes (cellulases, cutinases, and pectinases) can cause degradation of the plant cell wall. Many genes encode these hydrolytic enzymes.

COLONIZATION AND ALTERATION OF HOST PHYSIOLOGY

The pathogen may produce some molecules and thus modify the host environment to colonize the host plant. *Sclerotinia sclerotiorum* produces oxalic acid, which affects the opening and closing of stomata and suppresses the defense responses of the host; it also regulates polygalacturonases production and sclerotia formation (Cessna et al., 2000). Oxylipins from *Aspergillus* spp. have a role in developing interaction between fungus and host (Brodhagen and Keller, 2006). The *Aspergillus nidulans* mutant with a mutation in oxylipins is less able to colonize the seeds (Tsitsigiannis and Keller, 2006). Fungal oxylipins can stimulate mycotoxin production and sporulation (Walters et al., 2006). Some fungi can alter the host's physiology for colonization by inducing oxidative burst cell death. Necrotrophic fungi weaken the host's defense responses for colonization (van, 2006). Fungi protect themselves from the rapid oxidative defense of plants by producing enzymes (superoxide dismutase) that can cause degradation of reactive oxygen species, as the invaded host generates hydrogen peroxide. It inhibits the growth of the pathogen in plants. Disruption of superoxide dismutase genes (present in the pathogen) can reduce virulence (Hwang et al., 2002). Proline activates a protection mechanism of fungi; it also protects the pathogen from oxidative burst (Chen and Dickman, 2005). The fungi may secrete toxins, which act as virulence factors that kill the host plant's tissues. Toxins causing diseases in plants are categorized as non-host specific and host specific; some toxins are secondary metabolites with a variable structure range and low molecular weight (Chen and Dickman, 2005).

REPRODUCTION

Mitotically, reproduction of plant pathogenic fungi occurs during the disease cycle. Conidia production is common in the life cycle of plant pathogenic fungi, which are usually airborne or waterborne. Conidia formation control by genetic factors. The nutrients, pH, and light may trigger these pathways. In *Aspergillus fumigatus*, Ras protein interferes with the mitotic reproduction cycle, disturbing conidia and conidiophore formation (Fortwendel et al., 2004). Some proteins are involved in signaling pathways for asexual development and growth through the nutrient pathway (Panepinto et al., 2003). Hydrophobins are important for the dispersal and survival of spores and conidia (Whiteford and Spanu, 2001). The members of ascomycetes exhibit meiosis at some stages of the life cycle and produce resting spores for the next active season. Plant pathogens undergo either homothallism or heterothallism; heterothallism requires opposite mating types and may generate new genotypes. Genes of mating types of ascomycetes have been characterized (Poeggeler, 2001).

SEXUAL REPRODUCTION AND MATING SYSTEMS

The sexual reproduction process drives the genetic recombination in eukaryotes. It includes the fundamental processes of mating-type recognition, zygote formation by the fusion of cells, and gamete generation by meiosis. Sexual reproduction helps fungal organisms in adopting to the new environment. Fungal sexual reproduction has unique characters than other eukaryotic organisms. The nuclear membrane of the fungal cell remains intact (does not pass through two steps—dissolution and reformation of the nuclear membrane—as happens in plants, animals, and protists); some species exhibit gaps in the nuclear membrane. Fungal sexual reproduction generally consists of three main stages. The diploid nucleus divides into two cells (haploid state), with each cell having a single, complete chromosome set. In plasmogamy, two protoplasts of haploid cells fuse; in karyogamy, haploid nuclei fuse to form a diploid nucleus (zygote). The dikaryotic condition is highly prominent in fungal reproduction. After karyogamy, meiosis occurs in the cell to restore the haploid state. Fungal sexual reproduction includes mating systems and sex determinants. In fungi, opposite matting types are designated by *A* and *a* or + and –.

In sexual reproduction, recognition of a mate is the primary step. The fungi (yeast, filamentous) have a well-organized system for detecting mating partners using specific pheromones and receptors. The presence of the receptors of mating-type pheromones on the cell surface can sense and activate the MAP kinase (mitogen-activated proteins) to send pathways of mating types. In fungi, the recognition of mates by pheromone sensing is widespread. Recently, a pheromone-like protein has been identified in *Neurospora crassa* (Kim et al., 2002). Self-fertilization in *Aspergillus nidulans* also requires pheromone receptors. Basidiomycetes exhibit pheromones and receptors encoded by the genes present on the MAT locus. In *Cryptococcus neoformans* (a human pathogen), the pheromone system plays a vital role in same- and opposite-sex mating (Stanton et al., 2010; Lin et al., 2005). Pheromones and receptors are also important for pathogenesis (Daniels et al., 2006). First recognition of mates through pheromone sensing, the compatible cells undergo fusion and formation of dikaryon occur, afterwards karyogamy and meiosis take place.

The fusion occurs between hyphal partners in filamentous fungi through the pheromone receptor system. The male conidium of *N. Crassa* attracts female hyphae by releasing pheromones; thereafter, a fusion between male and female cells occurs (Kim and Borkovich, 2006), and then, all processes of division occur. The yeast cells fuse by making conjugation tubes in response to pheromones (Idnurm et al., 2005; McClelland et al., 2004).

MATING SYSTEMS

HOMOTHALLISM AND HETEROTHALLISM

Fungi have sexual systems, i.e., homothallism and heterothallism. In homothallism, the single thallus has both sex organs, which are self-compatible. In heterothallism, the sex organs produce by two different thalli (two compatible mating types reside on two different thalli).

Homthallism and heterothallism can be observed in different species or the same species of the same genus (Lin and Heitman, 2007; Heitman, 2006). Heterothallic fungi need two partners with compatible MAT (identity of sex in fungi is regulated by the MAT locus) that control different reproduction events (cell fusion, zygote formation, meiosis). Homothallic fungi do not require formatting of different genes. The homothallic fungus *Cochliobolus* spp. has both of the MAT idiomorphs located on the same chromosome, while heterothallic species carry one MAT on each isolate (Yun et al., 1999). That homothallism is derived from heterothallism is supported by the study of the structural arrangement of MAT in the *Cochliobolus* genus.

RESPONSE TO STRESS AND SENSING

The fungal cell can respond to the environmental (external) stresses by signal transduction as mitogen-activated protein kinases (Kultz, 1998). Three Mitogen-activated protein kinases (MAPKs) are involved in forming the core of MAPK. The specific receptors sense the signals, and the module is triggered directly; transcription factors are final effectors. The MAPK pathway regulates vegetative growth in response to stresses. MAPK pathways also control the virulence factor in pathogenic fungi. The production of heat shock protein in the cell is one of the cellular reactions of organisms to environmental stress. The gene expression is regulated by stress level as more heat shock proteins are produced in response to a high-stress condition. The production of heat shock protein in stressful conditions is a universal stress response by all organisms. This was well characterized in yeast fungi in the 1980s (Plesofsky, 2004). Cells also require these proteins for metabolic processes during development and growth in normal conditions. The molecular chaperons are required in protein synthesis and also in folding and assembling of proteins.

Hsp60 and Hsp70 perform specific functions. The forms of chaperon Hsp70 present in the mitochondria, cytoplasm, and endoplasmic reticulum are involved in the synthesis of proteins and transmembrane movement, while Hsp60 is present only in the mitochondria.

QUORUM SENSING

Cell-to-cell communication and expression of gene regulation of microorganisms are achieved by a phenomenon known as "quorum sensing." In the 1960s, it was revealed that microorganisms cannot not live alone, and signaling exists, in which cells communicate with other cells to develop coordination (Nealson et al., 1970). This was first studied in bacterial populations (Miller and Bassler, 2001; Fuqua et al., 1994). Molecules control the production of microbial cells' signaling (autoinducer) (Williams et al., 2012). The microbial cells can sense the inducer (QSM) concentration. The sensing processes will be initiated when QSM reaches a specific concentration (Lu et al., 2014). The population density of microbes coordinates with gene expression through the quorum-sensing mechanism. The roles of processes involved are proposed from the mechanisms of evolution (Miller and Bassler, 2001; Wuster and Babu, 2010). Quorum sensing regulates many important nutrient uptakes, symbiosis, morphological differentiation, pathogenesis, and production of secondary metabolites

(Bandara et al., 2012). The quorum-sensing mechanism has been well described for *Candida albicans* (Hornby et al., 2001). The quorum-sensing mechanism is not restricted to the same species; it exists between different species of bacteria, and even between bacteria and fungi, bacteria and the plant host (Barriuso et al., 2008), and bacteria and the human host (Camara et al., 2002). The phenomenon in which organisms disrupt the quorum-sensing signaling of other organisms is known as "quorum quenching" (Zhu and Kaufmann, 2013). Quorum quenching prevents the accumulation of the quorum-sensing molecules. *Candida albicans* was used as a model to study quorum-sensing mechanisms. *C. Albicans* exhibited the dimorphic behavior in which a fungus can switch from yeast growth to hyphal growth; this transition is controlled by signaling cascades (Lo et al., 1997) and quorum sensing mechanisms. Two molecules, tyrosol and farnesol, have the opposite effect of quorum- sensing molecules (autoinducer) (Hornby et al., 2001; Chen et al., 2004). It has been studied that the morphological transition and inoculum size are governed by the quorum-sensing mechanism in *Ophiostoma flocculosus, Penicillium isariaeforme, Histoplasma capsulatum, Cryptococcus neoformans,* and *Mucor rouxii* (Nickerson et al., 2006).

REFERENCES

Agnihotri, S.A., Mallikarjuna, N.N. and Aminabhavi, T.M. 2004. Recent advances on chitosan-based micro-and nanoparticles in drug delivery. *Journal of Controlled Release* 100(1):5–28.

Alexopoulos, C.J., Mims, C.W. and Blackwell, M. 1996. *Introductory mycology* (No. Ed. 4). John Wiley and Sons, New York.

Araujo-Bazan, L., Penalva, M.A. and Espeso, E.A. 2008. Preferential localization of the endocytic internalization machinery to the hyphal tips underlies polarization of the actin cytoskeleton in Aspergillus nidulans. *Molecular Microbiology* 67:891–905.

Bailey, A., Mueller, E. and Bowyer, P. 2000. Ornithine decarboxylase of Stagonospora (Septoria) nodorum is required for virulence toward wheat. *Journal of Biological Chemistry* 275:14242–14247.

Balhadere, P.V., Foster, A.J. and Talbot, N.J. 1999. Identification of pathogenicity mutants in the rice blast fungus *Magnaporthe grisea* by insertional mutagenesis. *Molecular Plant-Microbe Interactions* 12:129–142.

Balhadere, P.V. and Talbot, N.J. 2001. *PDE1* encodes a P-type ATPase involved in appressorium-mediated plant infection by the rice blast fungus *Magnaporthe grisea. Plant Cell* 13:1987–2004.

Bandara, H.M.H.N., Lam, O.L.T., Jin, L.J., et al. 2012. Microbial chemical signaling: A current perspective. *Critical Reviews in Microbiology* 38:217–249.

Barriuso, J., Solano, B.R., Fray, R.G., et al. (2008) Transgenic tomato plants alter quorum sensing inplant growth-promoting rhizobacteria. *Plant Biotechnology Journal* 6:442–452.

Bartnicki-Garcia, S. 1973. Fundamental aspects of hyphal morphogenesis. In: Ashworth, J.O., and Smith, I.E. (eds) *Microbial differentiation. Symposium of the Society for General Microbiology, vol 23*. Cambridge University Press, Cambridge, p. 245.

Bartnicki-Garcia, S., Hergert, F. and Gierz, G. 1989. Computer simulation of fungal morphogenesis and the mathematical basis for hyphal tip growth. *Protoplasma* 153:46e57.

Bassilana, M., Blyth, J. and Arkowitz, R.A. 2003. Cdc24, the GDP-GTP exchange factor for Cdc42, is required for invasive hyphal growth of *Candida albicans. Eukaryotic Cell* 2:9–18.

Bassilana, M., Hopkins, J. and Arkowitz, R.A. 2005. Regulation of the Cdc42/Cdc24 GTPase module during *Candida albicans* hyphal growth. *Eukaryotic Cell* 4:588–603.

Beadle, G.W. and Tatum, E.L. (1945) Neurospora II. Methods of producing and detecting mutations concerned with nutritional requirements. *American Journal of Botany* 32:678–686.

Belkhadir, Y., Subramaniam, R. and Dangl, J.L. 2004. Plant disease resistance protein signaling: NBS-LRR proteins and their partners. *Current Opinion in Plant Biology* 7:391–399.

Biemelt, S. and Sonnewald, U. 2006. Plant-microbe interactions to probe regulation of plant carbon metabolism. *Journal of Plant Physiology* 163:307–318.

Blackwell, M. 2011. The Fungi: 1, 2, 3 . . . 5.1 million species? *American Journal of Botany* 98(3):426–438.

Bolton, M.D., Thomma, B. and Nelson, B.D. 2006. *Sclerotinia sclerotiorum* (Lib.) de Bary: Biology and molecular traits of a cosmopolitan pathogen. *Molecular Plant Pathology* 7:1–16.

Borges-Walmsley, M.I., Chen, D., Shu, X. and Walmsley, A.R. 2002. The pathobiology of *Paracoccidioides brasiliensis*. *Trends in Microbiology* 10:80–87.

Brodhagen, M. and Keller, N. 2006. Signalling pathways connecting mycotoxins production and sporulation. *MolecularPlant Pathology* 7:285–301.

Calvo, A.M., Wilson, R.A., Bok, J.W. and Keller, N.P. 2002. Relationship between secondary metabolism and fungal development. *Microbiology and Molecular Biology Reviews* 66(3):447–459.

Camara, M., Williams, P. and Hardman, A. 2002. Controlling infection by tuning in and turning down the volume of bacterial small-talk. *Lancet Infectious Diseases* 2:667–676.

Casadevall, A. 2008. Evolution of intracellular pathogens. *Annual Review Microbiology* 62:19e33.

Casadevall, A. and Pirofski, L.A. 2003. The damage-response framework of microbial pathogenesis. *Nature Reviews Microbiology* 1:17–24.

Casadevall, A. and Pirofski, L.A. 2006. Polysaccharide-containing conjugate vaccines for fungal diseases. *Trends in Molecular Medicine* 12:6–9.

Cessna, S., Sears, V., Dickman, M. and Low, P. 2000. Oxalic acid, a Pathogenicity factor for *Sclerotinia sclerotiorum*, suppresses the oxidative burst of the host plant. *Plant Cell* 12:2191–2200.

Chen, C. and Dickman, M.B. 2005. Proline suppresses apoptosis in the fungal pathogen *Colletotrichum trifolii*. *Proceedings of the National Academy of Sciences USA* 102:3459–3464.

Chen, C., Ha, Y.-S., Min, J.-Y., Memmott, S.D. and Dickman, M.B. 2006. Cdc42 is required for proper growth and development in the fungal pathogen *Colletotrichum trifolii*. *Eukaryotic Cell* 5:155–166.

Chen, H., Fujita, M., Feng, Q., et al. 2004. Tyrosol is a quorum sensing molecule in *Candida albicans*. *Proceedings of the National Academy of Sciences USA* 101:5048–5052.

Christensen, M.J., Bennett, R.J., Ansari, H.A., Koga, H., Johnson, R.D., et al. 2008. Epichloe endophytes grow by intercalary hyphal extension in elongating grass leaves. *Fungal Genetics and Biology* 45:84–93.

Chu, J.H., Feudtner, C., Heydon, K., Walsh, T.J. and Zaoutis, T.E. 2006. Hospitalizations for endemic mycoses: A population-based national study. *Clinical Infectious Diseases* 42:822–825.

Clergeot, P.H., Gourgues, M., Cots, J., Laurans, F., Latorse, M.P., Pepin, R., Tharreau, D., Notteghem, J.L. and Lebrun, M.-H. 2001. *PLS1*, a gene encoding a tetraspanin-like protein, is required for penetration of rice leaf by the fungal pathogen *Magnaporthe grisea*. *Proceedings of the National Academy of Sciences USA* 98:6963–6968.

Cole, G.T. and Samson, R.A. 1979. *Patterns of Development in Conidial Fungi*. Pitman, London.

Crampin, H., Finley, K., Gerami-Nejad, M., Court, H., Gale, C., et al. 2005. Candida albicans hyphae have a Spitzenkorper that is distinct from the polarisome found in yeast and pseudohyphae. *Journal of Cell Science* 118:2935e2947.

Daniels, K.J., Srikantha, T., Lockhart, S.R., Pujol, C. and Soll, D.R. 2006. Opaque cells signal white cells to form biofilms in *Candida albicans*. *EMBO Journal* 25:2240–52.

Ding, C.C., Teng, S.H. and Pan, H. 2012. In-situ generation of chitosan/hydroxyapatite composite microspheres for biomedical application. *Materials Letters* 79:72–74.

Doehlemann, G., Berndt, P. and Hahn, M. 2006. Different signalling pathways involving a Gal-pha protein, cAMP and a MAP kinase control germination of *Botrytis cinerea* conidia. *Molecular Microbiology* 59:821–835.

Dunham, D. 2012. Holy bananas: Your favorite fruit may not be vegan anymore. www.blis-stree.com/2012/08/23/food/bananas-may-not-be-vegananymore- 707.

Elorza, V., Murgui, A. and Sentendreu, R. 1985. Dimorphism on *Candida albicans*: Contribution of mannoproteins to the architecture of yeast and mycelial cell walls. *Journal of General Microbiology* 131:2209–2216.

Elsabee, M.Z. and Abdou, E.S. 2013. Chitosan based edible films and coatings: A review. *Materials Science and Engineering: C* 33(4):1819–1841.

Flor, H.H. 1971. Current status of the gene-for-gene concept. *Annual Review of Phytopathology* 9:275–296.

Fortwendel, J.R., Panepinto, J., Seitz, A.E., Askew, D.S. and Rhodes, J.C. 2004. *Aspergillus fumigatus* rasA and rasB regulate the timing and morphology of asexual development. *Fungal Genetics and Biology* 41:129–139.

Fuqua, W.C., Winans, S.C. and Greenberg, E.P. 1994. Quorum sensing in bacteria—the Luxr-Luxi family of cell density-responsive transcriptional regulators. *Journal Bacteriology* 176:269–275.

Gantner, B., Simmons, R. and Underhill, D. 2005. Dectin-1 mediates macrophage recognition of *Candida albicans* yeast but not filaments. *European Molecular Biology Organization* 24:1277–1286.

Garnjobst, L. and Tatum, E.L. 1967. A survey of new morphological mutants in *Neurospora crassa*. *Genetics* 57:579–604.

Girbardt, M. 1957. Der Spitzenkorper von Polystictus versicolor. *Planta* 50:47–59.

Griffin, D.H. 1994. Spore development. In: *Fungal physiology*, 2nd edn. Wiley-Liss, Inc, New York, pp. 338–342.

Guldener, U., Seong, K.Y., Boddu, J. et al. (2006) Development of a Fusarium graminearum Affymetrix GeneChip for profiling fungal gene expression in vitro and in planta. *Fungal Genetics and Biology* 43:316–325.

Hamer, J.E. and Holden, D.W. 1997. Linking approaches in the study of fungal pathogenesis: A commentary. *Fungal Genetics and Biology* 21:11–16.

Harris, S.D. 2010. Hyphal growth and polarity. In: Borkovich, K.A. and Ebbole, D.J. (eds) *Cellular and molecular biology of filamentous fungi*. ASM Press, Washington, DC, pp. 238–259.

Harris, S.D., Read, N.D., Roberson, R.W., Shaw, B., Seiler, S., et al. 2005. Polarisome meets Spitzenkorper: Microscopy, genetics, and genomics converge. *Eukaryotic Cell* 4:225e229.

Hawksworth, D.L. 2012. Global species numbers of fungi: Are tropical studies and molecular approaches contributing to a more robust estimate? *Biodiversity and Conservation* 21(9):2425–2433.

Heitman J. 2006. Sexual reproduction and the evolution of microbial pathogens. *Current Biology* 16:R711–25.

Herrero, E., Sanz, P. and Sentendreu, R. 1987. Cell wall proteins liberated from zymolyase from several ascomycetous and imperfect yeasts. *Journal of General Microbiology* 133:2895–2903.

Hoog, G.S., Guarro, J., Gene, J. and Figueras, M.J. 2000. *Atlas of clinical fungi*, 2nd edn. Centraalbureau voor Schimmelcultures, Netherlands.

Hornby, J.M., Jensen, E.C., Lisec, A.D., et al. 2001. Quorum sensing in the dimorphic fungus *Candida albicans* is mediated by farnesol. *Applied Environmental Microbiology* 67:2982–2992.

Hwang, C.S., Rhie, G.E., Oh, J.H., Huh, W.K., Yim, H.S. and Kang, S.O. 2002. Copper- and zinc-containing superoxide dismutase (Cu/ZnSOD) is required for the protection of *Candida albicans* against oxidative stresses and the expression of its full virulence. *Microbiology* 148:3705–3713.

Icenhour, C.R., Kottom, T.J. and Limper, A.H. 2006. Pneumocystis melanins confer enhanced organism viability. *Eukaryotic Cell* 5:916–923.

Idnurm, A., Bahn, Y.S., Nielsen, K., Lin, X., Fraser, J.A. and Heitman, J. 2005. Deciphering the model pathogenic fungus *Cryptococcus neoformans*. *Nature Reviews Microbiology* 3:753–764.

Jones, L.A. and Sudbery, P.E. 2010. Spitzenkorper, exocyst, and polarisome ncomponents in *Candida albicans* hyphae show different patterns of localization and have distinct dynamic properties. *Eukaryotic Cell* 9:1455e1465.

Kendrick, B. 2003. Analysis of morphogenesis in hyphomycetes: New characters derived from considering some conidiophores and conidia as condensed hyphal systems. *Canadian Journal of Botany* 81:75–100.

Keon, J., Antoniw, J., Rudd, J., Skinner, W., Hargreaves, J. and Hammond- Kosack, K. 2005. Analysis of expressed sequence tags from the wheat leaf blotch pathogen *Mycosphaerella graminicola* (*anamorph Septoria tritici*). *Fungal Genetic and Biology* 42:376–389.

Khale, A. and Deshpande, M.V. 1992. Dimorphism in *Benjaminiella poitrasii*: Cell wall chemistry of parent and two stable yeast mutants. *Antonie van Leeuwenhoek* 62(4):299–307.

Kim, H. and Borkovich, K.A. 2006. Pheromones are essential for male fertility and sufficient to direct chemotropic polarized growth of trichogynes during mating in *Neurospora crassa*. *Eukaryotic Cell* 5:544–54.

Kim, H., Metzenberg, R.L. and Nelson, M.A. 2002. Multiple functions of *mfa-1*, a putative pheromone precursor gene of *Neurospora crassa*. *Eukaryotic Cell* 1:987–999.

Kim, S., Ahn, I., Rho, H. and Lee, Y. 2005. *MHP1*, a *Magnaporthe grisea* hydrophobin gene, is required for fungal development and plant colonization. *Molecular Microbiology* 57:1224–1237.

Kirk, P.M., Cannon, P.F., Minter, D.W. and Stalpers, J.A., eds. 2008. *Dictionary of the fungi*. 10th edn. CAB International, Wallingford, UK.

Klein, B.S. and Tebbets, B. 2007. Dimorphism and virulence in fungi. *Current Opinion in Microbiology* 10:314e319.

Kultz, D. 1998. Phylogenetic and functional classification of mitogen- and stress-activated protein kinases. *Journal of Molecular Evolution* 46:571–588.

Lewis, D.H. 1973. Concepts in fungal nutrition and the origin of biotrophy. *Biological Reviews* 48(2):261–277.

Liao, D.I., Basarab, G., Gatenby, A.A. and Jordan, D.B. 2000. Selection of a potent inhibitor of trihydroxynaphthalene reductase by sorting disease control data. *Bioorganic & Medicinal Chemistry Letters* 10:491–494.

Lin, X. and Heitman, J. 2007. Mechanisms of homothallism in fungi and transitions between heterothallism and homothallism. See Ref. 56, pp. 35–57.

Lin, X., Hull, C.M. and Heitman, J. 2005. Sexual reproduction between partners of the same mating type in *Cryptococcus neoformans*. *Nature* 434:1017–21.

Linden, J.C., Stoner, R.J., Knutson, K.W. and Gardner-Hughes, C.A. 2000. Organic disease control elicitors. *Agro Food Industry Hi-tech* 11(5):32–34.

Lo, H.J., Köhler, J.R., DiDomenico, B., et al. 1997. Nonfilamentous *C. albicans* mutants are avirulent. *Cell* 90:939–949.

Lu, Y., Su, C., Unoje, O., et al. 2014. Quorum sensing controls hyphal initiation in *Candida albicans* through Ubr1-mediated protein degradation. *National Academy of Sciences USA* 111:1975–1980.

Marzluf, G.A. 2004. Regulation of nitrogen metabolism in mycelia fungi. In: Brambl, R. and Marzluf, G.A. (eds) *The mycota III: Biochemistry and molecular biology*, 2nd edn. Springer, Berlin, pp. 357–368.

Maubon, D., Park, S., Tanguy, M., Huerre, M., Schmitt, C., Prevost, M.C., Perlin, D.S., Latge, J.P. and Beauvais, A. 2006. *AGS3*, an alpha(1–3) glucan synthase gene family member of *Aspergillus fumigatus*, modulates mycelium growth in the lung of experimentally infected mice. *Fungal Genetics and Biology* 43:366–375.

McClelland, C.M., Chang, Y.C., Varma, A. and Kwon-Chung, K.J. 2004. Uniqueness of the mating system in *Cryptococcus neoformans*. *Trends in Microbiology* 12:208–212.

Mendgen, K. and Hahn, M. 2002. Plant infection and the establishment of fungal biotrophy. *Trends in Plant Science* 7(8):352–356.

Miller, M.B. and Bassler, B.L. 2001. Quorum sensing in bacteria. *Annual Reviews of Microbiology* 55:165–199.

Mishra, N.C. 1977 Genetics and biochemistry of morphogenesis in Neurospora. *Advanced Genetics* 19:341–405.

Mur, L.A., Carver, T. and Prats, E. 2006. NO way to live; the various roles of nitric oxide in plant-pathogen interactions. *Journal of Experimental Botany* 57:489–505.

Muzzarelli, R.A., Ilari, P., Tarsi, R., Dubini, B. and Xia, W. 1994. Chitosan from Absidia coerulea. *Carbohydrate Polymers* 25(1):45–50.

Nealson, K.H., Platt, T. and Hastings, J.W. 1970. Cellular control of the synthesis and activity of the bacterial luminescent system. *Journal of Bacteriology* 104:313–322.

Nickerson, K.W., Atkin, A.L. and Hornby, J.M. 2006. Quorum sensing in dimorphic fungi: Farnesol and beyond. *Applied Environmental Microbiology* 72:3805–3813.

Oliver, R.P. and Solomon, P.S. 2004. Does the oxidative stress used by plants for defence provide a source of nutrients for pathogenic fungi? *Trends in Plant Science* 9(10):472–473.

Ortoneda, M., Guarro, J., Madrid, M.P., Caracuel, Z., Roncero, M.I.G., Mayayo, E. and Di Pietro, A. 2004. *Fusarium oxysporum* as a multihost model for the genetic dissection of fungal virulence in plants and mammals. *Infectious Immuninity* 72:1760–1766.

Osherov, N. and May, G. 2001. The molecular mechanisms of conidial germination. *Federation of European Microbiological Society Letter* 199:153–160.

Panepinto, J.C., Oliver, B.G., Fortwendel, J.R., Smith, D.L.H., Askew, D.S. and Rhodes, J.C. 2003. Deletion of the *Aspergillus fumigatus* gene encoding the Ras-related protein RhbA reduces virulence in a model of invasive pulmonary aspergillosis. *Infectious Immuninity* 71:2819–2826.

Park, J.-H., Choi, G.J., Jang, K.S., Lim, H.K., Kim, H.T., Cho, K.Y. and Kim, J.-C. 2005. Antifungal activity against plant pathogenic fungi of chaetoviridins isolated from *Chaetomium globosum*. *Federation of European Microbiological Society Letter* 252:309–313.

Pastor, F.I.J., Herrero, E. and Sentendreu, R. 1984. Structure of the *Saccharomyces cerevisiae* cell wall: Mannoproteins released by Zymolyase and their contribution to wall architecture. *Biochimica et Biophysica Acta* 802:292–300.

Paterson, P.J., Seaton, S., Yeghen, T., McHugh, T.D., McLaughlin, J., Hoffbrand, A.V. and Kibbler, C.C. 2005. Molecular confirmation of invasive infection caused by *Chaetomium globosum*. *Journal of Clinical Pathology* 58:334.

Perfect, S.E. and Green, J.R. 2001. Infection structures of biotrophic and hemibiotrophic fungal plant pathogens. *Molecular Plant Pathology* 2(2):101–108.

Pieterse, C.M. and Van Loon, L. 2004. NPR1: The spider in the web of induced resistance signaling pathways. *Current Opinion in Plant Biology* 7:456–464.

Plesofsky, N. 2004. Heat shock proteins and the stress response. In: Brambl, R. and Marzluf, G.A. (eds) *The mycota III: Biochemistry and molecular biology*, 2nd edn. Springer, Berlin, pp. 143–174.

Poeggeler, S. 2001. Mating-type genes for classical strain improvements of ascomycetes. *Applied Microbiology and Biotechnology* 56:589–601.

Read, N.D., Fleibner, A., Roca, M.G. and Glass, N.L. 2010. Hyphal fusion. In: Borkovich, K.A. and Ebbole, D.J. (eds) *Cellular and molecular biology of filamentous fungi*. ASM Press, Washington, DC, pp. 260–273.

Regenstein, J.M., Chaudry, M.M. and Regenstein, C.E. 2003. The kosher and halal food laws. *Comprehensive Reviews in Food Science and Food Safety* 2(3):111–127.

Riquelme, M., Bartnicki-Garcia, S., Gonzalez-Prieto, J.M., Sanchez-Leon, E., Verdin-Ramos, J.A., et al. 2007. Spitzenkorper localization and intracellular traffic of green fluorescent protein-labeled CHS-3 and CHS-6 chitin synthases in living hyphae of Neurospora crassa. *Eukaryotic Cell* 6:1853e1864.

Roozendaal, R. and Carroll, M. 2006. Emerging patterns in complement mediated pathogen recognition. *Cell* 125:29–32.

Scott, A.W. 1976. Biochemical genetics of morphogenesis in *Neurospora*. *Annual Reviews Microbiology* 30:85–104.

Scott, A.W., Mishra, N.C. and Tatum, E.L. 1973. Biochemical genetics of morphogenesis in *Neurospora*. *Brookhaven Symposium Biol*ogy 25:1–18.

Seong, K., Hou, Z., Tracy, M., Kistler, H.C. and Xu, J.R. 2005. Random insertional mutagenesis identifies genes associated with virulence in the wheat scab fungus Fusarium graminearum. *Phytopathology* 95:744–750.

Shea, J.M. and Del Poeta, M. 2006. Lipid signaling in pathogenic fungi. *Current Opinion in Microbiology* 9:352–358.

Shinya, T., Menard, R., Kozone, I., Matsuoka, H., Shibuya, N., Kauffmann, S., Matsuoka, K. and Saito, M. 2006. Novel beta-1,3-,1,6-oligoglucan elicitor from *Alternaria alternata* 102 for defense responses in tobacco. *FEBS Journal* 273:2421–2431.

Solomon, P.S. and Oliver, R.P. 2001. The nitrogen content of the tomato leaf apoplast increases during infection by Cladosporium fulvum. *Planta* 213(2):241–249.

Speth, C., Rambach, G., Lass-Florl, C., Dierich, M.P. and Wurzner, R. 2004. The role of complement in invasive fungal infections. *Mycoses* 47:93–103.

Stanton, B.C., Giles, S.S., Staudt, M.W., Kruzel, E.K. and Hull, C.M. 2010. Allelic exchange of pheromones and their receptors reprograms sexual identity in *Cryptococcus neoformans*. *PLoS Genetics* 6:e1000860.

Steinberg, G. 2007. Hyphal growth: A tale of motors, lipids, and the Spitzenkorper. *Eukaryotic Cell* 6:351e360.

St. Leger, R.J., Screen, S.E. and Shams-Pirzadeh, B. 2000. Lack of host specialization in *Aspergillus flavus*. *Applied Environmental. Microbiology* 66:320–324.

Sweigard, J.A., Carroll, A.M., Farrall, L., Chumley, F.G. and Valent, B. 1998. Magnaporthe grisea pathogenicity genes obtained through insertional mutagenesis. *Molecular Plant-Microbe Interaction* 11:404–412.

Taheri-Talesh, N., Horio, T., Araujo-Bazan, L., Dou, X., Espeso, E.A., et al. 2008. The tip growth apparatus of Aspergillus nidulans. *Molecular Biology of the Cell* 19:1439e1449.

Taylor, J.W. and Ellison, C.E. 2010. Mushrooms: Morphological complexity in the fungi. *Proceedings of the National Academy of Sciences of the United States of America* 107:11655e11656.

Thevelein, J.M. 1984. Regulation of trehalose mobilization in fungi. *Microbiological Reviews* 48(1):42–59.

Thines, E., Weber, R. and Talbot, N.J. 2000. MAP kinase and protein kinase A-dependent mobilization of triacylglycerol and glycogen during appressorium turgor generation by *Magnaporthe grisea*. *Plant Cell* 12(9):1703–1718.

Tsitsigiannis, D.I. and Keller, N.P. 2006. Oxylipins act as determinants of natural product biosynthesis and seed colonization in Aspergillus nidulans. *Molecular Microbiology* 59:882–892.

Upadhyay, S. and Shaw, B. 2008. The role of actin, fimbrin and endocytosis in growth of hyphae in Aspergillus nidulans. *Molecular Microbiology* 68:690–705.

Valentine, B.P. and Bainbridge, B.W. 1978. The relevance of a study of a temperature-sensitive ballooning mutant of *Aspergillus nidulans* defective in mannose metabolism to our understanding of mannose as a wall component and carbon/energy source. *Journal of General Microbiology* 109:155–168.

van Kan, J.A. 2006. Licensed to kill: The lifestyle of a necrotrophic plant pathogen. *Trends in Plant Science* 11:247–253.

Verdin, J., Bartnicki-Garcia, S. and Riquelme, M. 2009. Functional stratification of the Spitzen-korper of Neurospora crassa. *Molecular Microbiology* 74:1044e1053.

Virag, A. and Harris, S.D. 2006. The Spitzenkörper: A molecular perspective. *Mycological Research* 110(1):4–13.

Voegele, R.T. 2006. Uromyces fabae: development, metabolism, and interactions with its host Vicia faba. *Federation of European Microbiological Societies* 259:165–173.

Walters, D. 2003. Resistance to plant pathogens: Possible roles for free polyamines and poly-amine catabolism. *New Phytologist* 159:109–115.

Walters, D.R., Cowley, T. and Weber, H. 2006. Rapid accumulation of trihydroxy oxylipins and resistance to the bean rust pathogen *Uromyces fabae* following wounding in *Vicia faba. Ann. Bot.* 97:779–784.

Webster, J. and Weber, R. 2007. *Introduction to fungi.* Cambridge University Press, New York.

Wei, Y., Shen, W., Dauk, M., Wang, F., Selvaraj, G. and Zou, J. 2004. Targeted gene disruption of glycerol-3-phosphate dehydrogenase in *Colletotrichum gloeosporioides* reveals evi-dence that glycerol is a significant transferred nutrient from host plant to fungal patho-gen. *Journal of Biological Chemistry* 279:429–435.

White, S.A., Farina, P.R. and Fulton, I. 1979 Production and isolation of chitosan from *Mucor rouxii. Appliedl Environmental Microbiology* 38:323–330.

Whiteford, J.R. and Spanu, P.D. 2001. The hydrophobin HCf-1 of *Cladosporium fulvum* is required for efficient water-mediated dispersal of conidia. *Fungal Genetics and Biology* 32:159–168.

Williams, H.E., Steele, J.C., Clements, M.O., et al. 2012. *Gamma*-Heptalactone is an endoge-nously produced quorum-sensing molecule regulating growth and secondary metabolite production by *Aspergillus nidulans. Applied Microbiology and Biotechnology* 96:773–781.

Winnenburg, R., Baldwin, T.K., Urban, M., Rawlings, C., Köhler, J. and Hammond-Kosack, K.E. 2006. PHI-base: A new database for pathogen host interactions. *Nucleic Acids Reserach* 34:D459–D464.

Wu, T., Zivanovic, S., Draughon, F.A. and Sams, C.E. 2004. Chitin and chitosan value added products from mushroom waste. *Journal of Agricultural and Food Chemistry* 52:7905–7910.

Wuster, A. and Babu, M.M. 2010. Transcriptional control of the quorum sensing response in yeast. *Molecular BioSystems* 6:134–141.

Xia, W., Liu, P., Zhang, J. and Chen, J. 2011. Biological activities of chitosan and chitooligo-saccharides. *Food Hydrocolloids* 25(2):170–179.

Yun, S.H., Berbee, M.L., Yoder, O.C. and Turgeon, B.G. 1999. Evolution of the fungal self-fer-tile reproductive lifestyle from self-sterile ancestors. *Proceedings of the National Acad-emy of Sciences USA* 96:5592–5597.

Zhu, J. and Kaufmann, G.F. 2013. Quo vadis quorum quenching? *Current Opinion in Phar-macology* 13:688–698.

3 Fungal Genome and Genomics

Barbaros Çetinel, Zemran Mustafa,
Ayşe Andaç Özketen, Ahmet Çağlar Özketen
and Ömür Baysal

CONTENTS

INTRODUCTION

Fungi are universal microorganisms with a significant effect on agriculture, human beings, and the animal kingdom. Fungi have the second largest group of eukaryotic organisms accounted for—5.1 million species—on the earth (Prakash et al. 2017). Among plant pathogens, the most devastating diseases are caused by fungal pathogens (Fisher et al. 2012). These pathogenic fungi have developed mechanisms to suppress or evade host immunity and grow specialized structures, such as haustoria and intracellular or epicuticular hyphae, for virulence factor deliverance and feeding (Mukhtar et al. 2011, Green et al. 1995). Pathogenic and non-pathogenic fungi have been generally diversified according to standard techniques based on morphological, phenotypic properties and biochemical differences (Panackal 2011). For example, on the basis of feeding and colonizing strategies, plant pathogens are classified as biotrophs, necrotrophs, and hemibiotrophs. Biotrophic fungi have a parasitic life cycle, feeding on living plant tissues without killing them. Necrotrophic fungi kill the plant cells by forming necrotic lesions and feed by extracting nutrients from these necrotic plant tissues.

On the other hand, hemibiotrophic fungi have a facultative life cycle, starting the infection as biotrophs and changing to the necrotrophic lifestyle in subsequent infection stages (Vleeshouwers and Oliver 2014). Moreover, the genomic size, number of chromosomes, repeating content, and pathogenic fungi genes differ considerably. To date (September 2020), 1753 fungal genomes are listed in the MycoCosm database (https://mycocosm.jgi.doe.gov). Genomic sizes of plant pathogenic fungi vary from 19 Mb genomes of the smut fungi *Ustilago maydis* to 240 Mb genomes of *Phytophthora infestans* (Schirawski et al. 2010; Haas et al. 2009). While the powdery mildew pathogen, *Blumeria graminis,* has only around 5,854 protein-coding genes, the yellow rust pathogen, *Puccinia striiformis,* was found to have around 22,815 genes. Repetitive DNA can also constitute up to 74% of the genome in the example of *Phytophthora infestans* (Raffaele and Kamoun, 2012).

It has recently been demonstrated that fungi classification based on their morphological, phenotypic properties and biochemical differences is insufficient and results in outputs mistakes (Rizzato et al. 2015). Molecular biology is a tool for fungal phylogeny and precisely identifies the fungi belonging to a different genus, but it needs vast amounts of sequence data provided by molecular studies. This creates a new challenge, but large volumes of data can easily be stored with computational technology and bioinformatics tools annotated and accessed by the end-user. Thus, databases for fungal studies require detailed knowledge and scrutiny screening for choosing the right tool for each user. The fungal database comprises sequence information related to ribosomal DNA genes and genes encoding proteins as well as other associated metadata. To find the exact database for any study related to fungi and to precisely identify the bioinformatic tools, the following steps are necessary. These steps are also required for further aims, including storing, annotating, and retrieving data.

DNA BARCODE, THE IMPORTANCE OF HOUSE-KEEPING GENES FOR IDENTIFICATION OF FUNGI

DNA barcoding is an effective way for the rapid identification of fungal specimens and refers to biodiversity. It bases the technique on internal transcribed spacer (ITS) barcodes to address a variety of microorganisms. To identify the fungi at species level, using molecular techniques, short DNA sequences, called "DNA barcodes," are used. These barcodes constitute short and conserved regions of around 800 bp (Table 3.1). DNA barcode provides well-identified reference collections of each species' DNA sequences in the database (Frézal and Leblois 2008). However, the absence of a universal gene's confidential cases to accurately describe and identify fungal species remains a key constraint for applications (Wu et al. 2019). As the primary barcode for fungi, the ITS and rDNA sequences (Stielow et al. 2015) are mainly used for molecular identification and phylogenetic studies. These are obtained from public databases known as the International Nucleotide Sequence Database Collaboration (www.insdc.org/). Quality-controlled databases, such as National Center for Biotechnology Information (NCBI), Reference Sequence

TABLE 3.1.
Databases and links (Aja et al. 2017)[a]

Name of the Database	URL	Region Utilized
Barcode of Life Database, BOLD	http://www.boldsystems.org/index.php/IDS_OpenIdEngine	ITS
CBS-KNAW	http://www.cbs.knaw.nl/Collections/BioloMICSSequences.aspx	ITS
FUSARIUM-ID	http://isolate.fusariumdb.org	ITS, *tef1*, *RPB1*, *RPB2*, *tub2*
Fungal Barcoding	http://www.fungalbarcoding.org	ITS
Fungal MLST database Q-Bank	http://www.q-bank.eu/Fungi/	partial actin, *tub2*, *RPB1*. *RPB2*, *tef1* among others
ISHAM, The International Society for Human and Animal Mycology	http://its.mycologylab.org	ITS
Naïve Bayesian Classifier	http://rdp.cme.msu.edu/classifier/classifier.jsp	28S, ITS
RefSeq Target Loci (RTL)	http://www.ncbi.nlm.nih.gov/refseq/targetedloci/	ITS, 18S, 28S
International Subcommision on Hypocrea and Trichoderma (ISHT) TrichoKey and TrichoBLAST (Trichoderma)	http://www.isth.info/tools/blast/	ITS and *tef1*, *RPB2*
UNITE, User-friendly Nordic ITS Ectomycorrhiza Database	https://unite.ut.ee/	ITS

[a]For an exhaustive list, see Robert et al. (2013, 2015).

(RefSeq), The Barcode of Life Data (BOLD) System, User-friendly Nordic ITS Ectomycorrhiza (UNITE), and the International Society for Human and Animal Mycology Barcoding Internal Transcribed Spacer (ISHAM-ITS) (Irinyi et al. 2015). Artificial intelligence network–based open-source algorithm "Mycofier," consisting of a large pool of ITS1 sequence data, can also classify most fungi genus with an accuracy level that is comparable to the basic local alignment search tool (BLAST) algorithm (Delgado-Serrano et al. 2016). The exact identification of species based on sequences of housekeeping genes is needed, considering the β-tubulin (βTUB) gene for *Aspergillus* spp., the translation elongation factor 1α (TEF1α) for *Fusarium* spp., and the intergenic spacer 1 (IGS1), which are also valid for different genus (Short et al. 2013). For instance, the Aspergillus Genome Database (AspGD) is a genome-based database that collects biologically important information of genomic data, proteins, subcellular metabolites, and functions of different Aspergillus species (Cerqueira et al. 2014). However, researchers have no consensus on these barcodes for studies related to many taxa; barcodes are all still genus-dependent sequences. Various protein-coding genes from fungi for finding an appropriate candidate for fungal DNA barcode (Frézal and Leblois 2008) and TEF1α gene sequence showed correspondence with phylogenetic mappings (Robert et al. 2015).

Various strains of the fungi/ bacteria species can be classified with the sequence analysis of housekeeping genes. This comparison of nucleotide polymorphisms on sequence differentiation within selected specific regions of up to seven genes is called multi-locus sequence typing (MLST).

A new database called "PhytoPath" (www.phytopathdb.org) is a genomic and phenotypic data resource for plant pathogen species. The resource contains fungi and oomycetes plant pathogens with sequenced and annotated genomes. The PhytoPath database consists of 135 genomic sequences of 87 plant pathogen species, and 1364 genes play a role in pathogenicity. In recent years, the ten most important fungal plant pathogens added to this database according to global scientific importance caused a remarkable expansion in PhytoPath database genomic sequences (Pedro et al. 2016). Other pathogens concern global food and feed issues such as *Puccinia graminis* f. sp. *tritici* and the oomycete *Phytophthora infestans* (agent of potato late blight disease). Different species of Fusarium (agent of root rot/crown root rot disease) have also been introduced.

ONLINE DATABASES FOR TAXONOMY AND IDENTIFICATION

Many online databases, ranging from morphological, biochemical to gene and genome sequence–based databases, are now available (*https://nt.ars-grin.gov/fungaldatabases/*). On the basis of diagnostic criteria, microorganisms can be broadly grouped depending on (i) biochemical, morphological, and taxonomical properties; (ii) specific gene-based differences; (iii) locus-typing diversity; and (iv) genome-based classifications. Some of the databases provide images and describe the morphological characteristics of fungal species across the globe. Index Fungorum is an open-access database for fungal nomenclatures.

The most commonly used public database (Table 3.2), GenBank of the NCBI, contains sequences, annotations, and species definitions stemming from erroneous or mismatched sequences, which are accounted for 10% flawed ITS information (Benson et al. 2014). The RTL database, established by the NCBI (Table 3.2), contains fully annotated sequences from the type and verified data to solve this issue. It also provides a versatile interface, including multiple-sequence markers and whole-genome searches (Celio et al. 2006). Assembling the Fungal Tree of Life is another online resource for fungi with data matrices and phylogenetic mapping, combined with character illustrations and molecular data (Kauff et al. 2007). BOLD represents the first barcode database system incorporated within the ISHAM-ITS database (Irinyi et al. 2015).

Furthermore, the sequences of the most extensive living collection of fungal strains with more than 80,000 strains belonging to 15,000 different species are hosted by the The Centraalbureau voor Schimmelcultures Institute of the Royal Netherlands Academy of Arts and Sciences (CBS-KNAW) database. All available ITS and large subunit (LSU) loci sequences of the strains, including pathogenic species, render the possibility to perform pairwise DNA sequence alignments, besides physiological and morphological properties supported with molecular characterization for several different fungal groups (Fusarium, Penicillium, Phaeoacremonium, and yeasts). DNA alignments performed using several reference databases are combined into a single matching list using pairwise alignment to find the best results for MycoBank, ISHAM-ITS, International

TABLE 3.2.
Database links according to collection institutes Source: (Schoch et al. 2014)

Acronym	Collection Institute	Database link
ACBR	Austrian Center of Biological Resources and Applied Mycology	www.acbr-database.at/BioloMICS.aspx
ATCC	American Type Culture Collection	www.atcc.org/Products/Cells_and_Microorganisms/Fungi_and_Yeast.aspx
BCRC	Bioresource Collection and Research Center	https://catalog.bcrc.firdi.org.tw/BSAS_cart/controller?event=WELCOME
BPI	US National Fungus Collections, Systematic Botany and Mycology Laboratory	http://nt.ars-grin.gov/fungaldatabases/specimens/Specimens.cfm
CBS	Centraalbureau voor Schimmelcultures, Fungal and Yeast Collection	www.cbs.knaw.nl/Collections/Biolomics.aspx?Table=CBS%20 strain%20database
CFMR	Center for Forest Mycology Research	www.fpl.fs.fed.us/search/mycology_request.php
DSM	Deutsche Sammlung von Mikroorganismen und Zellkulturen GmbH	www.dsmz.de/catalogues/catalogue-microorganisms.html
FRR	Food Science Australia, Ryde	www.foodscience.csiro.au/fcc/search.htm
ICMP	International Collection of Microorganisms from Plants	http://scd.landcareresearch.co.nz/Search/Search/ICMP
JCM	Japan Collection of Microorganisms	www.jcm.riken.jp/JCM/catalogue.shtml
MA	Real Jardín Botánico de Madrid Herbarium	www.rjb.csic.es/jardinbotanico/jardin/index.php?Cab=109&len=es
MAFF	MAFF Genebank, Ministry of Agriculture, Forestry and Fisheries	www.gene.affrc.go.jp/databases-micro_search_en.php
MICH	University of Michigan	http://quod.lib.umich.edu/h/herb4ic?page=search
MTCC	Microbial Type Culture Collection and Gene Bank	http://mtcc.imtech.res.in/catalogue.php
MUCL	Mycotheque de l'Universite Catholique de Louvain	http://bccm.belspo.be/db/mucl_search_form.php
NBRC	NITE Biological Resource Center	www.nbrc.nite.go.jp/NBRC2/NBRCDispSearchServlet?lang=en
NRRL	Agricultural Research Service Culture Collection	http://nrrl.ncaur.usda.gov/cgi-bin/usda/index.html
PDD	New Zealand Fungal and Plant Disease Herbarium	http://nzfungi2.landcareresearch.co.nz
PYCC	Portuguese Yeast Culture Collection	http://pycc.bio-aware.com/BioloMICS.aspx?Table=PYCC%20strains
SAG	Sammlung von Algenkulturen at Universitat Gottingen	http://sagdb.uni-goettingen.de/
UAMH	University of Alberta Microfungus Collection and Herbarium	https://secure.devonian.ualberta.ca/uamh/searchcatalogue.php

Society for Human and Animal Mycology Barcoding Database Multilocus sequence typing, Institut Pasteur-FungiBank (IP-FungiBank); all use the same algorithm. The EzFungi database contains selected and verified ITS sequences for identifying fungal pathogens hosted by Seoul National University and ChunLab, Inc.

For pathogenic fungi, the French National Reference Center for Invasive Mycoses and Antifungals hosts the IP-FungiBank (http://fungibank.pasteur.fr/), which allows pairwise gene sequence alignments for important yeasts and molds. The sequence database systematically identified the strains on the basis of morphology using the MALDI-TOF MS and DNA sequences. The IP-FungiBank provides updated DNA sequence information using ITS sequences for species-specific identification, and it redesigns the nomenclature of Fusarium and Aspergillus. TEF1 gene, TUB, and IGS1, respectively, are housekeeping genes, and their sequences are useful for molecular identification. The taxonomy of the genus Fusarium is genetically very complex (Gieser et al. 2004), and providing reliable identification by using databases for the Fusarium community is vital due to its pathogenic plant species. The Fusarium ID Database, Pennsylvania State University, and the Fusarium MLST database are very strong web-based information banks for Fusarium species–related research (Park et al. 2011).

The Fusarium MLST database helps single-sequence and multi-sequence alignments for unknown sequence queries. This database linked with the GenBank (www.ncbi.nlm.nih.gov/genbank/) (Dennis et al. 2011) and the CBSKNAW sequence databases is among the most extensively used databases for Fusarium. The ISHAM-ITS database for human and animal pathogenic fungi was constructed in 2011 under the ISHAM international working group studying the "DNA barcoding of human and animal pathogenic fungi." Most species can be identified by ITS sequences of Aspergillus, Fusarium, Penicillium, and pathogenic yeasts. These fungal databases in the genomics era are priceless tools. These automated systems give rise to taxonomical novelties in online databases (Robert et al. 2013). Bioinformatics and metagenomics rely on data mining to understand macro/microorganisms (Gustavo et al. 2014). The concept of MLST purposed for epidemiological studies is changing from using only a couple of loci to whole-genome analysis and comparison with the massive database (Vu et al. 2014).

FUNGAL PATHOGEN GENOMES IN THE LIGHT OF CO-EVOLUTION WITH HOST PLANTS

Both biotic and abiotic factors affect the evolution of plants and their pathogens. However, the dominant driving force of the evolution in this pathosystem comes from the selection pressure of these organisms exerting on each other (Thrall et al. 2012). Plants and pathogens are in a constant fight for resistance and pathogenicity against each other, respectively. The struggle between them affects both sides' biology and drives a co-evolution in shared habitats. Furthermore, this conflict influences allele frequencies, genomic arrangement, and diversification in these populations (Ma and Guttman 2008; McDonald and Linde 2002b).

With the intervention of humans and artificial selection, selection pressure has been skewed toward agriculturally desired traits, causing unintentional dilution of natural pathogen resistance genes. Furthermore, introducing different plants to new environments has caused diseases to reemerge due to a lack of resistance in these unadapted cultivars (Möller and Stukenbrock 2017). In addition, the reduction of genetic diversity (as decrease in genetic diversity causes pathogen to overcome the resistance barrier developed by the species) and single-culture practices permit widespread pathogen outbreaks, causing significant yield losses (McDonald and Mundt 2016).

Agricultural environments favor the emergence, dispersal, and rapid evolution of plant pathogens due to reduced population diversity and growing monoculture plants over and over again. The lack of diversity, combined with densely grown plants in these ecosystems, induces new virulent pathogen strains. The dense planting practices create environments where disease transmission becomes easier, with a chance of multiple infections. The physical proximity of heterogeneous pathogens on the same plant allows occasional horizontal gene transfer between different pathogens and the transfer of virulence factors. Furthermore, plants' clonal nature favors the successful pathogens' unrestricted proliferation (Alizon et al. et al. 2013; Friesen et al. 2006; McDonald and Mundt 2016; Read 1994).

There are two main scenarios of host–pathogen co-evolution: (i) trench warfare and (ii) arms race in the gene-for-gene model (Flor 1971). The trench warfare scenario occurs mostly in heterogeneous natural ecosystems, and it exerts positive selection pressure on maintaining multiple alleles in host and pathogen populations. Alleles at these co-evolving loci remain at balanced, intermediate frequencies around a stable equilibrium. On the other hand, the arms race scenario is more prominent in homogenous agricultural environments, causing positive directional selection to fix beneficial alleles in host and pathogen. Alleles that are not fixated are lost from the local gene pool, lowering the overall variation. The lost allele can later be reintroduced from other geographically distant populations, or occasionally a new beneficial allele can emerge naturally by mutations and become fixated, replacing the old allele (Brown 2011; Tellier et al. 2014).

EVOLUTION MECHANISMS OF PATHOGENIC FUNGI

Mutations are permanent changes of genomic DNA sequence that can occur naturally and lead to different consequences, depending on the genome's location of occurrence.

The newly emerged mutation can decrease, be neutral, or increase the fitness of the organism. The fate of the mutation is decided by two important natural forces: natural selection, which tends to eliminate deleterious mutations and keep the beneficial ones, and genetic drift, where the fate of the mutation is decided by chance instead of its effect on the organism's survival odds. The relative effects of genetic drift over natural selection depend on sufficient population (N_e) size. Adequate population size is generally smaller than the actual population size and is determined on the basis of how much of the parental alleles are transmitted to the next generation. The smaller the N_e size, the more the genetic drift shapes the population. Because the selection in genetic drift is a matter of chance, it is possible to fix deleterious mutations and loss of beneficial mutations in contrast to the effects of natural selection (Lynch 2007).

Many pathogenic fungi have a high rate of mutations and can adapt quickly to new environments. The low genetic variance in the host due to monoculture agricultural practices leads to the pathogen being spread over large areas once the resistance is overcome (Möller and Stukenbrock 2017; Frantzeskakis et al. 2019).

Genome plasticity helps fungal pathogens adapt to new environments, especially during the co-evolutionary struggle between host and pathogen. Many filamentous fungal pathogens harbor "two-speed" genomes, which have two different genomic arrangements. The housekeeping core genes are repeat poor; genes reach regions with lower evolution rates, while virulence factors are positioned in repeat reach genomic regions with sparsely coding regions and high mutation rates (Raffaele and Kamoun 2012).

Transposable elements or transposons (TEs) are essential features in the pathogenic fungi genome that provide genome plasticity. TEs can change their position within the genome, causing frequent mutations due to double-strand DNA breakage during excision and being a hotspot of DNA recombination. These TE-rich genomic regions show accelerated evolution rates. Effectors and different virulence factors tend to be located in these regions' vicinity as the rearrangements and mutations are beneficial in creating new peptides, where every so often successful proteins arise, escaping host recognition (Faino et al. 2016; Seidl and Thomma 2014).

Chromosomal rearrangements are large-scale genomic modifications with mostly deleterious effects. However, they also play an active role in the evolution and speciation of organisms. These rearrangements include duplications, deletions, inversions, and translocations. The effects of such large-scale changes in the chromosome arrangements can change crossing-over rates that cause infertility. In addition, change in the length of chromosomes can produce chromosome-length polymorphism within the population (Plissonneau et al. 2016; Houben et al. 2014; Seidl and Thomma 2014; Stukenbrock 2013).

Accessory chromosomes are observed in some fungal species with characteristics of being dispensable and showing non-Mendelian inheritance. These chromosomes do not contain essential genes, and their presence may vary in a population. Accessory chromosomes generally contain more repetitive DNA and a few virulence coding genes. They have higher mutation rates, as the selection pressure is lower than in core chromosomes, and they provide an excellent diversification mechanism for the pathogen. In contrast, the core chromosomes of these fungi contain housekeeping genes necessary for proper metabolic functions, with lower repetitive DNA and higher gene density, showing classical Mendelian inheritance (Goodwin et al. 2011; Ma et al. 2010; Haas et al. 2009).

Horizontal gene transfer refers to gene transfer between two different species. It can occur through hybridization, hyphal transfer, or uptake of vectors such as plasmids, viruses, or bacteria. In some cases, the whole chromosome can be transferred across species, and this is termed "horizontal chromosome transfer" (Seidl and Thomma 2014; Ma et al. 2010; Coleman et al. 2009). The introduction of new DNA material can increase the overall fitness of the fungus and enable colonization of new hosts and adaptation to new niches.

Hybridization is the combination of two, generally distinct, genomes in one cell that can happen by sexual mating or asexual fusion of two distinct fungal species. The hybrids generated by hybridization generally have genomic incompatibility and are inferior to either parent but constitute introducing new genes across species. The adverse effects of hybridization are typically eliminated with introgressive hybrid-ization, where hybrids backcross with one of the parents several times, diluting the negative effects of hybridization while retaining some of the genes new to the species (Stukenbrock 2016; Baack and Rieseberg 2007).

Sexual recombination plays an integral part in fungal evolution and increases the adaptation rate in both host and pathogen (McDonald and Linde 2002a). It also pro-vides a means to retain the genomic variation and introduce new mutations during the crossing-over stage (Charlesworth et al. 1993; Duret and Galtier 2009). During sexual reproduction, recombined parental genomes bring together alleles otherwise residing in different organisms, enabling novel combinations with a higher chance of survival.

Understanding the evolution mechanisms of phytopathogenic fungi in natural and agriculture can help avoid the practices that accelerate the evolution rate and facili-tate new contagious pathogens. Multicultural planting environments, crop rotations, and strict quarantine monitoring are necessary actions to slow down the evolution and possible emergence of new devastating pathogens, thus enabling more sustain-able agricultural systems (McDonald and Mundt 2016).

FUNGAL EFFECTOROMES

The plant–pathogen interaction relies on the effectors and their contribution to vir-ulence or resistance if they are recognized by the innate host immunity (Jones and Dangl 2006). The effectors are generally defined as proteinaceous factors released from the pathogen to the host cells' apoplastic or cytosolic environment to jam or manipulate defense signals, assist pathogen fitness, and so on (De Jonge et al. 2011). Plant pathogenic fungi use specialized structures such as haustoria, specialized inva-sive hyphae, or biotrophic interfacial complex to deliver these effectors (Koeck et al. 2011). The delivery mechanism of bacterial pathogens is superior to the delivery of fungal effectors, for instance, the type 3 secretion system (T3SS) of bacterial effec-tors (Dodds and Rathjen 2010; De Jonge et al. 2011). There are a significant number of reports suggesting promising pathogen-independent translocation to the host cyto-plasm (Rafiqi et al. 2010; Kale et al. 2010; Andac et al. 2020). The uptake mechanism is still controversial, whether it is pathogen dependent or not (Petre and Kamoun 2014). Specific oomycete effectors bear the RxLR motif in their N-terminus region right after the signal peptide sequence (secretion signal), such as Avr3a and Avr1b of *P. infestans* and *P. sojae*, respectively, is demonstrated to localize the host cell (Dou et al. 2008). Other motifs are the RGD sequence of ToxA effector of wheat fungal pathogens *Pyrenophora tritici repentis* and *Stagonospora nodorum* and the LFLAK sequence in the N terminus region of crinkler (CRN) effectors discovered in *P. infes-tans* (Manning et al. 2008; Haas et al. 2009; Torto et al. 2003).

The discovery of conserved motifs among fungal effectorome is an intriguing research subject that will significantly reduce the efforts to pinpoint effector candi-dates and elucidate their functions. For that purpose, the [F/Y/W]xC motif is reported to be conserved among the expressed sequence tags predicted to possess N-terminal signal peptide sequences of powdery mildew and rust disease (Godfrey et al. 2010).

The YxSL[RK] motif is identified dominantly in the *Pythium ultimum* genome and transcriptome sequences rather than the RxLR or CRN effectors (Levesque et al. 2010). The [LI]xAR motif is generated in the genome sequence of *Magnaporthe oryzae* of rice blast disease (Yoshida et al. 2009). The CHxC motif is 80 amino acid long, and it is located in N-terminal regions of the predicted secretome repertoire of the biotrophic oomycete *Albugo laibachii*. The N-terminus of candidate CHxC9 effector was further demonstrated to translocate into the host cell as a fusion with the C-terminus of Avr3a, using the pEDV6 vector system of Pst (*Pseudomonas syringae* pv. Tomato) DC3000 (Kemen et al. 2011). [R/K]VY[L/I]R was a conserved motif in the powdery mildew pathogen, *Blumeria graminis* f.sp. *hordei* (Rideout et al. 2006).

In Silico Strategies to Identify Candidate Effectors

Current plant-fungal disease strategies exploit advances in next-generation sequencing (NGS) and bioinformatics to catalog candidate effectorome or secretome sets from large fungal genomes. As the cost-effectivity of NGS increases, more and more fungal genomeand transcriptome sets are becoming available. Approximately, 1000 sequence information of the fungal species present in plant's Bio project database is accessible on the NCBI web server. Various public omics databases are also useful for obtaining fungal genomes to compare and analyze sequence sets (Table 3.3). These large datasets are scrutinized to identify promising secreted proteins (secretomes) or effectors. The presence of an N-terminus signal peptide, absence of transmembrane domains, shorter amino acid length (e.g., <300 a.a), presence of previously studied motifs (e.g., RxLR), and abundance of cysteine residues are among the selection criteria and applied by several research groups (Yin et al. 2009; Duplessis et al. 2011; Hacquard et al. 2012; Saunders et al. 2012). The *in silico* prediction and characterization of the candidate secretome or effectorome is preferred to assist in the promising candidate effector search. These strategies employ new approaches like deep machine learning rather than setting a particular pipeline to filter proteins befitting the parameters. Sperschneider et al. (2018a) developed the online accessible EffectorP program that enables the advantage of machine learning algorithms; later, the authors developed EffectorP 2.0 (Sperschneider et al. 2016, 2018a). Other subcellular localization programs, such as Localizer and ApoplastP, are advantageous to characterize the filtered subset of candidate effectors (Sperschneider et al. 2017, 2018b). The databases for proteases, carbohydrate-active enzymes (CAZymes), lipases, and peroxidases are valuable to annotate and characterize the pool of candidate effectors (Kim et al. 2016; Ozketen et al. 2020). These enzymatic functions could undoubtedly aid in suppressing or manipulating host defense for an effector to contribute virulence. Table 3.4 lists individual databases constructed for analysis of such function prediction. Of course, predicted function should be validated using *in vivo* assays in planta to prove the effectors' significance. Genome/transcriptome-wide association studies (GWAS/TWAS), quantitative trait locus (QTL) mapping, and correlation analysis are yet other strategies that have proved to be convenient to study effectors from fungal genomes (Zhong et al. 2017; Xia et al. 2017; Sanchez-Vallet et al. 2018). AvrStb6 is a small, cysteine-rich, apoplastic effector of *Zymoseptoria tritici* discovered as an avirulence gene using GWAS/QTL combinations (Zhong et al. 2017). In summary,

TABLE 3.3

A variety of online databases serve as a resource for fungal genomes.

Database	Description	Website	Reference
NCBI	Curated non-redundant sequence of genomes, transcriptomes, and proteomes	www.ncbi.nlm.nih.gov/genome/	Pruit et al. 2005
EnsemblFungi	Assemblies and genomes, transcriptomes and proteomes	https://fungi.ensembl.org/	Howe et al. 2020
Fungidb	Genomic database resources (genome, transcriptome, proteome) for data mining and functional genomics	https://fungidb.org/	Basenko et al. 2018
JGI	Joint Genome Institute genome databases along with analytical tools	https://genome.jgi.doe.gov/portal/	Nordberg et al. 2014
FGI	Fungal genome initiative of Broad Institute providing databases for fungal genomes	www.broadinstitute.org/fungal-genome-initiative/	Cuomo and Birren 2010, Haas et al. 2011

TABLE 3.4

Examples of certain secretome/effectorome prediction, annotation, or characterization programs

Webtool	Function/Description	Resource Link	References
SignalP 5.0/4.1	Prediction of N-terminal secretion signal	www.cbs.dtu.dk/services/SignalP/ www.cbs.dtu.dk/services/SignalP-4.1/	Armenteros et al. 2019 Petersen et al. 2011
SecretomeP 2.0	Prediction of non-classical protein secretion using ab initio modeling	www.cbs.dtu.dk/services/SecretomeP	Bendtsen er al. 2004
DeepLoc 1.0	Subcellular localization prediction for eukaryotes	www.cbs.dtu.dk/services/DeepLoc-1.0/	Armenteros et al. 2017
TargetP 2.0	Subcellular localization prediction for eukaryotes	www.cbs.dtu.dk/services/TargetP/	Armenteros et al. 2019
Secretool	Prediction and characterization of fungal- secreted proteins	genomics.cicbiogune.es/SECRETOOL/	Cortazar et al. 2014
WolfPSORT	Prediction of subcellular localization	www.genscript.com/wolf-psort.html https://wolfpsort.hgc.jp/	Horton et al. 2007
TMHMM	Prediction of transmembrane helices	www.cbs.dtu.dk/services/TMHMM/	Krogh et al. 2001
EffectorP 1.0/2.0	Fungal effector prediction via machine learning method	http://effectorp.csiro.au/	Sperschneider et al. 2018a
ApoplastP	Prediction of proteins localized into apoplast region via machine learning method	http://apoplastp.csiro.au/	Sperschneider et al. 2018b

(Continued)

TABLE 3.4 (CONTINUED)

Webtool	Function/Description	Resource Link	References
Localizer	Subcellular localization prediction for eukaryotes via machine learning	localizer.csiro.au/	Sperschneider et al. 2017
PHI-base	Database for Pathogen–Host Interaction	www.phi-base.org/	Urban et al. 2015 Urban et al. 2017
MEME Suite	Motif discovery, enrichment, and analysis tools	meme-suite.org/	Bailey and Elkan 1995 Bailey et al. 2009
FSD	Fungal secretome database	fsd.snu.ac.kr/	Choi et al. 2010
fPoxDB	Fungal peroxidase database	peroxidase.riceblast.snu.ac.kr/	Choi et al. 2014
FunSecKB	Database of Fungal Secretome KnowledgeBase	bioinformatics.ysu.edu/secretomes/fungi.php	Lum and Min 2011
MEROPS	Database for peptidases, inhibitors, and substrates of peptidases	www.ebi.ac.uk/merops/	Rawlings et al. 2014 Rawlings et al. 2018
dbCAN	Database of carbohydrate-active enzymes (CAZymes)	cys.bios.niu.edu/dbCAN/	Yin et al. 2012
CAZy	Database of CAZymes that degrade, modify, or create glycosidic bonds	www.cazy.org/	Lombard et al. 2014
LED	Lipase Engineering Database for lipases and lipase-associated proteins	www.led.uni-stuttgart.de/	Fischer and Pleiss 2003

data mining of Big Data generated by NGS on fungal genomes/transcriptomes is a game-changing methodology to narrow down the ever-expanding pool, hunt the most promising candidates down, and give pace to effector detection to understand host–pathogen interaction.

FUNCTIONAL ANALYSIS IN FUNGAL PLANT PATHOLOGY

Effector functionality and effector host targets have been investigated to understand the relationship between the host plant and the pathogen. Hence, the molecular mechanism of effector virulence and the recognition process by the plant immune system cell can be comprehend (Gouveia et al. 2017). To date, only a few pathogen species have been identified. Nevertheless, there are many ongoing studies about fungal genomes that will broaden our knowledge about plant–pathogen interaction.

The accessibility of the pathogen genomes enables the characterization of the putative effectors of fungus. Bioinformatically identified effectors could be functionally characterized in many ways, such as through molecular or cellular approaches. In this part of the review, we discuss the functional analysis methods that enlarge our knowledge about effector biology.

In Planta Expression System

Once a candidate effector gene is expressed in planta, numerous assays can be performed to discover its function and virulence. The most widely used and well-established method is Agrobacterium-mediated gene expression under the control of the cauliflower mosaic virus 35S promoter in the model plant *N. benthamiana* (Bozkurt et al. 2014). Despite the success of this method in dicots, this expression method cannot work efficiently in some plant species, such as monocots (e.g., cereals). The *Pseudomonas* species–mediated T3SS is used to overcome this limitation (Upadhyaya et al. 2014). It has been shown that many oomycete effectors from *P. infestans* and *Hyaloperonospora arabidopsis* were characterized by T3SS (Whisson et al. 2007; Fabro et al. 2011). The T3SS has also been successfully applied to characterize effectors from stem rust and bean rust; there are still problems in fungal effector screening and application in cereals, though (Upadhyaya et al. 2014; Qi et al. 2019; Saur et al. 2019). This difficulty may arise from unfolding and refolding the effectors before the translocation process, particularly in cysteine-rich effectors that form cysteine bridges (Kanja and Hammond-Kosack 2020).

To overcome the problems in Agrobacterium- and Pseudomonas-mediated gene transfer system in several cereal species, the virus-mediated heterologous protein expression has been established as a viral overexpression (VOX) (Lee et al. 2012). The barley stripe mosaic virus (BSMV) was developed for use in the overexpression of effectors, and it was first used to determine the function of ToxA (Manning et al. 2010). However, the BSVM–VOX expression system has limitations, such as it can be used for the expression of proteins that have at most 150 amino acids in size, and this system is also inappropriate for screening many effectors due to having a tripartite RNA genome. All three subgenomes should be successfully combined to express the protein (Bouton et al. 2018). Then, the foxtail mosaic virus (FoMV) has been improved for use in VOX systems in cereals, and it overcomes the problems faced in the BSMV system (Bouton et al. 2018). It has a monopartite RNA genome, and it can be expressed in proteins up to 600 amino acids in length. Despite the effectiveness of the FoMV–VOX system, many problems arise from overexpression of the effector proteins. First, the expressed protein levels may be much higher from the expression of the fungus itself. This may lead to hypomorphic or neophormic phenotypes. In addition, overexpression of the effector may affect its localization, and its proper localization cannot be determined (Dalio et al. 2018).

After expressing the plant's effector, regardless of the heterologous expression systems indicated, the first assay that can be performed to elucidate the effector's effect on plant cells is monitoring the hypersensitive response (HR) () activity (Whisson et al. 2007). This HR activity can be produced by possible Avr–R gene interaction or the toxin-induced by the effector (Qutob et al. 2002). This method should be preferred to be performed in the natural host of the pathogen because HR formation may come from the nonhost resistance induced by the effector instead of the potential Avr–R gene interaction (Kettles et al. 2017). The expression of the effector genes in their natural host will eliminate this outcome and give the most accurate information about the characterization.

In addition, by using another HR-inducing elicitor, the suppression activity of the effector can be tested. In this method, a very well-known HR-induced *P. infestans,*

INF1, is used to assay whether the candidate effector can suppress this HR formation (Kamoun et al. 1998). Thus, it is an easy and effective way to assess the function of the effector.

Subcellular localization is the second method to determine the effector's function after expression by the Agrobacterium-mediated planta method. It effectively reveals the effector's function in host cell compartments to modulate virulence (Bozkurt et al. 2014; Whisson et al. 2007). The heterologous expression of the fluorescently tagged effectors can be visualized by confocal microscopy to detect their subcellular localization. Numerous effectors were found to target diverse cellular compartments, such as a nucleus, membrane, vacuole, and recently, it was found that they localize to the most important organelles, i.e., chloroplast and mitochondria (Mach 2017; Petre, Lorrain, et al. 2016). Determining the host's subcellular localization is very important for obtaining information about the effector function and intervene in its virulence by inhibiting its translocation (Dalio et al. 2018).

Host-Induced Gene Silencing

RNA silencing mechanisms have become an effective method to investigate the function of the effector genes. Host-induced gene silencing (HIGS) is used in the RNA silencing mechanism by delivering silencing constructs of antisense RNA. HIGS is developed from virus-induced gene silencing, which permits the silencing of genes in plant pathogens. HIGS is an RNA-interference (RNAi)-based method that is post-transcriptional gene silencing, and it is used for expression of double-stranded RNA in the host plant to silence the desired gene of plant pathogens and following gene expression reduction in this pathogen gene (Huang et al. 2006; Qi et al. 2019; Sang and Kim 2020). Subsequently, the pathogen effector is degraded by RNAi, and the plant is protected from pathogen invasion.

In many plants, especially in wheat and barley, HIGS was used to reduce the effect of pathogen virulence. For instance, the HIGS method was used to protect wheat from three major *Puccinia* species: *Puccinia striiformis* f. sp. *tritici*, *Puccinia graminis* f. sp. *tritici*, and *Puccinia triticina* (Panwar et al. 2018; Yin et al. 2011; Zhu et al. 2017). Furthermore, many effectors of *Blumeria graminis* f. sp. *hordei* were also silenced in barley by the HIGS method, and a decrease in infection was observed. Other than wheat and barley, the HIGS method has been applied to maize and rice, and a successful reduction in pathogens' virulence specialized to each crop has been observed (Pliego et al., 2013; Sang and Kim 2020).

Despite its effectiveness, this method still cannot be applied to many crops as it lacks appropriate transformation processes. In this case, the developing method of clustered regularly interspaced short palindrome repeats (CRISPR)/CRISPR-associated protein 9 (Cas9) system is more advantageous than the HIGS method because it is specific and manageable. In addition, combining HIGS/Cas9 will improve many plants' resistance against several fungal pathogens (Jang and Joung 2019; Sang and Kim 2020).

CRISPR/Cas-Mediated Genome Editing

Deletion/disruption techniques are used in effector biology to determine an effector involved in disease formation. CRISPR/Cas9 is a powerful and effective method for genome editing to disrupt the virulence mechanism of pathogens. The CRISPR/Cas9 method is acquired from the bacterial and archaeal immune systems, and it consists of two elements: a single guide RNA (sgRNA) and an endonuclease enzyme, Cas9. The sgRNA identifies a sequence named "protospacer." Then, it binds to the targeted DNA region and interacts with Cas9 to form a Cas9–sgRNA complex. After recognition is accomplished, double-stranded DNA is cleaved by Cas9, which produces a double-stranded break and causes DNA repair activation (Doudna and Charpentier 2014; Khan et al. 2020).

In functional characterization studies for effectors, many reports show successful results in generation mutation in several fungus species. For example, Fang and Tyler could disrupt and replace an effector, an AVR gene, in the oomycete *P. sojae* (Fang and Tyler, 2016). Many CRISPR/Cas9-mediated gene disruption approaches was carried out in several filamentous fungi, such as *Trichoderma reesei*, *Neurospora crassa*, *Magnaporthe oryzae*, *Fusarium oxysporum*, and Aspergillus species (Muñoz et al. 2019).

Furthermore, by the advantage of CRISPR/Cas9, which allows the editing of more than one genomic site, related genes can be disrupted by using a single sgRNA. Thus, the related genes can be deleted with this sgRNA, which recognizes them according to their conserved region. This system can be applied to effectors that share conserved regions, such as cysteine-rich effectors. *CRISPR/Cas9* successfully edited *Palmivora* with a single sgRNA (Gumtow et al., 2018). *Ustilago maydis* is another fungus in which CRISPR /Cas9 works successfully, and two paralogous effector genes were disrupted by the CRISPR/Cas9 multiplexing–mediated approach (Schuster et al. 2018).

The CRISPR/Cas9 method provides a time-saving and efficient genome editing technology to determine the functional characterization of effectors for single and multiple genes. Targeting related genes will improve our understanding of effector function because of the chance of resulting conclusive phenotype (Selin et al. 2016). It is thought that the non-virulent CRISPR-mutant strains of pathogens may be the potential competitors of plant pathogens in the case of nutrition and space competition between them (Muñoz et al. 2019).

CONCLUSION

Future research aimed to characterize important fungi causing economic losses in cultivation areas needs high quality and maximum-utility databases for these subjects attracting researchers' attention. A vast expanse of biodiversity should be investigated, and new fungal species should be defined using fungal databases. Correspondingly, it will help in better understanding and improve new strategies for discrimination of the fungi and accurately identify major fungal pathogen species causing diseases.

REFERENCES

Aja, H., Andrew, M., Cedric, P., & Oberlies, N. (2017). Fungal identification using molecular tools: A primer for the natural products research community. *Journal of Natural Products,* 80. doi:10.1021/acs.jnatprod.6b01085.

Alizon, S., de Roode, J. C., & Michalakis, Y. (2013). Multiple infections and the evolution of virulence. *Ecology Letters*, 16(4), 556–567.

Almagro Armenteros, J. J., Sønderby, C. K., Sønderby, S. K., Nielsen, H., & Winther, O. (2017). DeepLoc: Prediction of protein subcellular localization using deep learning. *Bioinformatics*, 33(21), 3387–3395.

Andac, A., Ozketen, A. C., Dagvadorj, B., & Akkaya, M. S. (2020). An effector of *Puccinia striiformis* f. sp. *tritici* targets chloroplasts with a novel and robust targeting signal. *European Journal of Plant Pathology*, 157(4), 751–765.

Armenteros, J. J. A., Salvatore, M., Emanuelsson, O., Winther, O., Von Heijne, G., Elofsson, A., & Nielsen, H. (2019). Detecting sequence signals in targeting peptides using deep learning. *Life Science Alliance*, 2(5).

Baack, E. J., & Rieseberg, L. H. (2007). A genomic view of introgression and hybrid speciation. *Current Opinion in Genetics & Development*, 17(6), 513–518.

Bailey, T. L., Boden, M., Buske, F. A., Frith, M., Grant, C. E., Clementi, L., . . . Noble, W. S. (2009). MEME SUITE: Tools for motif discovery and searching. *Nucleic Acids Research,* 37(suppl_2), W202–W208.

Bailey, T. L., & Elkan, C. (1995, July). The value of prior knowledge in discovering motifs with MEME. In *Ismb* (Vol. 3, pp. 21–29).

Basenko, E. Y., Pulman, J. A., Shanmugasundram, A., Harb, O. S., Crouch, K., Starns, D., . . . Roos, D. S. (2018). FungiDB: An integrated bioinformatic resource for fungi and oomycetes. *Journal of Fungi*, 4(1), 39.

Bendtsen, J. D., Jensen, L. J., Blom, N., Von Heijne, G., & Brunak, S. (2004). Feature-based prediction of non-classical and leaderless protein secretion. *Protein Engineering Design and Selection*, 17(4), 349–356.

Benson, D. A., Clark, K., Karsch-Mizrachi, I., Lipman, D. J., Ostell, J., & Sayers, E. W. (2014). GenBank. *Nucleic Acids Research,* 42, 32–37.

Benson, D. A., Karsch-Mizrachi, I., Lipman, D. J., Ostell, J., & Sayers, E. W. (2011). GenBank. *Nucleic Acids Research*, 39(suppl_1), 32–37.

Bouton, C., King, R. C., Chen, H., Azhakanandam, K., Bieri, S., Hammond-Kosack, K. E., & Kanyuka, K. (2018). Foxtail mosaic virus: A viral vector for protein expression in cereals. *Plant Physiology*, 177(4), 1352–1367.

Bozkurt, T. O., Richardson, A., Dagdas, Y. F., Mongrand, S., Kamoun, S., & Raffaele, S. (2014). The plant membrane-associated REMORIN1.3 accumulates in discrete perihaustorial domains and enhances susceptibility to *Phytophthora infestans*. *Plant Physiology,* 165(3), 1005–1018.

Brown, G. D. (2011). Innate antifungal immunity: The key role of phagocytes. *Annual Review of Immunology*. 29, 1–21. doi:10.1146/annurev-immunol-030409-101229

Celio, G. J., Padamsee, M., Dentinger, B. T., Bauer, R., & McLaughlin, D. J. (2006). Assembling the Fungal Tree of Life: Constructing the structural and biochemical database. *Mycologia*, 98, 850–859.

Cerqueira, G. C., Arnaud, M. B., Inglis, D. O., Skrzypek, M. S., Binkley, G., Simison, M., . . Wortman, J. R. (2014). The aspergillus genome database: Multispecies curation and incorporation of RNA-Seq data to improve structural gene annotations. *Nucleic Acids Research*, 42(1), 705–710.

Charlesworth, B., Morgan, M. T., & Charlesworth, D. (1993). The effect of deleterious mutations on neutral molecular variation. *Genetics*, 134(4), 1289–1303.

Choi, J., Détry, N., Kim, K. T., Asiegbu, F. O., Valkonen, J. P., & Lee, Y. H. (2014). fPoxDB: Fungal peroxidase database for comparative genomics. *BMC Microbiology*, 14(1), 117.

Choi, J., Park, J., Kim, D., Jung, K., Kang, S., & Lee, Y. H. (2010). Fungal secretome database: Integrated platform for annotation of fungal secretomes. *BMC Genomics*, 11(1), 105.

Coleman, J. J., Rounsley, S. D., Rodriguez-Carres, M., Kuo, A., Wasmann, C. C., Grimwood, J., . . . Schwartz, D. C. (2009). The genome of *Nectria haematococca*: Contribution of supernumerary chromosomes to gene expansion. *PLoS Genetics*, 5(8), e1000618.

Cortázar, A. R., Aransay, A. M., Alfaro, M., Oguiza, J. A., & Lavín, J. L. (2014). SECRE-TOOL: Integrated secretome analysis tool for fungi. *Amino Acids*, 46(2), 471–473.

Cuomo, C. A., & Birren, B. W. (2010). The fungal genome initiative and lessons learned from genome sequencing. In *Methods in enzymology* (Vol. 470, pp. 833–855). Academic Press.

Dalio, R. J. D., Herlihy, J., Oliveira, T. S., McDowell, J. M., & Machado, M. (2018). Effector biology in focus: A primer for computational prediction and functional characterization. *Molecular Plant-Microbe Interactions*, 31(1), 22–33.

De Jonge, R., Bolton, M. D., & Thomma, B. P. (2011). How filamentous pathogens co-opt plants: The ins and outs of fungal effectors. *Current Opinion in Plant Biology*, 14(4), 400–406.

Delgado-Serrano, L., Restrepo, S., Bustos, J. R., Zambrano, M. M., & Anzola, J. M. (2016). Mycofier: A new machine learning-based classifier for fungal ITS sequences. *BMC Research Notes*, 9(1), 1–8. https://doi.org/10.1186/s13104-016-2203-3

Dodds, P. N., & Rathjen, J. P. (2010). Plant immunity: Towards an integrated view of plant—pathogen interactions. *Nature Reviews Genetics*, 11(8), 539–548.

Dou, D., Kale, S. D., Wang, X., Jiang, R. H., Bruce, N. A., Arredondo, F. D., . . . Tyler, B. M. (2008). RXLR-mediated entry of Phytophthora sojae effector Avr1b into soybean cells does not require pathogen-encoded machinery. *The Plant Cell*, 20(7), 1930–1947.

Doudna, J. A., & Charpentier, E. (2014). The new frontier of genome engineering with CRIS-PR-Cas9. *Science*, 346(6213), 1258096.

Duplessis, S., Cuomo, C. A., Lin, Y. C., Aerts, A., Tisserant, E., Veneault-Fourrey, C., . . . Chiu, R. (2011). Obligate biotrophy features unraveled by the genomic analysis of rust fungi. *Proceedings of the National Academy of Sciences*, 108(22), 9166–9171.

Duret, L., & Galtier, N. (2009). Biased gene conversion and the evolution of mammalian genomic landscapes. *Annual Review of Genomics and Human Genetics*, 10, 285–311.

Fabro, G., Steinbrenner, J., Coates, M., Ishaque, N., Baxter, L., Studholme, D. J., Körner, E., Allen, R. L., Piquerez, S. J., Rougon-Cardoso, A., & Greenshields, D. (2011). Multiple candidate effectors from the oomycete pathogen *Hyaloperonospora arabidopsidis* suppress host plant immunity. *PLoS Pathogens*, 7(11), e1002348.

Faino, L., Seidl, M. F., Shi-Kunne, X., Pauper, M., van den Berg, G. C., Wittenberg, A. H., & Thomma, B. P. (2016). Transposons passively and actively contribute to evolution of the two-speed genome of a fungal pathogen. *Genome Research*, 26(8), 1091–1100.

Fang, Y., & Tyler, B. M. (2016). Efficient disruption and replacement of an effector gene in the oomycete P hytophthora sojae using CRISPR/C as9. *Molecular Plant Pathology*, 17(1), 127–139. doi:10.1111/mpp.12318

Fischer, M., & Pleiss, J. (2003). The lipase engineering database: A navigation and analysis tool for protein families. *Nucleic Acids Research*, 31(1), 319–321.

Fisher, M., Henk, D., & Briggs, C. (2012). Emerging fungal threats to animal, plant and ecosystem health. *Nature*, 484, 186–194.

Flor, H. H. (1971). Current status of the gene-for-gene concept. *Annual Review of Phytopathology*, 9(1), 275–296.

Frantzeskakis, L., Kusch, S., & Panstruga, R. (2019). The need for speed: Compartmentalized genome evolution in filamentous phytopathogens. *Molecular Plant Pathology*, 20(1), 3–7.

Frézal, L., & Leblois, R. (2008). Four years of DNA barcoding: Current advances and prospects. *Infection, Genetics and Evolution*, 8, 727–736.

Friesen, T. L., Stukenbrock, E. H., Liu, Z., Meinhardt, S., Ling, H., Faris, J. D., Rasmussen, J. B., Solomon, P. S., McDonald, B. A., & Oliver, R. P. (2006). Emergence of a new disease as a result of interspecific virulence gene transfer. *Nature Genetics*, 38(8), 953–956. doi:10.1038/ng1839

Geiser, D. M., del Mar Jiménez-Gasco, M., Kang, S., Makalowska, I., Veeraraghavan, N., Ward, T. J., Zhang, N., Kuldau, G. A., & O'Donnell, K. (2004). FUSARIUM-ID v. 1.0: A DNA sequence database for identifying Fusarium. *European Journal of Plant Pathology*, 110, 473–479.

Godfrey, D., Böhlenius, H., Pedersen, C., Zhang, Z., Emmersen, J., & Thordal-Christensen, H. (2010). Powdery mildew fungal effector candidates share N-terminal Y/F/WxC-motif. *BMC Genomics*, 11(1), 1–13.

Goodwin, S. B., M'barek, S. B., Dhillon, B., Wittenberg, A. H., Crane, C. F., Hane, J. K., & Antoniw, J. (2011). Finished genome of the fungal wheat pathogen Mycosphaerella graminicola reveals dispensome structure, chromosome plasticity, and stealth pathogenesis. *PLoS Genet*, 7(6), e1002070.

Gouveia, B. C., Calil, I. P., Machado, J. P. B., Santos, A. A., & Fontes, E. P. (2017). Immune receptors and co-receptors in antiviral innate immunity in plants. *Frontiers in Microbiology*, 7, 2139. doi:10.3389/fmicb.2016.02139

Green, J. R., Pain, N. A., Cannell, M. E., Leckie, C. P., McCready, S., Mitchell, A. J., & Mendgen, K. (1995). Analysis of differentiation and development of the specialized infection structures formed by biotrophic fungal plant pathogens using monoclonal antibodies. *Canadian Journal of Botany*, 73(S1), 408–417.

Gumtow, R., Wu, D., Uchida, J., & Tian, M. (2018). A Phytophthora palmivora extracellular cystatin-like protease inhibitor targets papain to contribute to virulence on papaya. *Molecular Plant-Microbe Interactions*, 31(3), 363–373.

Haas, B. J., Kamoun, S., Zody, M. C., Jiang, R. H., Handsaker, R. E., Cano, L. M., Grabherr, M., Kodira, C. D., Raffaele, S., Torto-Alalibo, T., & Bozkurt, T. O. (2009). Genome sequence and analysis of the Irish potato famine pathogen Phytophthora infestans. *Nature*, 461(7262), 393–398. https://doi.org/10.1038/nature08358

Haas, B. J., Zeng, Q., Pearson, M. D., Cuomo, C. A., & Wortman, J. R. (2011). Approaches to fungal genome annotation. *Mycology*, 2(3), 118–141.

Hacquard, S., Joly, D. L., Lin, Y. C., Tisserant, E., Feau, N., Delaruelle, C., & Frey, P. (2012). A comprehensive analysis of genes encoding small secreted proteins identifies candidate effectors in *Melampsora larici-populina* (poplar leaf rust). *Molecular Plant-Microbe Interactions*, 25(3), 279–293.

Horton, P., Park, K. J., Obayashi, T., Fujita, N., Harada, H., Adams-Collier, C. J., & Nakai, K. (2007). WoLF PSORT: Protein localization predictor. *Nucleic Acids Research*, 35(suppl_2), W585–W587.

Houben, A., Banaei-Moghaddam, A. M., Klemme, S., & Timmis, J. N. (2014). Evolution and biology of supernumerary B chromosomes. *Cellular and Molecular Life Sciences*, 71(3), 467–478. doi: https://doi.org/10.1007/s00018-013-1437-7

Howe, K. L., Contreras-Moreira, B., De Silva, N., Maslen, G., Akanni, W., Allen, J., . . . Carbajo, M. (2020). Ensembl genomes 2020—enabling non-vertebrate genomic research. *Nucleic Acids Research*, 48(D1), D689–D695.

Huang, G., Allen, R., Davis, E. L., Baum, T. J., & Hussey, R. S. (2006). Sciences of the USA. *PNAS*, 103, 14302–14306.

Irinyi, L., Serena, C., Garcia-Hermoso, D., Arabatzis, M., Desnos-Ollivier, M., Vu, D., et al. (2015). International Society of Human and Animal Mycology (ISHAM)- ITS reference DNA barcoding database—the quality controlled standard tool for routine identification of human and animal pathogenic fungi. *Medical Mycology*, 53, 313–337.

Jang, G., & Joung, Y. H. (2019). CRISPR/Cas-mediated genome editing for crop improvement: Current applications and future prospects. *Plant Biotechnology Reports*, 13(1), 1–10.

Jones, J., & Dangl, J. L. (2006). The plant immune system. *Nature*, 444, 323–329. https://doi.org/10.1038/nature05286

Kale, S. D., Gu, B., Capelluto, D. G., Dou, D., Feldman, E., Rumore, A., & Lawrence, C. B. (2010). External lipid PI3P mediates entry of eukaryotic pathogen effectors into plant and animal host cells. *Cell*, 142(2), 284–295.

Kamoun, S., Van West, P., Vleeshouwers, V. G. A. A., De Groot, K. E., & Govers, F. (1998). Resistance of Nicotiana benthamiana to *Phytophthora infestans* is mediated by the recognition of the elicitor protein INF1. *Plant Cell*, 10(9), 1413–1425.

Kanja, C., & Hammond-Kosack, K. E. (2020). Proteinaceous effector discovery and characterization in filamentous plant pathogens. *Molecular Plant Pathology*, (July), 1–24.

Kauff, F., Cox, C. J., & Lutzoni, F. (2007). WASABI: An automated sequence processing system for multigene phylogenies. *Systematic Biology*, 56(3), 523–531.

Kemen, E., Gardiner, A., Schultz-Larsen, T., Kemen, A. C., Balmuth, A. L., Robert-Seilaniantz, A., & Jones, J. D. (2011). Gene gain and loss during evolution of obligate parasitism in the white rust pathogen of Arabidopsis thaliana. *PLoS Biology*, 9(7), e1001094.

Kettles, G. J., Bayon, C., Canning, G., Rudd, J. J., & Kanyuka, K. (2017). Apoplastic recognition of multiple candidate effectors from the wheat pathogen Zymoseptoria tritici in the nonhost plant *Nicotiana benthamiana*. *New Phytologist*, 213(1), 338–350.

Khan, H., McDonald, M. C., Williams, S. J., & Solomon, P. S. (2020). Assessing the efficacy of CRISPR/Cas9 genome editing in the wheat pathogen *Parastagonspora nodorum*. *Fungal Biology and Biotechnology*, 7(1), 1–8.

Kim, K. T., Jeon, J., Choi, J., Cheong, K., Song, H., Choi, G., . . . Lee, Y. H. (2016). Kingdom-wide analysis of fungal small secreted proteins (SSPs) reveals their potential role in host association. *Frontiers in Plant Science*, 7, 186.

Koeck, M., Hardham, A. R., & Dodds, P. N. (2011). The role of effectors of biotrophic and hemibiotrophic fungi in infection. *Cellular Microbiology*, 13(12), 1849–1857.

Krogh, A., Larsson, B., Von Heijne, G., & Sonnhammer, E. L. (2001). Predicting transmembrane protein topology with a hidden Markov model: Application to complete genomes. *Journal of Molecular Biology*, 305(3), 567–580.

Lee, W. S., Hammond-Kosack, K. E., & Kanyuka, K. (2012). Barley stripe mosaic virus-mediated tools for investigating gene function in cereal plants and their pathogens: Virus-induced gene silencing, host-mediated gene silencing, and virus-mediated overexpression of heterologous protein. *Plant Physiology*, 160(2), 582–590.

Lévesque, C. A., Brouwer, H., Cano, L., Hamilton, J. P., Holt, C., Huitema, E., . . . Zerillo, M. M. (2010). Genome sequence of the necrotrophic plant pathogen Pythium ultimum reveals original pathogenicity mechanisms and effector repertoire. *Genome Biology*, 11(7), R73.

Lombard, V., Golaconda Ramulu, H., Drula, E., Coutinho, P. M., & Henrissat, B. (2014). The carbohydrate-active enzymes database (CAZy) in 2013. *Nucleic Acids Research*, 42(D1), D490–D495.

Lum, G., & Min, X. J. (2011). FunSecKB: The fungal secretome knowledgebase. *Database*. https://doi.org/10.1093/database/bar001

Lynch, M. (2007). The evolution of genetic networks by non-adaptive processes. *Nature Reviews Genetics*, 8(10), 803–813. doi:10.1038/nrg2192

Ma, L. J., Van Der Does, H. C., Borkovich, K. A., Coleman, J. J., Daboussi, M. J., Di Pietro, A. & Houterman, P. M. (2010). Comparative genomics reveals mobile pathogenicity chromosomes in Fusarium. *Nature*, 464(7287), 367–373.

Ma, W., & Guttman, D. S. (2008). Evolution of prokaryotic and eukaryotic virulence effectors. *Current Opinion in Plant Biology*, 11(4), 412–419.

Mach, J. (2017). Tracking the bacterial type III secretion system: Visualization of effector delivery using split fluorescent proteins. *Plant Cell*, 29(7), 1547–1548.

Manning, V. A, Chu, A. L., Scofield, S. R., & Ciuffetti, L. M. (2010). Intracellular expression of a host-selective toxin, ToxA, in diverse plants phenocopies silencing of a ToxA-interacting protein, ToxABP1. *New Phytologist*, 187(4), 1034–1047.

Manning, V. A., Hamilton, S. M., Karplus, P. A., & Ciuffetti, L. M. (2008). The Arg-Gly-Asp—containing, solvent-exposed loop of Ptr ToxA is required for internalization. *Molecular Plant-microbe Interactions*, 21(3), 315–325.

McDonald, B. A., & Linde, C. (2002a). Pathogen population genetics, evolutionary potential, and durable resistance. *Annual Review of Phytopathology*, 40(1), 349–379.

McDonald, B. A., & Linde C. (2002b). The population genetics of plant pathogens and breeding strategies for durable resistance. *Euphytica,* 124(2), 163–180. doi: 10.1023/A:1015678432355

McDonald, B. A., & Mundt, C. C. (2016). How knowledge of pathogen population biology informs management of Septoria tritici blotch. *Phytopathology,* 106(9), 948–955. doi:10.1094/PHYTO-03-16-0131-RVW

Möller, M., & Stukenbrock, E. H. (2017). Evolution and genome architecture in fungal plant pathogens. *Nature Reviews Microbiology,* 15(12), 756–771. doi:10.1038/nrmicro.2017.76

Mukhtar, M. S., Carvunis, A. R., Dreze, M., Epple, P., Steinbrenner, J., Moore, J., Tasan, M., Galli, M., Hao, T., Nishimura, M. T., & Pevzner, S. J. (2011). Independently evolved virulence effectors converge onto hubs in a plant immune system network. *Science,* 333(6042), 596–601. doi:10.1126/science.1203659

Muñoz, I. V., Sarrocco, S., Malfatti, L., Baroncelli, R., & Vannacci, G. (2019). CRISPR-CAS for fungal genome editing: A new tool for the management of plant diseases. *Frontiers in Plant Science,* 10(February), 1–5.

Nordberg, H., Cantor, M., Dusheyko, S., Hua, S., Poliakov, A., Shabalov, I., . . . Dubchak, I. (2014). The genome portal of the Department of Energy Joint Genome Institute: 2014 updates. *Nucleic Acids Research,* 42(D1), D26–D31.

Ozketen, A. C., Andac, A., Dagvadorj, B., Demiralay, B., & Akkaya, M. S. (2020). In-depth secretome analysis of *Puccinia striiformis* f. sp. *tritici* in infected wheat uncovers effector functions. *Biosci Rep.* 2020 Dec, 23; 40(12): BSR20201188. doi: 10.1042/BSR20201188. PMID: 33275764; PMCID: PMC7724613.

Panackal, A. A. (2011). Global climate change and infectious diseases: Invasive mycoses. *Journal of Earth Science and Climatic Change,* 1, 108.

Panwar, V., Jordan, M., McCallum, B., & Bakkeren, G. (2018). Host-induced silencing of essential genes in *Puccinia triticina* through transgenic expression of RNAi sequences reduces severity of leaf rust infection in wheat. *Plant Biotechnology Journal,* 16(5), 1013–1023.

Park, B., Park, J., Cheong, K. C., Choi, J., Jung, K., Kim, D., . . Kang, S. (2011). Cyber infrastructure for Fusarium: Three integrated platforms supporting strain identification, phylogenetics, comparative genomics and knowledge sharing. *Nucleic Acids Research,* 39, 640–646.

Pedro, H., Maheswari, U., Urban, M., Irvine, A. G., Cuzick, A., Mcdowall, M. D., . . Kersey, P. J. (2016). PhytoPath: An integrative resource for plant pathogen genomics". *Nucleic Acids Research,* 44, 688–639. doi:10.1093/nar/gkv1052. PMC 4702788. PMID 26476449

Petersen, T. N., Brunak, S., Von Heijne, G., & Nielsen, H. (2011). SignalP 4.0: Discriminating signal peptides from transmembrane regions. *Nature Methods,* 8(10), 785–786.

Petre, B., & Kamoun, S. (2014). How do filamentous pathogens deliver effector proteins into plant cells? *PLoS Biology,* 12(2), e1001801.

Petre, B., Lorrain, C., Saunders, D. G. O., Win, J., Sklenar, J., Duplessis, S., & Kamoun, S. (2016). Rust fungal effectors mimic host transit peptides to translocate into chloroplasts. *Cellular Microbiology,* 18(4), 453–465.

Pliego, C., Nowara, D., Bonciani, G., Gheorghe, D. M., Xu, R., Surana, P., et al. (2013). Host-induced gene silencing in barley powdery mildew reveals a class of ribonuclease-like effectors. *Molecular Plant-Microbe Interactions,* 26(6), 633–642.

Plissonneau, C., Stürchler, A., & Croll, D. (2016). The evolution of orphan regions in genomes of a fungal pathogen of wheat. *MBio,* 7(5).

Prakash, P. Y., Irinyi, L., Halliday, C., Chen, S., Robert, V., & Meyer, W. (2017). Online databases for taxonomy and identification of pathogenic fungi and proposal for a cloud-based dynamic data network platform. *Journal of Clinical Microbiology,* 55, 1011–1024.

Pruitt, K. D., Tatusova, T., & Maglott, D. R. (2005). NCBI Reference Sequence (RefSeq): A curated non-redundant sequence database of genomes, transcripts and proteins. *Nucleic Acids Research*, 33(suppl_1), D501–D504.

Qi, T., Guo, J., Peng, H., Liu, P., Kang, Z., & Guo, J. (2019). Host-induced gene silencing: A powerful strategy to control diseases of wheat and barley. *International Journal of Molecular Sciences*, 20(1).

Qutob, D., Kamoun, S., & Gijzen, M. (2002). Expression of a Phytophthora sojae necrosis-inducing protein occurs during transition from biotrophy to necrotrophy. *The Plant Journal*, 32(3):361–373.

Raffaele, S., & Kamoun, S. (2012). Genome evolution in filamentous plant pathogens: Why bigger can be better. *Nature Reviews Microbiology*, 10(6), 417–430.

Rafiqi, M., Gan, P. H., Ravensdale, M., Lawrence, G. J., Ellis, J. G., Jones, D. A., . . . Dodds, P. N. (2010). Internalization of flax rust avirulence proteins into flax and tobacco cells can occur in the absence of the pathogen. *The Plant Cell*, 22(6), 2017–2032.

Rawlings, N. D., Barrett, A. J., Thomas, P. D., Huang, X., Bateman, A., & Finn, R. D. (2018). The MEROPS database of proteolytic enzymes, their substrates and inhibitors in 2017 and a comparison with peptidases in the PANTHER database. *Nucleic Acids Research*, 46(D1), D624–D632.

Rawlings, N. D., Waller, M., Barrett, A. J., & Bateman, A. (2014). MEROPS: The database of proteolytic enzymes, their substrates and inhibitors. *Nucleic Acids Research*, 42(D1), D503–D509.

Read, A. F. (1994). The evolution of virulence. *Trends in Microbiology*, 2(3), 73–76.

Ridout, C. J., Skamnioti, P., Porritt, O., Sacristan, S., Jones, J. D., & Brown, J. K. (2006). Multiple avirulence paralogues in cereal powdery mildew fungi may contribute to parasite fitness and defeat of plant resistance. *The Plant Cell*, 18(9), 2402–2414.

Rizzato, C., Lombardi, L., Zoppo, M., Lupetti, A., & Tavanti, A. (2015). Pushing the Limits of MALDI-TOF mass spectrometry: Beyond fungal species identification. *Journal of Fungi*, 367. doi:10.3390/jof1030367

Robert, V., Cardinali, G., Stielow, B., Vu, T., dos Santos, F. B., Meyer, W., & Schoch, C. (2015). Fungal DNA barcoding. *Molecular Biology of Food and Water Borne Mycotoxigenic and Mycotic Fungi*, 37–56.

Robert, V., Vu, D., Amor, A. B., van de Wiele, N., Brouwer, C., Jabas, B., . . Crous, P. W. (2013). MycoBank gearing up for new horizons. *IMA Fungus*, 4, 371–379.

Sánchez-Vallet, A., Hartmann, F. E., Marcel, T. C., & Croll, D. (2018). Nature's genetic screens: Using genome-wide association studies for effector discovery. *Molecular Plant Pathology*, 19(1), 3.

Sang, H., & Kim, J. Il. (2020). Advanced strategies to control plant pathogenic fungi by host-induced gene silencing (HIGS) and spray-induced gene silencing (SIGS). *Plant Biotechnology Reports*, 14(1), 1–8.

Saunders, D. G., Win, J., Cano, L. M., Szabo, L. J., Kamoun, S., & Raffaele, S. (2012). Using hierarchical clustering of secreted protein families to classify and rank candidate effectors of rust fungi. *PLoS One*, 7(1), e29847.

Saur, I. M., Bauer, S., Kracher, B., Lu, X., Franzeskakis, L., Müller, M. C., Sabelleck, B., Kümmel, F., Panstruga, R., Maekawa, T., & Schulze-Lefert, P. (2019). Multiple pairs of allelic MLA immune receptor-powdery mildew AVRA effectors argue for a direct recognition mechanism. *Elife*, 8, e44471.

Schirawski, J., Mannhaupt, G., Münch, K., Brefort, T., Schipper, K., Doehlemann, G., . . . Müller, O. (2010). Pathogenicity determinants in smut fungi revealed by genome comparison. *Science*, 330(6010), 1546–1548.

Schoch, C. L., Robbertse, B., Robert, V., Vu, D., Cardinali, G., Irinyi, L., . . Federhen, S. (2014). Finding needles in haystacks: Linking scientific names, reference specimens and molecular data for Fungi. *Database: The Journal of Biological Databases and Curation*, 2014, bau061.

Schuster, M., Schweizer, G., & Kahmann, R. (2018). Comparative analyses of secreted proteins in plant pathogenic smut fungi and related basidiomycetes. *Fungal Genetics and Biology*, 112, 21–30.

Seidl, M. F., & Thomma, B. P. (2014). Sex or no sex: Evolutionary adaptation occurs regardless. *Bioessays*, 36(4), 335–345. doi:10.1002/bies.201300155

Selin, C., de Kievit, T. R., Belmonte, M. F., & Fernando, W. G. D. (2016). Elucidating the role of effectors in plant-fungal interactions: Progress and challenges. *Frontiers in Microbiology*, 7(April), 1–21.

Short, D. P., O'Donnell, K., Thrane, U., Nielsen, K. F., Zhang, N., Juba, J. H., & Geiser, D. M. (2013). Phylogenetic relationships among members of the Fusarium solani species complex in human infections and the descriptions of *F. keratoplasticum* sp. nov. and *F. petroliphilum* stat. nov. *Fungal Genetics and Biology*, 53, 59–70.

Sperschneider, J., Catanzariti, A. M., DeBoer, K., Petre, B., Gardiner, D. M., Singh, K. B., . . . Taylor, J. M. (2017). LOCALIZER: Subcellular localization prediction of both plant and effector proteins in the plant cell. *Scientific Reports*, 7(1), 1–14.

Sperschneider, J., Dodds, P. N., Gardiner, D. M., Singh, K. B., & Taylor, J. M. (2018a). Improved prediction of fungal effector proteins from secretomes with EffectorP 2.0. *Molecular Plant Pathology*, 19(9), 2094–2110.

Sperschneider, J., Dodds, P. N., Singh, K. B., & Taylor, J. M. (2018b). ApoplastP: Prediction of effectors and plant proteins in the apoplast using machine learning. *New Phytologist*, 217(4), 1764–1778.

Sperschneider, J., Gardiner, D. M., Dodds, P. N., Tini, F., Covarelli, L., Singh, K. B., . . . Taylor, J. M. (2016). EffectorP: Predicting fungal effector proteins from secretomes using machine learning. *New Phytologist*, 210(2), 743–761.

Stielow, J., Levesque, C., Seifert, K., Meyer, W., Irinyi, L., Smits, D., . . Chaduli, D. (2015). One fungus, which genes? Development and assessment of universal primers for potential secondary fungal DNA barcodes. *Persoonia*, 35, 242–263.

Stukenbrock, E. H. (2013). Evolution, selection and isolation: A genomic view of speciation in fungal plant pathogens. *New Phytologist*, 199(4), 895–907.

Stukenbrock, E. H. (2016). The role of hybridization in the evolution and emergence of new fungal plant pathogens. *Phytopathology*, 106(2), 104–112.

Tellier, A., Moreno-Gámez, S., & Stephan, W. (2014). Speed of adaptation and genomic footprints of host—parasite co-evolution under arms race and trench warfare dynamics. *Evolution*, 68(8), 2211–2224.

Thrall, P. H., Laine, A. L., Ravensdale, M., Nemri, A., Dodds, P. N., Barrett, L. G., & Burdon, J. J. (2012). Rapid genetic change underpins antagonistic co-evolution in a natural host-pathogen metapopulation. *Ecology Letters*, 15(5), 425–435. doi:10.1111/j.1461-0248.2012.01749.x

Torto, T. A., Li, S., Styer, A., Huitema, E., Testa, A., Gow, N. A., & Kamoun, S. (2003). EST mining and functional expression assays identify extracellular effector proteins from the plant pathogen Phytophthora. *Genome Research*, 13(7), 1675–1685.

Upadhyaya, N. M., Mago, R., Staskawicz, B. J., Ayliffe, M. A., Ellis, J. G., & Dodds, P. N. (2014). A bacterial type III secretion assay for delivery of fungal effector proteins into wheat. *Molecular Plant-Microbe Interactions*, 27(3), 255–264.

Urban, M., Cuzick, A., Rutherford, K., Irvine, A., Pedro, H., Pant, R., & Hammond-Kosack, K. E. (2017). PHI-base: A new interface and further additions for the multi-species pathogen—host interactions database. *Nucleic Acids Research*, 45(D1), D604–D610.

Urban, M., Pant, R., Raghunath, A., Irvine, A. G., Pedro, H., & Hammond-Kosack, K. E. (2015). The Pathogen-Host Interactions database (PHI-base): Additions and future developments. *Nucleic Acids Research*, 43(D1), D645–D655.

Vleeshouwers, V. G., & Oliver, R. P. (2014). Effectors as tools in disease resistance breeding against biotrophic, hemibiotrophic, and necrotrophic plant pathogens. *Molecular Plant-microbe Interactions*, 27(3), 196–206. http://dx.doi.org/10.1094/MPMI-10-13-0313-IA

Vu, D., Szoke, S., Wiwie, C., Baumbach, J., Cardinali, G., Rottger, R., & Robert, V. (2014). Massive fungal biodiversity data re-annotation with multi-level clustering. *Scientific Reports*, 4, 6837.

Whisson, S. C., Boevink, P. C., Moleleki, L., Avrova, A. O., Morales, J. G., Gilroy, E. M., . . . Hein, I. (2007). A translocation signal for delivery of oomycete effector proteins into host plant cells. *Nature*, 450(7166), 115–118.

Wu, B., Hussain, M., Zhang, W., Stadler, M., Liu, X., & Xiang, M. (2019). Current insights into fungal species diversity and perspective on naming the environmental DNA sequences of fungi. *Mycology*, 10(3), 127–140. doi:10.1080/21501203.2019.1614106

Xia, C., Wang, M., Cornejo, O. E., Jiwan, D. A., See, D. R., & Chen, X. (2017). Secretome characterization and correlation analysis reveal putative pathogenicity mechanisms and identify candidate avirulence genes in the wheat stripe rust fungus *Puccinia striiformis* f. sp. *tritici*. *Frontiers in Microbiology*, 8, 2394.

Yin, C., Chen, X., Wang, X., Han, Q., Kang, Z., & Hulbert, S. H. (2009). Generation and analysis of expression sequence tags from haustoria of the wheat stripe rust fungus *Puccinia striiformis* f. sp. *tritici*. *BMC Genomics*, 10(1), 1–9.

Yin, C., Jurgenson, J. E., & Hulbert, S. H. (2011). Development of a host-induced RNAi system in the wheat stripe rust fungus *Puccinia striiformis* f. sp. *tritici*. *Molecular Plant-Microbe Interactions*, 24(5), 554–561.

Yin, Y., Mao, X., Yang, J., Chen, X., Mao, F., & Xu, Y. (2012). dbCAN: A web resource for automated carbohydrate-active enzyme annotation. *Nucleic Acids Research*, 40(W1), W445–W451.

Yoshida, K., Saitoh, H., Fujisawa, S., Kanzaki, H., Matsumura, H., Yoshida, K., & Kamoun, S. (2009). Association genetics reveals three novel avirulence genes from the rice blast fungal pathogen *Magnaporthe oryzae*. *The Plant Cell*, 21(5), 1573–1591.

Zhong, Z., et al. (2017). A small secreted protein in *Zymoseptoria tritici* is responsible for avirulence on wheat cultivars carrying the Stb6 resistance gene. *New Phytologist*, 214(2), 619–631.

Zhu, X., Qi, T., Yang, Q., He, F., Tan, C., & Ma, W (2017). Host-induced gene silencing of the MAPKK gene PsFUZ7 confers stable resistance to wheat stripe rust. *Plant Physiology*, 175(4), 1853–1863. doi:10.1104/pp.17.01223

4 Pathogenicity Genes

*Muhammad Kaleem Sarwar, Imran Ul Haq,
Siddra Ijaz, Iqrar Ahmad Khan
and Aşkım Hediye Sekmen Çetinel*

CONTENTS

PATHOGENICITY GENES

The pathogens for tomato wilt, apple scab, and black rust of wheat are *Fusarium oxysporum* f. sp. *lycopersici, Venturia inaequalis*, and *Puccinia graminis* f. sp. *tritici*, respectively. It is worth mentioning that specific pathogens mainly develop diseases in particular host pants; thus, *Venturia inaequalis* attacks only the apple tree, not the other two plants. The pathogen has a set of pathogenicity genes that help identify the host and enable consequent disease development. Pathogenicity and virulence genes of a pathogen are specific to a few related host plants; likewise, susceptibility genes are present only in a few plants, making them susceptible to the particular pathogen. The phenomenon that enables one particular pathogen to attack only the host rather than all plants and the ability of plants to induce defenses and resistance against most microbes, except a few they are susceptible to, is called non-host resistance. Contrarily, necrotrhophs, due to less gene specificity, can attack a vast number of hosts compared with specialized plant pathogens. Field crops and fruit trees are grown over vast lands, simultaneously and in close vicinity. However, wide disease spread or epidemics are often not because of complex disease development, induced resistance against pathogenicity genes, and advanced breeding and management strategies. Among all the strategies and technologies, identifying pathogenicity genes and discovering resistance genes against them, usually from wild plant relatives, is the most effective and reliable strategy.

Pathogenicity genes are crucial for establishing a causal relationship between pathogen and host. These are responsible for host recognition, germination of the pathogen propagative structures on the host surface, and subsequent penetration and colonization of host tissue. Cell wall detoxifying enzymes, crucial for cell wall penetration, and pathogen toxins, such as victorin, aflatoxins, and HC toxin, produced during plant–pathogen interactions are indispensable for disease development. They are known as pathogenicity factors produced by pathogenicity genes. In some cases, these compounds are produced by pathogens during interaction with the plant but are not essential for disease development; instead, they increase the disease severity and are known as virulence factors produced by virulence genes. These compounds, however, impact the pathogen propagation in the plant. Pathogenic microbes employ ingenious strategies to infect a plant host; they utilize various genes during this process and quite frequently turn them on and off according to local conditions. Instead of considerable debate, we can define pathogenicity genes as genes responsible for making a particular microorganism a potential pathogen of specific host. Their deletion/disruption results in a complete or drastic reduction in the disease-causing ability of a pathogen.

PATHOGENICITY GENES OF PHYTOPATHOGENIC FUNGI

Although most plant pathogens belong to kingdom fungi, most fungi are non-pathogens and are also limited in phylogenetic distribution. This limited distribution of pathogenicity genes across the fungal kingdom and high mutation rates make it tricky to recognize common ancestral genes, orthologs. Disease pressure evolved the host defense proteins, whereas pathogenicity genes evolved to attack the updated defense in the host. Horizontal gene transfer is another crucial step in fungal pathogenicity genes, as introducing a new pathogenicity gene or a cluster of genes can produce novel pathogens or make a recipient attack new hosts (Jones and Dangl, 2006).

Rhynchosporium secalis, a barley pathogen, has three sets of necrosis genes: NIP1, NIP2, and NIP3. Analysis shows that most isolates contain either NIP1 or NIP2, and gene deletion is independent of one another and does not reduce pathogenicity significantly. However, NIP3 is present in almost all isolates, and its deletion significantly reduces pathogenicity. *Pyrenophora tritici* is host specific and induces necrosis in wheat. Necrosis-producing proteins are encoded by ToxA, which converts non-pathogenic isolates to pathogens of wheat (Ciuffetti et al., 1997; Schürch et al., 2004; van der Does and Rep, 2007).

HOST SURFACE RECOGNITION AND APPRESSORIUM FORMATION PATHOGENICITY GENES

The amenity of genomics, particularly tagged mutagenesis, has made it possible to understand better the diverse strategies that fungi employ to infect host plants. Before the wide availability of molecular techniques, natural isolates or chemical-induced mutants (non-pathogenic) were used to identify pathogenicity genes by comparison with pathogenic isolates (Andrivon, 1993). Pathogenicity genes are required by the pathogen to develop a disease but are not indispensable to complete its life cycle. The type of pathogenicity genes in a fungus is defined by the infection process; some pathogens may degrade the cell wall or the appressorium, while others use natural openings or wounds to penetrate the host plant epidermis. On the basis of nutrient absorption from the host after colonization, fungi can be classified into biotrophs, hemi-biotrophs, and necrotrophs.

The appressorium, due to its high turgor pressure, punctures the host epidermis. This high pressure, as in *Magnaporthe grisea*, is due to the high glycerol concentration maintained by the presence of melanin, which prevents glycerol from leaking out. Hence, mutants with modified melanin genes cannot hold the pressure and subsequent host penetration, as observed in *Colletotrichum lagenarium,* in which at least three genes control melanin production (Rasmussen and Hanau, 1989). Apart from melanin-producing genes, other genes, such as hydrophobin (mpg1), are involved in appressorium development. Mutants of this gene reduced pathogenicity and conidiation. It was found through the cis-trans test that three regulatory genes control mpg1 transcription (Lau and Hamer, 1996). Mutations in the spore specificity gene of *M. grisea*, acropetal, result in defective conidiation and significant loss in pathogenicity. Similarly, mutations in the pth11 and ORP1 pathogenicity genes alter pathogen phenotype, which cannot recognize host surface and normal appressorium formation, respectively.

CUTICLE, PECTIN, AND CELL WALL DEGRADATION

Cell wall and cuticle degradation enzymes, usually encoded by multi-genes, also play a significant role in pathogenicity to a specific host. As several unrelated genes are involved in their production, this redundancy enables pathogens to mask the inactivity of a certain gene because other alternate genes can compensate for the disrupted gene. Apart from producing enzymes to degrade cell walls, cuticles, and Pectin, these genes play an essential role in suppressing plant defenses and eliciting defense responses. This feature could be used to breed resistant varieties. For instance, xylanase mutation in *Trichoderma reesei*, despite reducing its activity, could elicit a defense response in tomato plants (Enkerli et al., 1999). Different heterologous expression and antibody inhibition studies demonstrate that the disruption of the cutinase gene in *Fusarium solani* f. sp. *pisi* results in non-pathogenic mutants, proving their role as pathogenicity factors.

Pectin degradation enzymes—pectin methylesterase, pectate lyase, and polygalacturonase—degrade Pectin in host plants' cell walls and middle lamellae, playing an important role in plant cell penetration and infection. Significant loss of pathogenicity in *C. gloeosporioides* against avocado fruits was observed after mutation in pelBgene, a pectate lyase encoder. Contrarily, mutation of *F. solani* f. sp. *pisi* at pelA or pelD genes does not significantly affect strain pathogenicity. However, a mutated strain, with both genes disrupted negatively, impacted the fungal pathogenicity in pea. This multi-gene family produces fragments of pectin degradation enzymes; hence, disruption of one gene is covered by other genes due to redundancy. An *Aspergillus flavus* strain, lacking in endo-polygalacturanase production, increased disease severity and lesion size on cotton bolls after induction and expression of the responsible gene. Alternatively, disruption of the endo-polygalacturonase gene Bcpg 1 in *Botrytis cinerea* reduced symptom severity in tomato and apple. The method of fungal infection might govern whether a particular gene expression is required for pathogenicity or not; disruption of the endo-polygalacturonase gene in *A. citri* (responsible for post-harvest diseases in citrus) results in reduced pathogenicity, whereas disruption of the homologous gene in *A. alternata* (responsible for a brown spot in rough lemon) has no impact on pathogenicity. (Shieh et al., 1997; Have et al., 1998; Rogers et al., 2000; Isshiki et al., 2001).

HOST ENVIRONMENT AND PATHOGENICITY GENES

Pathogens use various mechanisms (avoidance, degradation, or altered physiology) to overcome plant defense secondary metabolites, i.e., phytoalexins and phytoanticipins. AvenacinA-1, a saponon belonging to membrane-disrupting class of phytoanticipin, present in oat root epidermis, is defeated by avenacinase of *Gaeumannomyces graminis*. Targeted mutation of the avenacinA-1 gene in oat plants prevents "take all fungus" from attacking and developing the disease. The Phytoalexin pisatin detoxifying gene PDA1 encodes pisatin demethylase in pea fungus *Nectria haematococca*, located at 1.6 Mb dispensable chromosome. Disruption of the PDA1 gene alone results in a slight loss, whereas loss of chromosome results in a more significant loss of pathogenicity. Characterization revealed that three other putative pathogenicity genes are located in that cluster, which is usually the case in pathogenicity islands of bacteria (Wasmann and VanEtten, 1996; Han et al., 2001).

In *M. grisea*, a mutation in the ATP-binding cassette transporter gene ABC1 resulted in non-pathogenic isolates, whereas disruption of the transporter gene BxatrB in *Botrytis cinerea* increased fungicide and phytoalexin sensitivity, resulting in just a slight reduction in pathogenicity (Schoonbeek et al., 2001). When mutated, pathogenicity genes produce auxotrophic strains of a particular pathogen, leading to the proposition that the nutritional content of fungal cells has a significant impact on host–pathogen interaction. Alteration in the enzyme responsible to carryout 6th step of histidine biosynthesis pathway of *Magnaporthe grisea* results in mutant with compromised pathogenicity, similar reduced pathogenicity results were observed in auxotrophic mutatnt of *Ustilago maydis*. A non-pathogenic arginine auxotroph was identified in *F. oxysporum*, as well. Similarly, a mutant of *Parastagononspora nodorum* with a disrupted polyamine biosynthesis pathway produced non-pathogenic strains. Fungi under unfavorable environmental conditions might increase expression of pathogenicity genes; for example *Cladosporium fulvum* produces Aox1 and Aldh1 under stress conditions disruption which results in reduced growth and sporulation of pathogen, *C. fulvum,* upon host plant colonization (Idnurm and Howlett, 2001; Segers et al., 2001).

PATHOGENICITY AND FUNGAL TOXINS

Various plant fungi produce toxins, host specific and nonspecific, which can alter/ disable normal cell functioning in, or kill, host cells. Host-specific fungal toxins and the defining host range also determine the pathogenicity. Genes responsible for the biosynthesis of mycotoxins are usually clustered, although not all genes sharing the same cluster are toxins producing. *Cochliobolus carbonum* produces HC-toxin by involving six gene repeats, spanning 600kb; most of these are part of toxin biosynthesis. HTS1, through peptide synthetase, assembles four residue amino acids without involving ribosome, producing the tetrapeptide HC toxin; additionally, encode the beta sub-units of the following enzymes, fatty acid synthase, and branches – chain amino acid transaminase, respectively. These all are essential for fungal pathogenicity. Whereas TOXG and TOXEp encode alanine racemase and regulate toxin secretion, except HTS1, respectively; alteration in these results in toxins changes composition, but the pathogen retains its pathogenic ability intact (Cheng et al., 1999; Cheng and Walton, 2000).

PATHOGENICITY AND SIGNALING GENES

Fungi use signaling genes, mitogen-activated and AMP-dependent protein kinase, and heterotrimeric g proteins to respond to external stimuli. Mutation in these genes can reduce fungal pathogenicity and impact vital cell functions, conidiation, and toxin production. Signaling pathways and their interactions vary considerably among different fungi, disruption of which may cause multiple effects. Although disruption of the signaling pathway gene PMK1 in *M. grisea* produced normal mycelium and conidia, it affected appressorium formation, which rendered these mutants non-pathogenic even when inoculated into the wound. Similarly, *C. lagenaium* mutants with disrupted CMK1 reduced pathogenicity and impaired appressorial melanization, conidiation, and conidia germination (Takano et al., 2000).

PATHOGENICITY GENES IN COLLETOTRICHUM TRUNCATUM

Colletotrichum truncatum, the causal organism of chili anthracnose, causes infection by following an asymptomatic and short endophytic lifestyle different from the hemibiotrophic lifestyle adopted by the rest of the genus species. Genomic analysis revealed various pathogenicity genes encoding effectors, cell wall degradation enzymes, secondary metabolites, and secretory proteins, crucial in host–pathogen interactions. It was rich in pectin lyases, effectors, and flavin adenine dinucleotide oxidases, all of which play an important role in plant–pathogen interactions. CtNnudix is an effector expressed by the host-specific *C. lentis* during the necrotrophic phase. This gene has no homologue in biotrophs or necrotrophs (Bhadauria et al., 2013; Crouch et al., 2014).

CAZYMES

CAZyme analysis revealed lineage-specific expansion of pectinases and cutinases in fruit-rotting fungi *C. truncatum*, and *C. fructicola*. A group of genes in CAZyme families help fruit-rotting fungi degrade complex fruit cell walls, an evolutionary advantage. It plays a crucial role, especially during early infection, by releasing chitinases or chitin oligomers, Which prevent pathogen recognition by the host defense system. It also enables chili pathogens to employ stealth strategy or quiescent phase during the endophytic life phase (Amselem et al., 2011; Kombrink and Thomma, 2013; Ranathunge and Sandani, 2016).

CELL WALL DEGRADATION

Serine and metallo proteases, major cell wall degradation enzymes, are most important in the *Colletotrichum* species. A family of proteolytic enzymes, the aspartic protease, which helps digest fruit cell walls in an acidic environment, is also present in *Colletotrichum*. Interestingly, proteases and CAZymes are expanded in *Trichoderma*, probably due to its saprophytic nature, where, these enzymes help in degrading cell walls and other dead plant parts. The presence of metallo-proteases in *C. truncatum* indicates a link between necrotrophs and metallo-proteases. A single gene of fungalysin, zinc metallo-proteases, is present in most fungi of Sordariomycetes. Phylogeny studies suggest that gene duplication occurred even before the appearance of this fungal family, followed by loss of gene copies in most of its members, thus explaining the two copies of fugalysins in *C. truncatum* species (Sanz-Martín et al., 2016).

Most *Colletotrichum* species release ammonia to make the local environment alkaline, so that hydrolases, such as pentolytic enzymes, can perform efficiently, likewise function is performed by alkaline proteases of subtilisin family. It gives these species the ability to adjust/withstand pH of the encountered environment, either acidic or alkaline, at different life stages during infection. This pathogenicity gene is most likely transferred through Horizontal Gene Transfer (HGT) from the plant and called Colletotrichum plant-like subtilisins (Gan et al., 2013).

SECONDARY METABOLITES

Colletotrichum species have a specific set of secondary metabolite gene clusters, transporters, and transcription factors that regulate the transcription and affect virulence, nutrient uptake, and export of produced toxins. This cluster of 73 genes in *C. truncatum* explains their broad host range. It also has a unique set of polyketide and Non-Ribosomal peptide genes, with a greater ability of secondary metabolite production, which plays an important role in appressorium penetration and pathogenicity. PHI-base is a pathogen–host interactions database that serves as an authentic catalogue for pathogenicity genes (Rao and Nandineni, 2017).

PATHOGENICITY GENES OF PLANT BACTERIA

Bacterial pathogens cause various diseases in plants, such as blight, leaf spots, wilt, and canker. There is a distinct molecular mechanism of disease development in each case. Several secretory systems are important pathogenicity factors in plant pathogenic bacteria. *Xanthomonas*, a gram-negative bacterium, uses a T3SS to penetrate hosts (Palmieri et al., 2010). Bacteria deliver effector proteins directly into host cells using the secretory system, clustered on pathogenicity islands (Winstanley and Hart, 2001).

ADHESION ON THE PLANT SURFACE

Pathogenicity islands in *Xanthomonas* comprise avirulence genes and hypersensitive response and pathogenicity clusters. Hrp genes monitor the host–pathogen interaction, whereas Avr genes produce effectors and initiate hypersensitivity response in resistant hosts (Astua-Monge et al., 2005). The Hrp gene hrpG regulates the gene cluster cascade and gene functions. However, hpaA is an effector gene that disturbs the plant immune system and accelerates pathogenicity. It also plays a crucial role in pilus secretion and effector and translocon expression and transport to host; hence, it is an important component of the *Xynthomonas* secretary system (Lorenz et al., 2008).

Bacteria enter through natural openings or injuries in the host plant; hence, they require an adhesion mechanism rather than penetrating structures. *Agrobacterium* is a crown gall pathogen that attaches to the plant surface to transfer its T-DNA and cause symptom development. The initial attachment step involves three major components—glucan synthesis, cellulose synthesis, and *att* region—which contain genes for attachment. Additionally, it also has pilus biosynthesis genes. Others, such as *Xanthomonas, Xylella*, and *Pseudomonas*, have various gene homologs to pili genes (IV type), which is important in responding to environmental stresses and bacterial attachment in a turbulent xylem environment, where bacterial pathogen adheres in conjunction with polysaccharides (Rakhashiya et al., 2016).

BACTERIAL SECRETION SYSTEMS

Bacterial proteins and other essential chemicals are translocated outside with the help of secretion systems. These bacterial secretion systems are divided into five categories on the basis of the resultant protein. Type I-SS utilizes ATP hydrolysis energy for exporting plant pathogenic toxins as cyclolysin and hemolysins. Type II-SS is present in gram-negative bacteria and export proteins in two steps: secretory pathway in the inner membrane unfolds proteins to the periplasm; then, through a 12–14 proteins-based apparatus, folded proteins in the periplasm are secreted out of the outer membrane. *Ralstonia* and *Xanthomonas* have two, whereas *Xylell* and *Agrobacteria* have only one Type II-SS per cell, which secretes pathogenicity gene products such as pectinases and cellulases (Korotkov et al., 2012; Green and Mecsas, 2016).

Type III-SS transports effectors in the host plant cell. Protein component encoder genes have 75% similarity and are called hypersensitive response conserved (Hrc) genes. In contrast, transported protein encoder genes have a 35% similarity. Type IV-SS exports macromolecules into the plant cell. In *Agrobacterium tumefaciens*, the T-DNA strand is transported from pathogen to host sell with the help of a 11 proteins structure that extends from the inner bacterial membrane to the outer membrane and ends in a protruding pilus-like structure (Cascales and Christie, 2003; Troisfontaines and Cornelis, 2005).

CELL WALL DEGRADATION AND EXTRACELLULAR POLYSACCHARIDE GENES

Cell wall-degrading bacteria produce pectinolytic enzymes (most important pathogenicity enzymes), proteases, and cellulases. Pectinases (pectate lyase, pectinlyase, pectin methyl esterase, and poly-galacturonases) degrade host middle lamella components and are indirectly responsible for cell death. The expression of pectic enzymes and isozymes in Erwinia is sequential and controlled by separately regulated genes. The presence of quorum sensing, a regulatory system, maximizes the enzymatic activity of significant enzymes. Mutation at a single gene does not significantly affect the pathogenicity of bacteria, as several genes are encoding the cell maceration proteins (Chang et al., 2016).

When bacterial population and homo-serine lactose (HSL) levels during infection reach a critical point, extracellular enzymes are secreted, consequently developing maceration symptoms in the soft rot–affected host. HSL genes are critical for pathogenicity, and their disruption results in reduced pathogenicity. Bacterial propagation within host tissue is due to quorum sensing without triggering plant defense phytoalexins. Hence, bacterial cell wall degradation enzymes facilitate cell penetration and colonization; additionally, they are also responsible for typical symptom development. *X. campestris*, the causal organism of black rot in crucifers, has genes encoding two pectin esterases, whereas *X. campestris pv citri*, causing citrus canker, lacks in pectin esterases. As described earlier cell wall degradation enzymes are responsible for tissue maceration, and their absence explains the different symptomology of two diseases (Agrios, 2005).

Extracellular polysaccharides (EPS) directly intervene in host cells and avoid oxidative stress resistance to bacteria. EPS1 in *Ralstonia solanacearum* is produced by involving 12 genes and is produced in high concentrations, making it almost 90% of the total polysaccharides. Amylovoran is the major component of EPS in fire blight bacterium. Its biosynthesis involves various genes, and mutants with compromised amylovoran production are no more pathogenic. Regulatory system proteins in bacteria, VirA (membrane sensor) and VirG (cytoplasmic response regulator), sense wounds through plant exudations and respond to susceptible cells. Disruption of these genes results in a complete or significant loss of pathogenicity (Cho and Winans, 2005; Subramoni et al., 2014).

PATHOGENICITY GENES OF PLANT VIRUSES

Viruses have genes for coat proteins to protect nucleic acid, Nucleic acid replicates to make copies of the genome, movement genes to help it move during infection, and some might have additional genes for vector transmission. The virus only codes the proteins, whereas the host plant produces them. Furthermore, viruses use essential life steps such as transcription and movement by utilizing host proteins. Viruses need supportive host factors and escape from the host immunity system for effective infection. Virus pathogenesis occurs when infection alters the host's physiology, resulting in phenotypic growth abnormalities in plants or symptoms (Culver and Padmanabhan, 2007).

RNA REPLICASE GENES AND RELATED PROTEINS

RNA polymerase affects virus replication, so crucial in pathogenesis. Mutation strain, a single mutation in the sequence of RNA replicase of plum pox virus, drastically increases disease symptom in pea plants. The tobacco mosaic virus (TMV) showed symptom attenuation when a single amino acid mutation in its RNA replicase–related protein was made (Lewandowski and Dawson, 1993). Viral polymerases also break the resistance sources, such as a single mutation in Tomato Mosaic Virus (ToMV) allows it to escape the resistance gene tm-1. Similarly, a single-point mutation in the protein of PVX replicase enables the potato virus to escape the resistance gene JAX1 (Sugawara et al., 2013).

PATHOGENICITY GENES AND COAT PROTEINS

Virus assembly and disassembly are essential for its multiplication and pathogenicity. Coat proteins play a vital role by destabilizing coat protein RNA and releasing a few sub-units. Ribosomes find the exposed end and initiate RNA replicase translation. This translation removes coat protein sub-units. Coat protein (CP) interacts with movement proteins.

Viral coat proteins are multifunctional, involving encapsidation, viral replication, and movement. In some cases, coat proteins are necessary even for the cell-to cell-movements and infections. Single mutations in the coat proteins of TMV and Turnip Crinkle Virus (TCV) resulted in reduced pathogenicity against specific hosts. These coat proteins can also elicit host responses, e.g., a single mutation in TMV coat protein resulted in a hypersensitivity response in woodland tobacco (*Nicotiana sylvestris*) (Duff-Farrier et al., 2015).

TYPICAL PLANT VIRUS SYMPTOMS AND UNDERLYING MOLECULAR PHENOMENON

The most common viral infection symptoms are light and dark green mosaic on plant leaves, indicating an attack on the host's chloroplasts and the consequent impact on photosynthesis. The tobamo virus disturbs the oxygen-evolving complex in host plants by inhibiting electron transport in Photosystem II. Alfalfa mosaic virus (AMV) coat protein, upon interaction with PsbP, inhibits virus replication. AMV coat protein acts as an effector upon interaction and prevents PSII biogenesis, resulting in reactive oxygen species (ROS) and plant defense activation. In the case of ToMV, its coat protein interacts with IP-L, tomato thylakoid membrane protein. This interaction disturbs chloroplast functioning, resulting in chlorosis (Zhang et al., 2008).

Viruses alter the host plant cell cycle through proteins—Rep proteins in Gemini viruses—upon interaction with the retinoblastoma-related proteins of the host. This interaction inhibits the host's pRBR activity. This alters the host cell cycle and puts it in the S phase; resultantly, the virus assumes control of host replication machinery. The Faba bean necrotic yellows virus produces F-box proteins that interact with Skp-1 and pRBR of *Medicago sativa*, and its capacity to bind with these host proteins determines viral replication (Aronson et al., 2000; Kong et al., 2000).

CONCLUSION

Plant pathogens have evolved various genes for overcoming the plant immune system to ensure pathogenicity upon host–pathogen interaction. Mutations in these genes are common, resulting in more severe disease incidence or equipping a microbe to attack a new host. Several pathogenicity genes have been identified so far in fungi, bacteria, and viruses, but there is still a need to find many and explore their interaction with host cells in order to establish a causal relationship. Research on pathogenicity genes is crucial, as it helps incorporate new R-genes in host plants to develop resistant crop varieties. This chapter has discussed important pathogenicity genes and their function and possibilities upon interaction with the host.

REFERENCES

Agrios, G.N. 2005. *Introduction to plant pathology*. Elsevier, Burlington, MA.

Amselem, J., C.A. Cuomo, J.A. van Kan, M. Viaud, E.P. Benito, A. Couloux, P.M. Coutinho, R.P. de Vries, P.S. Dyer, and S. Fillinger. 2011. Genomic analysis of the necrotrophic fungal pathogens Sclerotinia sclerotiorum and Botrytis cinerea. *PLoS Genetics* 7:e1002230

Andrivon, D. 1993. Nomenclature for pathogenicity and virulence: The need for precision. *Phytopathology* 83:889–890

Aronson, M.N., A.D. Meyer, J.n. Györgyey, L. Katul, H.J. Vetten, B. Gronenborn and T. Timchenko. 2000. Clink, a nanovirus-encoded protein, binds both pRB and SKP1. *Journal of Virology* 74:2967–2972

Astua-Monge, G., J. Freitas-Astua, G. Bacocina, J. Roncoletta, S.A. Carvalho and M.A. Machado. 2005. Expression profiling of virulence and pathogenicity genes of Xanthomonas axonopodis pv. citri. *Journal of Bacteriology* 187:1201–1205

Bhadauria, V., S. Banniza, A. Vandenberg, G. Selvaraj and Y. Wei. 2013. Overexpression of a novel biotrophy-specific Colletotrichum truncatum effector, CtNUDIX, in hemibiotrophic fungal phytopathogens causes incompatibility with their host plants. *Eukaryotic Cell* 12:2–11

Cascales, E. and P.J. Christie. 2003. The versatile bacterial type IV secretion systems. *Nature Reviews Microbiology* 1:137–149

Chang, H.-X., C.R. Yendrek, G. Caetano-Anolles and G.L. Hartman. 2016. Genomic characterization of plant cell wall degrading enzymes and in silico analysis of xylanses and polygalacturonases of *Fusarium virguliforme*. *BMC Microbiology* 16:1–12

Cheng, Y.-Q., J.-H. Ahn and J.D. Walton. 1999. A putative branched-chain-amino-acid transaminase gene required for HC-toxin biosynthesis and pathogenicity in *Cochliobolus carbonum* The GenBank accession number for the nucleotide sequence reported in this paper is AF157629. *Microbiology* 145:3539–3546

Cheng, Y.-Q. and J.D. Walton. 2000. A eukaryotic alanine racemase gene involved in cyclic peptide biosynthesis. *Journal of Biological Chemistry* 275:4906–4911

Cho, H. and S.C. Winans. 2005. VirA and VirG activate the Ti plasmid repABC operon, elevating plasmid copy number in response to wound-released chemical signals. *Proceedings of the National Academy of Sciences* 102:14843–14848

Ciuffetti, L.M., R.P. Tuori and J.M. Gaventa. 1997. A single gene encodes a selective toxin causal to the development of tan spot of wheat. *The Plant Cell* 9:135–144

Crouch, J., R. O'Connell, P. Gan, E. Buiate, M.F. Torres, L. Beirn, K. Shirasu and L. Vaillancourt. 2014. *The genomics of Colletotrichum, Genomics of plant-associated fungi: Monocot pathogens*. Springer-Verlag, Berlin, Heidelberg. pp. 69–102.

Culver, J.N. and M.S. Padmanabhan. 2007. Virus-induced disease: Altering host physiology one interaction at a time. *Annual Review of Phytopathology* 45:221–243

Duff-Farrier, C.R., A.M. Bailey, N. Boonham and G.D. Foster. 2015. A pathogenicity determinant maps to the N-terminal coat protein region of the P epino mosaic virus genome. *Molecular Plant Pathology* 16:308–315

Enkerli, J.r., G. Felix and T. Boller. 1999. The enzymatic activity of fungal xylanase is not necessary for its elicitor activity. *Plant Physiology* 121:391–398

Gan, P., K. Ikeda, H. Irieda, M. Narusaka, R.J. O'Connell, Y. Narusaka, Y. Takano, Y. Kubo and K. Shirasu. 2013. Comparative genomic and transcriptomic analyses reveal the hemibiotrophic stage shift of Colletotrichum fungi. *New Phytologist* 197:1236–1249

Green, E.R. and J. Mecsas. 2016. Bacterial secretion systems: An overview. *Microbiology Spectrum* 4:4.1.13

Han, Y., X. Liu, U. Benny, H.C. Kistler and H.D. VanEtten. 2001. Genes determining pathogenicity to pea are clustered on a supernumerary chromosome in the fungal plant pathogen Nectria haematococca. *The Plant Journal* 25:305–314

Have, A.t., W. Mulder, J. Visser and J.A. van Kan. 1998. The endopolygalacturonase gene Bcpg1 is required for full virulence of Botrytis cinerea. *Molecular Plant-Microbe Interactions* 11:1009–1016

Idnurm, A. and B.J. Howlett. 2001. Pathogenicity genes of phytopathogenic fungi. *Molecular Plant Pathology* 2:241–255

Isshiki, A., K. Akimitsu, M. Yamamoto and H. Yamamoto. 2001. Endopolygalacturonase is essential for citrus black rot caused by *Alternaria citri* but not brown spot caused by Alternaria alternata. *Molecular Plant-Microbe Interactions* 14:749–757

Jones, J.D. and J.L. Dangl. 2006. The plant immune system. *Nature* 444:323–329

Kombrink, A. and B.P. Thomma. 2013. LysM effectors: Secreted proteins supporting fungal life. *PLoS Pathogens* 9:e1003769

Kong, L.J., B.M. Orozco, J.L. Roe, S. Nagar, S. Ou, H.S. Feiler, T. Durfee, A.B. Miller, W. Gruissem and D. Robertson. 2000. A geminivirus replication protein interacts with the retinoblastoma protein through a novel domain to determine symptoms and tissue specificity of infection in plants. *The EMBO Journal* 19:3485–3495

Korotkov, K.V., M. Sandkvist and W.G. Hol. 2012. The type II secretion system: Biogenesis, molecular architecture and mechanism. *Nature Reviews Microbiology* 10:336–351

Lau, G. and J.E. Hamer. 1996. Regulatory genes controlling MPG1 expression and pathogenicity in the rice blast fungus Magnaporthe grisea. *The Plant Cell* 8:771–781

Lewandowski, D.J. and W.O. Dawson. 1993. A single amino acid change in tobacco mosaic virus replicase prevents symptom production. *Molecular Plant Microbe Interactions* 6:157–157

Lorenz, C., O. Kirchner, M. Egler, J. Stuttmann, U. Bonas and D. Büttner. 2008. HpaA from Xanthomonas is a regulator of type III secretion. *Molecular Microbiology* 69:344–360

Palmieri, A.C.B., A.M.d. Amaral, R.A. Homem and M.A. Machado. 2010. Differential expression of pathogenicity-and virulence-related genes of Xanthomonas axonopodis pv. citri under copper stress. *Genetics Molecular Biology* 33:348–353

Rakhashiya, P.M., P.P. Patel, B.P. Sheth, J.G. Tank and V.S. Thaker. 2016. Detection of virulence and pathogenicity genes in selected phytopathovars. *Archives of Phytopathology Plant Protection* 49:64–73

Ranathunge, N.P. and H.B.P. Sandani. 2016. Deceptive behaviour of *Colletotrichum truncatum*: Strategic survival as an asymptomatic endophyte on non-host species. *Journal of Plant Protection Research* 56

Rao, S. and M.R. Nandineni. 2017. Genome sequencing and comparative genomics reveal a repertoire of putative pathogenicity genes in chilli anthracnose fungus *Colletotrichum truncatum*. *PLoS One* 12:e0183567

Rasmussen, J. and R. Hanau. 1989. Exogenous scytalone restores appressorial melanization and pathogenicity in albino mutants of *Colletotrichum graminicola*. *Canadian Journal of Plant Pathology* 11:349–352

Rogers, L.M., Y.-K. Kim, W. Guo, L. González-Candelas, D. Li and P.E. Kolattukudy. 2000. Requirement for either a host-or pectin-induced pectate lyase for infection of *Pisum sativum* by *Nectria hematococca. Proceedings of the National Academy of Sciences* 97:9813–9818

Sanz-Martín, J.M., J.R. Pacheco-Arjona, V. Bello-Rico, W.A. Vargas, M. Monod, J.M. Díaz-Mínguez, M.R. Thon and S.A. Sukno. 2016. A highly conserved metalloprotease effector enhances virulence in the maize anthracnose fungus *Colletotrichum graminicola. Molecular Plant Pathology* 17:1048–1062

Schoonbeek, H., G. Del Sorbo and M. De Waard. 2001. The ABC transporter BcatrB affects the sensitivity of *Botrytis cinerea* to the phytoalexin resveratrol and the fungicide fenpiclonil. *Molecular Plant-Microbe Interactions* 14:562–571

Schürch, S., C.C. Linde, W. Knogge, L.F. Jackson and B.A. McDonald. 2004. Molecular population genetic analysis differentiates two virulence mechanisms of the fungal avirulence gene NIP1. *Molecular Plant-Microbe Interactions* 17:1114–1125

Segers, G., N. Bradshaw, D. Archer, K. Blissett and R. Oliver. 2001. Alcohol oxidase is a novel pathogenicity factor for *Cladosporium fulvum*, but aldehyde dehydrogenase is dispensable. *Molecular Plant-microbe Interactions* 14:367–377

Shieh, M.-T., R.L. Brown, M.P. Whitehead, J.W. Cary, P.J. Cotty, T.E. Cleveland and R.A. Dean. 1997. Molecular genetic evidence for the involvement of a specific polygalacturonase, P2c, in the invasion and spread of *Aspergillus flavus* in cotton bolls. *Applied Environmental Microbiology* 63:3548–3552

Subramoni, S., N. Nathoo, E. Klimov and Z.-C. Yuan. 2014. *Agrobacterium tumefaciens* responses to plant-derived signaling molecules. *Frontiers in Plant Science* 5:322

Sugawara, K., T. Shiraishi, T. Yoshida, N. Fujita, O. Netsu, Y. Yamaji and S. Namba. 2013. A replicase of Potato virus X acts as the resistance-breaking determinant for JAX1-mediated resistance. *Molecular Plant-microbe Interactions* 26:1106–1112

Takano, Y., T. Kikuchi, Y. Kubo, J.E. Hamer, K. Mise and I. Furusawa. 2000. The *Colletotrichum lagenarium* MAP kinase gene CMK1 regulates diverse aspects of fungal pathogenesis. *Molecular Plant-Microbe Interactions* 13:374–383

Troisfontaines, P. and G.R. Cornelis. 2005. Type III secretion: More systems than you think. *Physiology* 20:326–339

van der Does, H.C. and M. Rep. 2007. Virulence genes and the evolution of host specificity in plant-pathogenic fungi. *Molecular Plant-Microbe Interactions* 20:1175–1182

Wasmann, C. and H. VanEtten. 1996. Transformation-mediated chromosome loss and disruption of a gene for pisatin demethylase decrease the virulence of *Nectria haematococca* on pea. *Molecular plant-microbe interactions: MPMI* 9: 793–823

Winstanley, C. and C.A. Hart. 2001. Type III secretion systems and pathogenicity islands. *Journal of Medical Microbiology* 50:116–126

Zhang, C., Y. Liu, X. Sun, W. Qian, D. Zhang and B. Qiu. 2008. Characterization of a specific interaction between IP-L, a tobacco protein localized in the thylakoid membranes, and Tomato mosaic virus coat protein. *Biochemical Biophysical Research Communications* 374:253–257

5 Molecular Mechanisms– Relayed Plant Defense Responses Against Fungal Pathogens

Aşkım Hediye Sekmen Çetinel, Barbaros Çetinel, Azime Gokce, Cemre Tatli and Erhan Erdik

CONTENTS

INTRODUCTION

Plants are the world's primary producers, who harvest the sun's rays and convert light energy into chemical energy. Furthermore, they constitute over half the organisms on earth (Windsor, 1998). Plants are repeatedly challenged by pathogens such as fungi, bacteria, viruses, and even other plants along their life span. Plant pathogens may affect the composition of plant populations and cause the local extinction of host

DOI: 10.1201/9781003162742-5

plant species and yield reductions in crops (Gilbert, 2002; Alexander et al., 2014). Plants have evolved complicated mechanisms to protect themselves from pathogens attacks. In a particular species or cultivar, a small number of pathogens can cause disease (compatible response), whereas most possible aggressors are blocked by plant defenses (incompatible response) in their penetration. Plants have physical defenses (cuticles and cell walls), defense compounds (antioxidants, defense peptides, secondary metabolites, and antimicrobial proteins), and a variety of biological and molecular mechanisms to prevent pathogen attacks (De Gara et al., 2003).

Pathogens have to overcome these host defense strategies to reach water and nutrients in their host plants. Pathogens such as protozoa, some bacteria, and most viruses are placed directly into host plant cells, while almost all fungi first come into contact with the host plant's external surface through their spores adhesion that results in penetration and colonization of the host plant. They encounter the plant cell wall, an exoskeleton around the cell protoplast, and a plant cuticle. The plant cell wall consists of highly integrated polysaccharides, such as pectin, cellulose, and hemicellulose—a dynamic structure remodeled in response to pathogen attack. To enter the host plant, pathogen fungi must break down enzymes and structural proteins in the host plant's cell wall by using mechanical force or releasing cell wall-degrading enzymes.

On the other hand, at sites of attempted pathogen penetration, plants often deposit callose-rich cell wall appositions and accumulate phenolic compounds and various toxins in the cell wall (Fig. 5.1). Furthermore, lignin-like polymers are also synthesized to strengthen the wall (Hématy et al., 2009). Therefore, it is suggested that one of the important protective structures of plant cells against pathogen attacks may be the cell wall.

From biotroph pathogens to necrotroph pathogens, various pathogen lifestyles affect interactions with the plant cell wall. Biotrophic fungi can only grow and reproduce in living hosts in nature, while necrotrophic fungi feed only on dead organic material. Furthermore, hemibiotrophic fungi combine both strategies. Necrotrophic fungi release abundant amounts of cell wall-degrading enzymes, while biotrophic fungi release only a few cell wall-degrading enzymes (Hématy et al., 2009). Necrotrophs secrete cell wall-disrupting enzymes that attempt to break down plant cells before they make an effective defense (Hématy et al., 2009). Among these, polygalacturonase (PG), a pectin-degrading enzyme, is one of the first proteins that the pathogen secretes when it enters the plant cell via penetrating its cell wall (Cooper and Wood, 1975). When pathogens secrete PG, it first attacks the complex pectin network in the host's cell wall, breaking down the primary cell wall. In turn, the plant activates polygalacturonase inhibitor proteins (PGIPs) in its cell walls and inhibits PG activity by preventing pathogen fungi penetration into the cell.

Furthermore, activating PGIPs can also result in the release of other cell wall components that are involved in various plant defense responses, e.g., when pathogens release plant degradative enzymes, such as plant chitinases, plants can also release pathogen-dependent molecular patterns (PAMPs). After sensing these PAMPS, plants activate various defense strategies, such as producing reactive oxygen species (ROS) and antimicrobial compounds, exporting these compounds, and

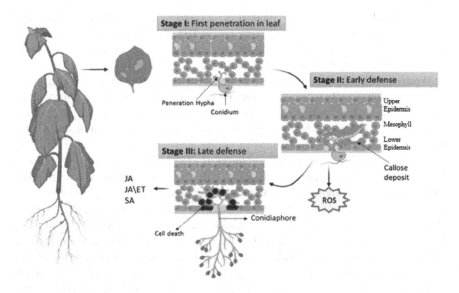

FIGURE 5.1 Penetration stages in Nicotiana tobaccum leaves.

strengthening cell wall. Besides, even if plants do not self-recognize PAMPs, they can directly detect the presence of pathogens and activate intracellular defenses, such as the salicylic acid (SA) and jasmonic acid (JA) pathways. This prepares the host plant for the next battle phase (Hématy et al., 2009). However, the nature of the plant's specific defense responses depends not only on the biotrophic or necrotrophic lifestyle of the pathogen but also on other factors, such as the timing of the pathogen attack, recognition timing, type of plant tissue in which the defense is expressed, and the activity of pathogen effectors (De Vos et al., 2005). This review is focused on the biochemical and molecular bases of plant resistance and defense responses to fungal pathogens.

PLANT DEFENSE RESPONSES TO FUNGAL PATHOGEN

The penetration of pathogen into plant tissues occurs in three stages. The first contact of the pathogen with the host involves the pathogen attacking the plant's cell wall (Stage I). As mentioned earlier, some fungi can directly penetrate the cell wall through specialized penetration structures, such as appressoria. In contrast, others use natural openings (stomata, lenticel, etc.), wounds, and insect-caused wounds (*Grosmannia clavigera* attack and *Ophiostomata ulmi* attack, etc.). After the plants recognize the pathogen, they quickly close their natural opening, stoma, and the defense begins. This defense type is pre-invasive defense.

On the other hand, pathogens face the second barrier of early post-invasive protection following successful penetration (Stage II). Cellular and biochemical reactions by the plants, such as papilla formation and rapid ROS accumulation, occur at the regions of pathogen attack, and Stage II is followed by transcriptomic and metabo-lomic reprogramming. In the final stage (Stage III), phytohormones responsible for defense (JA, ET [ethylene], SA, etc.) and intracellular/extracellular signaling mole-cules are produced to halt the pathogen's progression (Fig. 5.1).

Wounds: As explained earlier, some pathogen fungi disrupt the cell wall enzy-matically or physically to access water and nutrients in the plant protoplast. Thus, small wounds occur where the pathogen penetrates the plant cell wall. Once wounding occurs, plant defense responses are induced. It is demonstrated that wound-induced systemic resistance in young rice plants protects against infec-tion by the rice blast fungus (Schweizer et al., 1998). In the regions of pathogen attack and at the damaged sites, the plant cell wall is strengthened with papillae containing callose, lignin-like polymers, structural proteins, and antimicrobial proteins (Hématy et al., 2009). Callose is often accumulated around the edges of wounds and completely encloses attacked cells occasionally (Fig. 5.1 Callose deposit). Thus, this polysaccharide is known to serve as a physical reinforcement in the region of pathogen attack. First, this polysaccharide forms a physical bar-rier in the damaged areas. For example, the β-1, 3-d-glucan callose is rapidly synthesized and deposited at the damaged sites. Furthermore, it also participates in the retention of pathogen-derived elicitors at the infection sites of the glucan synthase.

In addition to the accumulation of callose, lignin, and extensins, hydrophobic polymer and hydroxyproline-rich protein, respectively, are deposited at the damaged cells to strengthen the wall. Therefore, damaged cell walls are transformed into more robust structures with the accumulation of lignin or extensins (Hématy et al., 2009). Cellulose is the most vital and abundant component of the plant cell wall. Therefore, when a fungus enters a plant cell, it first breaks down cellulose to obtain glucose. Deficiencies in the callose or cellulose directly activate various signaling pathways, such as JA, ET, and abscisic acid (ABA). For example, it was demonstrated that wounding causes an increase in JA and ET levels. They are important signal mol-ecules in wound-induced systemic resistance that results in activation of acidic PR genes (Wasternack and Parthier, 1997).

Stoma: Stoma is one of the most important cell types in plants. Most abiotic and biotic influences, including radiation and the plant hormone ethylene, control stomatal movements (Wilkinson and Davies, 2010). These microscopic pores in the epidermis allow plants to photosynthesize, sweat, and exchange gases. However, the large number of pores on the plant surface provide various pathogens opportunities to enter the plant. Rust fungi, such as *Uromyces* and *Puccinia,* start the infection by entering through plants' natural openings, such as stomata. These fungi directly generate appressoria on the stomatal guard cells. Plants can improve their pre-in-vasive resistance to penetration by quickly closing stomata upon microbe sensing, which occurs within one hour after a pathogen attack (Ton et al., 2009). Fungal toxin causes a decrease in stomatal conductance. Some fungal toxins promote a stomatal opening by stimulating H+-pump in the plasma membrane by triggering H^+-ATPase (Wang et al., 2014). Stomatal closure is seen as a defense in plants under a fungal attack (Dehgahi et al., 2015). Even stomatal closure is an initial response of the plant defense against pathogen attack.

ELICITOR MOLECULES IN PLANT DEFENSE RESPONSES

Elicitors are the chemicals bio-factors from diverse resources that can induce physiological and morphological responses and phytoalexin accumulation in the target living organisms (Gowthami, 2018). Depending on their origin and molecular structure, they are classified as biotic and abiotic elicitors. Elicitors of non-biological origin are called "Abiotic elicitors" (Baenas et al., 2014). Abiotic elicitors, such as heavy metal ions, UV light, and several metabolic inhibitors, trigger physiological stress responses and contribute to plant resistance. Biotic elicitor molecules are of pathogen or host origin and can stimulate defense responses (such as phytoalexin accumulation or hypersensitive response) in plant tissues. The first biotic elicitors were found in 1968.

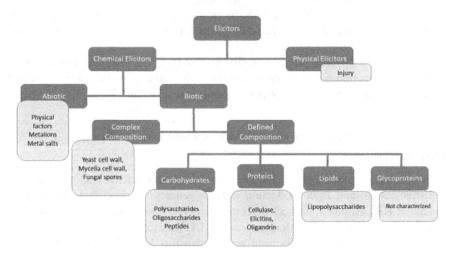

FIGURE 5.2   Classification of Elicitors

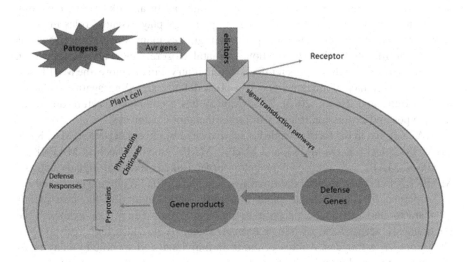

FIGURE 5.3   Elicitor-mediated defense responses in plants

Pattern recognition receptors (PRRs) containing Leucine-rich repeat preferentially bind proteins or peptides. When elicitors connect to the PRR receptors on the cell wall, signal transduction pathways get activated, resulting in defense gene production. These defense genes release several gene products that help in the primary immune response of plants, such as secondary metabolites (abiotic stress), pathogenesis-related (PR) proteins, phytoalexins, and chitinases (biotic stress) (Gowthami, 2018).

PR Proteins and Defense Responses in Plant

After the fungus comes into contact with the host plant, either infection occurs or the plant resists itself through a resistance mechanism. During the resistance, a series of responses occurs around the plant cells, including the oxidative stress that causes cell death. Many PR proteins are induced by the fast, transient production of enormous amounts of ROS during oxidative stress.

In addition to systemic acquired resistance (SAR), some plant hormones such as SA, JA, ET function as signaling molecules in the induced resistance process and control basal resistance against phytopathogens (Mengiste, 2012).

Mitogen-activated protein kinases (MAPKs) Signaling in Plant Defense Responses

When pathogens attack a plant, the plant activates various defense responses, including stomatal closure, ROS production, defense genes activation, phytoalexin accumulation, HR cell death, and cell wall modification. These defensive responses are controlled by a complex signaling network—MAPK cascades.

In the last few years, it has been well known that MAPKs play a central role in defense against pathogens in various plants, and MAPK controls the signaling events initiated by pathogens. Protein kinase (MAPK) activated by mitogens is a highly conserved general signal transduction pathway in all eukaryotes, including yeast, animals, and plants. MAPK signaling pathways play essential roles in plants rather than yeast and animals. This pathway regulates many stress stimuli, such as growth, proliferation, ROS production, stomatal closure, cell wall strengthening, programmed cell death, pathogen infection, and injury. Furthermore, the MAPK signaling pathway has significant impact in plant defense signaling against pathogen attack (Meng and Zhang, 2013). This pathway has a complex structure due to its many types and expression in many different metabolic pathways.

MAPKs exist in the form of triple kinase activated by phosphorylation: MAPKKK (MAP3K), MAPKK (MAP2K), and MAPK. MAPK enzymes such as MAPKKK–MAPKK–MAPK are top-down receptor–target linked. External signals activate MAPKKKs via the phosphorylation of two Ser/Thr motifs (Widmann et al., 1999).

In plants, MAPKs activate plant immunity while regulating the host plant's mechanical and enzymatic penetration into pathogenic fungi. For example, the continuously active form of MAPKK in *Arabidopsis* provides high resistance to *Phytophthora* pathogens. Appressorium is an individual cell specific to many types of fungi that is used to infect host plants.

Phytohormones play essential roles in the plant developmental process, such as in maturation, leaf and flower senescence, root development, and plant defense responses to abiotic and biotic stresses. In particular, ABA, ET), SA, and JA have vital functions in pathogen defense responses. In this part of the chapter, we review the role of these hormones in stress responses in plants.

ABA

ABA mediates diverse physiological and developmental processes in plants, such as seed germination, stomatal conductance, vegetative growth, and bud dormancy (Wasilewska, 2008); it is well known for its role in abiotic stress tolerance. ABA promotes resistance of some plants against pathogens, while it increases the sensitivity of others. For example, ABA-deficient tomato (*Solanum lycopersicum*) and ABA-deficient *Arabidopsis thaliana* mutants are susceptible to insect attack (Bodenhausen and Reymond 2007). Similarly, Boba et al. (2020) showed that the non-mevalonate signaling pathway in flax (*Linum usitatissimum*) is strongly activated within the first hours of pathogen infection, and metabolites are redirected to ABA synthesis. Besides, the increase in ABA synthesis increased the resistance in flax against *F. oxysporum*. On the other hand, ABA play opposite roles (negative or positive regulator) in different phases of plant defense (Fig 1. I, II, III Stage) on the basis of the type of pathogen, tissue type of host plant, route of entry of the pathogen into the host, and many other factors (Ton et al., 2009). As explained earlier, some pathogen fungi use pre-existing openings, such as stomata and wounds, to penetrate plant tissues. Plants withstand via closing stomata rapidly, i.e., pre-invasive penetration resistance (Fig 1. Stage I). ABA-deficient *aba3-1* plants infected by pathogenic bacteria fail to express rapid stomatal closure, limiting the pre-invasive penetration resistance of plants (Melotto et al., 2008). It is also known that stomatal closure can be inhibited by fungal virulence factors, such as fusicoccin and oxalate (Turner and Graniti, 1969; Guimaraes and Stotz, 2004). These results show that ABA induces plant pre-invasive defense against pathogen by mediating stomatal closure. However, ABA has negative or positive roles in early post-invasive penetration resistance (Fig 1. Stage II) against pathogen fungi. Asselbergh et al. (2007) found that the ABA-deficient tomato mutant *sitiens* is resistant to *Botrytis cinerea*, the necrotrophic fungus. This resistance is based on increased ROS accumulation by the fungus in the early stages of plant penetration. Contrary to these results, Abu Qamar et al. (2006) reported that the ABA-inducible myeloblastosis (MYB) transcription factor AIM1 (ABA-induced MYB1) regulates ABA sensitivity and basal resistance against *B. cinerea* in tomatoes via increased accumulation of root Na^+ ions. Similarly, it is reported that *aba1-3* and *abi1-1* mutants have high sensitivity against *Leptosphaeria maculans* due to decreased callose deposition. However, unlike *abi1-1*, the *abi 2-1* mutant is resistant to *L. maculans*. These results indicate that ABI1 and ABI2 play converse roles in the callose deposition against this pathogenic fungus. In addition to these studies, Merlot et al. (2001) reported that penetration resistance in ABA-inducible NAC transcription factor ATAF1 (*Arabidopsis* NAC domain–containing protein 1) mutants reduces against the non-host fungus *Blumeria graminis* f.sp. *hordei* (*Bgh*). Contrary to Stage

I and Stage II, during Stage III, ABA interacts with SA-, ET- and JA-dependent pathways. (Fig 1.) ABA inhibits this pathway-dependent resistance mechanism against pathogenic fungi (Mohr and Cahill, 2007; Ton et al., 2009).

ET

ET, a gaseous plant hormone, can be produced in all higher plants' organs. The precursor molecule of ET is methionine, and methionine is converted into S-adenosylmethionine and 1-aminocyclopropone-1-carboxylic acid (ACC) by the ACC synthase. Then, ACC is converted into ET by ACC oxidase. ET is an essential regulator of various plant physiological and developmental processes. Moreover, ET participates in the plant defense response by itself or by cross-talk with other phytohormones, such as JA.

Several research studies have shown that ET affects plant resistance against pathogens in different ways and ET induces the pathogen attacked–plant responses associated with disease resistance or disease susceptibility. When plants are treated with ET before pathogen inoculation, ET reduces disease development or does not affect it. However, when plants are treated with ET post infection, it increases disease development (Abeles et al., 1992; Loon et al., 2006). On the other hand, ethylene applications promote *Botrytis cinerea*–induced disease development (Elad, 1993).

Some studies have determined ET's effects on the symptom severity in the *Arabidopsis thaliana* ethylene-response mutants against pathogen fungi attack. In these studies, when the responses of *A. thaliana* ethylene-insensitive 2 (*ein2*) mutants were compared to different pathogenic fungi types, disease severity increased or decreased. For example, disease susceptibility in *A. thaliana ein2* increased against necrotrophic fungi such as *B. cinerea* (Berrocal et al., 2002), *Chalara elegans, Fusarium oxysporum* f sp. *lycopersici* and f sp. *Conglutinans*, and *Pythium* spp. (Geraats et al., 2002). In contrast, symptom severity in *A. thaliana* etr1–1 mutant decreased against the necrotrophic fungi *Verticillium dahliae*. These results indicate that ethylene reduces or stimulates diseases caused by various necrotrophic fungi. Therefore, ethylene's effect on disease development can depend on the timing of the exposure of plants to ethylene and plant–pathogen interactions.

Plants have two primary mechanisms for induced resistance against pathogens: SAR and induced systemic resistance (ISR). SAR is activated by plants only after they have been exposed to an elicitor from pathogenic fungi, bacteria, viruses, or non-pathogenic microbes. On the other hand, ISR is activated by infection. SA is necessary for SAR, while JA and ET are necessary for ISR.

For example, 1 mM ACC application caused a significant increase in the ISR of *A. thaliana* seedlings against *P. syringae* pv. *Tomato* (Pieterse et al., 2000). However, ET response mutants such as *ein2–1* and *etr1–1* developed severe symptoms after infection with *P. syringae* pv. *tomato*. Pieterse et al. (2000) showed that exogenous application of methyl jasmonate has no effect on the ISR generation in *etr1* or *ein2* mutants. These results indicated that generation of ISR in plants against pathogens is dependent on ET responsiveness.

SA

SA, the natural phenolic compound, is a critical signal transduction molecule in locally and systemically induced disease resistance responses against pathogenic fungi (Dempsey et al., 1999). It is required for symptom development (O'Donnell et al., 2003). When pathogens attack a plant, specific pathogen recognition mechanisms lead to a hypersensitive response (HR) in the infected plant's parts. Necrotic lesions in these areas occur as part of the HR or as a disease symptom caused by the pathogen. Meanwhile, other resistance responses, such as synthesis of phytoalexins, accumulation of antimicrobial compounds, and PR proteins are also induced in plants. One of the complex resistance responses in plants is SAR, of which SA is a crucial signaling component (Bakker et al., 2010). Gaffney et al. (1993) showed that transgenic NahG (bacterial salicylate hydroxylase gene) plants cannot accumulate SA and therefore cannot develop SAR.

Nevertheless, exogenous application of SA or its synthetic analog (BTH, benzo (1,2,3) thiadiazole-7-carbothioic acidS-methyl ester) causes an increase in the expression of PR genes and resistance to pathogens. These results indicate that SA accumulation is required for the SA-dependent defense responses, such as the activation of PR-gene expression and SAR (Gaffney et al., 1993). Preventing SA accumulation by degrading it to catechol or blocking SA synthesis inhibits SA-dependent defense responses.

In plants, SA can be synthesized in two ways: through the phenylpropanoid pathway or the isochorismate pathway. However, both pathways require the chemical chorismate derived from the shikimate pathway. Some bacteria have two enzymes that catalyze SA synthesis from the chorismate: isochorismate synthase (ICS) and isochorismate pyruvate lyase (IPL) (Serino et al., 1995). Similarly, plants also have analogous pathways. Recent studies indicated that overexpression of these two bacterial enzymes in transgenic plants caused a significant increase of SA accumulation (Serino et al., 1995; Verberne et al., 1999). Both *nahG* and *sid2* mutants exhibited reduced local and systemic resistance and were more susceptible to biotrophic pathogens due to SA biosynthesis dysfunction (Gaffney et al., 1993; Wildermuth et al., 2001).

SA works with other phytohormones, such as ET or JA (Vlot et al., 2009). Blocking the response to any of these can make plants more susceptible to pathogens or pests (Gaffney et al., 1993. These phytohormones cross-talk to determine the most appropriate response to pathogens under certain environmental conditions. In this cross-talk, the SA signal is mediated by at least two mechanisms: one is non-expressor of PR1 (NPR1) dependent, also called non-inducible immunity1 (NIM1), and the other is independent of NPR1 (Shah et al., 1997).

NPR1-Dependent SA Signaling

NPR1 is the primary regulator and SAR in plants through a pathway that leads to a massive increase of antimicrobial genes. NPR1 is also a negative regulator of ICS1 gene expression (Zhang et al., 2010). In *Arabidopsis*, at low SA levels, NPR1 is present in the cytoplasm as an oligomer. However, NPR3 and NPR4 bind residual NPR1in the nucleus, and NPR1 function is prevented. On the other hand, disulfide bonds in NPR1 oligomers are broken at high SA levels, and released monomeric NPR1 enters the nucleus. Here, it interacts with TGAs in SA-responsive promoters; defense responses are thereby activated.

Meanwhile, SA binds to NPR3 and NPR4. Furthermore, it blocks their transcriptional repression activity. Overexpression of NPR1 in plants such as *Arabidopsis* or rice increases plant resistance to the pathogen (M. Kinkema et al., 2000; Chern et al., 2001).

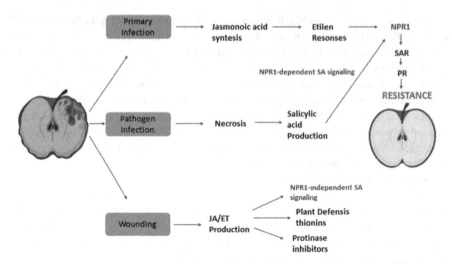

FIGURE 5.4 Pathogen infection and systemic signaling pathway after injury in apple. Primary infection: JA, ET, and SA phytohormones' production is always triggered during the pathogen attack. After NPR1, SAR is activated and resistance gained. SA-dependent SAR activation occur when tissues become necrotic as a result of pathogen attack. Activation of SAR can be induced by SA; thus, it is expressed in the PR environment, and resistance occurs. As a result of the injury, JA and ET production is triggered.

NPR1-Independent SA Signaling

Some plant–pathogen interactions show an SA-dependent defense mechanism independent of NPR1 (Fig. 4). NPR1-independent mechanisms have been elucidated in many studies in *A. thaliana* constitutive-defense-signaling mutants (Dong et al., 2001. For example, the SA-insensitivity2 (ssi2) gene provides acquired resistance to *Pseudomonas syringae* and *Peronospora parasitica* (Shah et al., 2001). The NPR1-independent, SA-dependent defense mechanism is activated in *Arabidopsis* ssi1 and the constitutive expression of PR genes (cpr5 and cpr 6) and HR-like lesions1 (hrl1) mutants. The cpr5 and cpr6 genes cross-talk with SA, JA, and ET signal molecules in defense against *Pseudomonas syringae* and *Peronospora parasitica* (Clarke et al., 2000). It is known that methyl salicylate (MeSA), which is formed by the methylation of SA, can be a systemic signal molecule for SAR.

JA

JA (3-oxo-2–2′-cis-pentenyl-cyclopentane-1-acetic acid) and its bioactive derivates, such as methyl jasmonate (MeJA), are cyclic fatty acid–derived regulators. Furthermore, JA is abundant in plant parts such as young leaves, ripe fruits, flowers, root tips, pericarp, and vegetative parts. When JA increases in plant tissues, it rearranges the expression of some genes involved in root growth, flower development, senescence, and secondary metabolite production. Additionally, JA functions as a signaling molecule in the induction of systemic defense responses. For example, when a pathogen attacks a plant, JA is rapidly produced during necrotizing infections

(Penninckx et al., 1996). Furthermore, exogenous application of JA induces some defense genes encoding plant defensins and thionin proteins, which have antimicrobial activity and cause induction of osmotic and phytoalexin synthesis (Wasternack, 1997). Similarly, the exogenous use of methyl jasmonate (MeJA) also activates the thionin gene *Thi2.1* and the plant defensin *PDF1.2* in *Arabidopsis* after infection (Penninckx et al., 1996 and Vignutelli et al., 1998). The PDF1.2 gene expression is blocked in *Arabidopsis* JA-insensitive mutant plants (Penninckx et al., 1996). Interestingly, induction in the PDF1.2 gene expression is associated with increased endogenous JA levels, but PR genes are correlated with SA (Bowling et al., 1997). Anthocyanins are responsible for multiple physiological functions in plants. Moreover, they protect plants from abiotic and biotic stress. Therefore, anthocyanin accumulation is one of the effective defense strategies in plants against pathogen infection. Several studies have shown that anthocyanin accumulation increases in the infected tissues of apple and maize attacked by cedar-apple rust (*Gymnosporangium yamadae Miyabe*) and biotrophic fungus (*Ustilago maydis*) (Liu et al., 2020). Liu et al. (2020) found that JA signaling mediates *Penicillium corylophilum* fungus-induced accumulation of anthocyanin in *Arabidopsis* (Liu et al., 2020). MeJA also increases anthocyanin accumulation in plants (Shan et al., 2009). JA has as a central role in the regulation of biosynthesis of many secondary metabolites.

Gibberellin

Like other plant growth hormones, gibberellin (GA) plays a significant role in plant immune responses by regulating host defense responses elicited by SA–JA–ET signaling systems (Pieterse et al., 2012). Overall, 126 GAs have been determined in plants, fungi, and bacteria. However, very few active GAs (GA1, GA3, GA4, GA7) have been documented (Yamaguchi, 2008). The most active forms in plants are GA4 and GA1 (Olszewski et al., 2002). Studies have shown that GAs regulate the JA and SA signaling systems as well as plant immune responses (Navarro et al., 2006; Yang et al., 2008). GAs create sensitivity or resistance to different pathogenic fungi and bacteria (De Vleesschauwer et al., 2012; Qin et al., 2013; Yang et al., 2008). Studies have shown that they increase SAR against pathogens (Xia et al., 2010).

Intracellular suppressors of GA signaling are called DELLA proteins. When gain-of-function mutations occur in DELLA genes, GA signal decreases. However, loss of function causes an increase in GA signaling. Besides, DELLAs can directly activate the transcription of downstream genes and suppress GA responses (Hirano et al., 2012).

GA enables the development of resistance or susceptibility to different pathogens that cause various diseases in plants. GA regulates resistance to plant pathogens by inducing the degradation of DELLA proteins (Navarro et al., 2008). A DELLA gene called SLENDER RICE1 (SLR1) has been found in rice. One study identified GA-deficient and GA-insensitive rice mutants that accumulate SLR1 excess (Ueguchi-Tanaka et al., 2008). Disruption in GA biosynthesis, insensitivity to GA resulted in greater sensitivity to *Pythium graminicola* compared with wild-type rice plants. In addition, improved sensitivity to *P. graminicola* was observed in two independent SLR1 gain-of-function mutants. Apart from this, the *slr1–1* mutant (a dysfunctional allele that exhibits a constructive GA response phenotype) showed increased resistance to this oomycete pathogen (De Vleesschauwer et al., 2012). These studies suggest that GA induces resistance by negatively regulating the DELLA protein. With increasing concentrations of GA in rice plants, *P. graminicola* resistance also increased, depending on the GA concentration. On the contrary, administration of uniconazole (GA biosynthesis inhibitor) encouraged disease susceptibility as it reduced GA levels (De Vleesschauwer et al., 2012).

GAs can trigger a susceptibility mechanism toward some pathogens. Elongated uppermost internode (EUI), which deactivates biologically active GAs, is a P450 monooxygenase (Zhu et al. 2006). If a mutation causes loss of function in these EUIs, large amounts of bioactive GA will accumulate (Xu et al. 2004; Yang et al. 2008; Zhu et al. 2006). *Eui* mutant rice, which causes high GA production, is more sensitive to the blast pathogen *M. oryzae* (Yang et al., 2008). Accordingly, GAs seem to effectively regulate rice disease resistance and play an important role in reducing susceptibility to diseases caused by bacterial and fungal pathogens (Yang et al., 2008). It has been found that, in rice plants where Eui is overexpressed, the level of SA decreases in increased disease resistance against bacterial blight and explosion pathogens, while the JA level is decreased in eui mutant plants with greater susceptibility to pathogens (Yang et al., 2008). In transgenic rice plants overexpressing the Eui gene, it was observed that enhanced resistance was provided against bacterial and fungal diseases (Yang et al., 2008). In another study, it was observed that plants became more susceptible to the necrotrophic fungal pathogen *Alternaria brassicicola* due to the mutation of DELLA proteins in *Arabidopsis* (Robert-Seilaniantz et al., 2007). Navarro et al. (2008) showed that DELLAs are involved in JA signaling. DELLA mutant *gai* (structurally active dominant) was sensitized for JA-response gene induction. High resistance against the necrotrophic fungal pathogen *Alternaria brassicicola* was observed to be weakened in this mutant (Navarro et al., 2008). A DELLA protein, RGA-LIKE3 (RGL3) protein, has been required to enhance JA-mediated responses in *Arabidopsis*. This RGL3 protein was positively regulated by JA-mediated resistance against the necrotrophic fungal pathogen *Botrytis cinerea* (Wild et al., 2012).

The interaction of GA with fungal pathogens has not been fully elucidated. How it acts in defense against pathogens is still being investigated. This complex mechanism will be the subject of many more studies.

Auxin

The plant hormone auxin, also known as indole-3-acetic acid (IAA), is a simple signaling molecule involved in regulating different signaling pathways (Peer et al., 2013; Tatsuki et al., 2013). Aux/IAA suppressors and auxin response factor (ARF) transcription factors are required to modulate the auxin signal. The specificity of gene expression regulated by auxin is ensured (Hayashi, 2012). Although it is known that auxins interact with signal transduction mechanisms, such as SA, JA, and ET, that lead to disease resistance, auxin exhibits a complex interaction with various defense signaling systems (Savatin et al., 2011). It is involved in the modulation of various defense mechanisms of the plant, such as the signaling of ROS, ethylene SA, JA, ABA, and cytokine (Peer et al., 2013; Tatsuki et al. 2013; Wang et al., 2007, Tabata et al., 2010; Rock and Sun, 2005; Naseem and Dandekar, 2012).

Auxin regulates plant defense responses negatively or positively, and it is known to increase either susceptibility to disease or resistance to disease (Vidhyasekaran, 2015).

It is known to synthesize the auxin hormone in plants against their pathogens, when they infecting the plants. Furthermore, auxin production in pathogens is known as a significant pathogenicity marker in plant–pathogen interactions. It was observed that the rice blast pathogen *Magnaporthe oryzae* produced auxin IAA during the biotrophic phase of the infection, and it was reported that this auxin could play an essential role in the pathogenesis of the early biotrophic stages of infection (Tanaka et al., 2011). However, here we will talk about auxin produced by plants against fungal pathogens.

It has been observed that the endogenous plant IAA levels increase at a high rate during the defense of plants against pathogen attacks. Accordingly, a five-fold increase in free IAA levels was reported in cotton roots infected with *F. oxysporum*, as in many other pathogens (Dowd et al., 2004).

In a study on soybeans, auxin was found to have an inverse effect on the inducibility of enzymes involved in induced defense responses and proteins associated with pathogenesis (Leguay and Jouanneau, 1987). It is known that auxin is effective in plant defense responses with other hormones. The study conducted in *Arabidopsis* on the interaction of auxin with JA in fungal pathogens. Plant interaction has also shown that *Arabidopsis* needs AXR1 (an auxin response gene that plays a vital role in auxin signaling) for defense against the opportunistic pathogen *Pythium irregular*. As AXR1 was seen to bind directly through both auxin and JA proteasome pathways, it was thought that auxin and JA signaling systems could both be involved in the defense response. In *Arabidopsis* axr1, *jar1, and jar1/axr1* mutants, a link between auxin and JA-mediated defense response as a result of the interaction of auxin and JA signaling systems against *Pythium irregulare* and JAR1 (Jasmonate resistant1) in triggering disease resistance (an important component in the JA signaling system) and shows that AXR1 affects (Tiryaki and Staswick, 2002).

In *Arabidopsis*, fungal and bacterial pathogens take over host auxin metabolism and promote the development of the disease by causing the accumulation of IAA–Asp, a conjugated form of the hormone. This IAA–Asp task is to increase pathogen development in the plant by regulating the transcription of virulence genes. It has been determined that *Arabidopsis* accumulates this conjugated form in response to *Botrytis cinerea* (González-Lamothe et al., 2012). These results suggest that GH3.2 (auxin response gene), mediated by IAA–Asp, plays a role in activating the expression of virulence genes in pathogens.

In addition to the studies mentioned, auxin transport contributes to auxin-induced sensitivity. This is because polar auxin transport plays a role in triggering sensitivity (Kidd et al., 2011). Pin-formed (PIN) proteins regulate auxin efflux. A mutation in Pin2 was found, that results in more plant resistance to *F. oxysporum*. As a result of it suppression using the inhibitor, the development of wilt disease caused by *F. oxysporum* in *Arabidopsis* was also reduced. Studies with all these auxin transport mutants show that auxin transport systems are important in plant susceptibility (Vidhyasekaran et al., 2015). Auxin, like other hormones, has a complex plant defense mechanism against fungal pathogens and will continue to be the subject of research in the future.

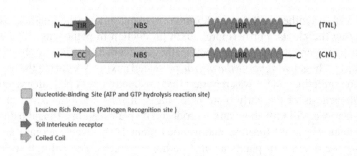

FIGURE 5.5 NBS–LRR classes and structures

NUCLEOTIDE-BINDING SITE LEUCINE-RICH REPEAT IN PLANT DEFENSE RESPONSES

Plants recognize avirulent genes of diverse pathogens through their R-genes (disease resistance genes). Furthermore, recognizing these genes activates numerous defense responses, HR, and PCD (programmed cell death) (Yang et al., 1997). Most of the R genes in plants encode NBS–LRR (nucleotide-binding site leucine-rich repeat) proteins characterized by NBS and LRR domains as well as variable amino and carboxy-terminal domains (McHale et al., 2006). NBS–LRR genes can be classified into two types: non-TIR and TIR (Toll/Interleukin-1 Receptor).

ANTIMICROBIAL COMPOUNDS IN PLANT DEFENSE RESPONSES

Plants can protect themselves from pathogen attacks with their inherent defense mechanisms, such as cell wall strengthening and cuticular wax as explained earlier. Alternatively, many plants produce secondary antifungal metabolites. Plants synthesize these compounds in response to pathogen attacks as part of their defense response. Antimicrobial compounds extracted from plants are of low molecular weight and inhibit the growth of fungi. In addition to the primary metabolites (sugars, amino acids, nucleic acids), which play a significant role in the protection of plant viability, growth, development, or reproduction, secondary metabolites (terpenes, polyphenols, quinones, alkaloids, peptides, etc.) serve as toxic chemicals or defense-related proteins (Freeman and Beattie, 2008). The separation between primary and secondary metabolism is uncertain because plants produce a diverse array of secondary metabolites (Tiku, 2018).

Phenolics are one of the secondary metabolites formed by plants. They play an important role in plant defense mechanisms against microbial threats (Mazid et al., 2011). Polyphenols are among the most important and numerous secondary metabolite groups, and a series of pigments with the quinone structure is responsible for the color of fruits and flowers (alizarin, purpurin, benzoquinone, juglone). Polyphenols have a role in adapting to abiotic and biotic stress and are involved in the defense against UV radiation, oxidizing agents, or some phytopathogenic agents (Cowan, 1999). Furthermore, they trap and scavenge free radicals, regulate nitric oxide, decrease leukocyte immobilization, induce apoptosis, inhibit cell proliferation and angiogenesis, and exhibit phytoestrogen activity in animal models and in vitro systems (Nijveldt et al., 2001; Yang et al., 2001). Terpenes are the biggest group of secondary metabolites present in all plants, and they are classified on the basis of the number of isoprene units used in their construction. They are an intricate part of the plant defense mechanism (Gershenzon and Croteau, 1991). Alkaloids are present mostly in herbaceous dicots and some gymnosperms and monocots, and they are also the most prominent family of bitter-tasting nitrogenous compounds (Hegnauer, 1988). The content of alkaloids depends on the plant, region, climate, and season (Ferdes, 2018). Pyrrolizidine alkaloids help defend against attacks and are somewhat toxic. Coniine, nicotine, tropane, atropine, cocaine, quinine, papaverine, morphine, codeine, strychnine, and caffeine are the most important alkaloids (Shin et al., 2018; Cowan, 1999). Several alkaloids have the ability to intercalate with DNA and modify the nucleotide sequences. Those are highly unsaturated planar quaternary alkaloids (Guil-Guerrero et al., 2016).

Fungal pathogens may counter the antimicrobial compounds produced by their host plants in different ways. Furthermore, they produce antimicrobial phtytoprotectants against antimicrobial compound (i) Some fungi may get around the host antimicrobial compounds (avoiding mechanism), (ii) some phytopathogenic fungi detoxify plant antimicrobial compounds by various mechanisms, and (iii) some fungi have an innate resistance to host plant antifungal compounds (Osbourn, 1999).

AUTOPHAGY IN PLANT–FUNGAL PATHOGEN INTERACTION

Autophagy is an evolutionarily protected process in eukaryotes. It is an important procedure for maintaining or restoring homeostasis under stress and then aiding plant endurance. In the meantime, some of the damaged proteins or organelles are suppressed by autophagic vesicles with bilayer structure and transported to lysosomes (creatures) or vacuoles (organisms and plants) for degradation and recycling (Liu et al., 2012).

Although autophagy can be broadly divided into two procedures that share some common aspects but ave very different cell capacities. Macrophagy, the first form of autophagy, describes the mass reuse of cell materials. This procedure begins with arranging a single cup-shaped membrane structure called the "phagophore," by suppressing organelles or potentially recyclable organelles or potentially the cytoplasm (Levine and Klionsky 2004; Mizushima, 2007). Conversely, specific autophagy is utilized to debase pexophagy (peroxisomes), reticulophagy (endoplasmic reticulum), or mitophagy (mitochondria) in an organelle-specific procedure that requires a particular arrangement of proteins, fundamental for cargo selection, take part in the cytoplasm to vacuole targeting (Cvt) on pathway that has been explained in yeast (Klionsky et al., 2007; Meijer et al., 2007). Besides these primary autophagy forms, plants are exposed to developmentally regulated microautophagy, where they accumulate storage proteins in seeds and mobile starch granules during seed germination (Bassham, 2007).

While basal degrees of autophagy serve the most part in cell homeostasis and quality control, expanded autophagy action permits adjustment to stress conditions brought about by an enormous assortment of developmental and environmental cues (Boya et al., 2013). Besides the critical commitment to cell and organismal endurance, autophagy has been alleged to regulate and execute PCD in different eukaryotic organisms (Anding and Baehrecke, 2015). In plants, autophagy is increasingly perceived for its significance being developed, metabolism, reproduction, senescence, and resistance to abiotic and biotic stress (Michaeli et al., 2016; Yang and Bassham, 2015). It seems that autophagy is an essential process in eukaryotic organisms for cellular differentiation and tissue remodeling during embryonic development, required for the cellular stress response as a main regulatory mechanism in metabolic control and host defense against pathogens (Klionsky, 2007; Kourtis and Tavernarakis, 2009; Levine and Klionsky, 2004; Mizushima, 2007). In recent decades, plant autophagy can gain either the host by participating in immune responses or the invading agent by contributing to contamination. Many studies have also demonstrated that autophagy assumes a significant role in the pathogenicity of plant pathogens (Hofius et al., 2017). Plants interface with autophagic forms will ultimately improve the control of the plant fungal diseases (Zhu et al., 2018).

Additionally, autophagy is not well understood, how it is manipulated by molecular mechanisms and adapted pathogens that support selective autophagy in plants. This is because pathogenic fungi utilize autophagy mechanisms while infecting plants and understanding autophagy through standard molecular mechanisms is difficult. However, several recent studies on plant autophagy machines using proteins produced by pathogens have revealed new autophagy-related defense components and shed light on the functioning of plant authophagy machines (Dagdas et al., 2016; Hafrén et al., 2017; Haxim et al., 2017).

Filamentous plant pathogens, such as fungi and oomycetes, pose a significant threat to global food safety. Many aggressive forms, including the rice blast pathogen, *Magnaporthe oryzae*, create close interactions with their hosts and are very useful in penetrating plant-created barriers. For example, *Magnaporthe oryzae* creates a dome-shaped cellular structure known as "appressorium" by crossing the host cuticle and subsequently disrupting the cell wall structure (Talbot, 2003). This stage is crucial because the pathogen penetrates the host cells and reaches the host's nutrient-rich part. Building blocks and energy (glycogen and lipids) are transported from neighboring conidia cells suffering from cell death due to autophagy (Wilson and Talbot, 2009). Therefore, autophagy mutants cannot produce suitable appressoria and cannot penetrate the host.

Similarly, Autophagy related 1 (ATG1) protein is induced in the fungal pathogen *Botrytis cinerea* during host colonization (Ren et al., 2017). This supports the view that appressorium formation is common in ATG1 mutants due to autophagy in fungi.

Recently, although there have been studies on the role of autophagy in the attachment of fungi to the plant, we will talk more about the role of autophagy in the plant's defense responses against fungal infection.

Autophagy functions as a response to biotic stress in many plants, including crop species, and manipulating autophagy leads to a change in resistance to various types of disease. For example, in bananas, the resistance of autophagy to *Fusarium oxysporum* f. sp. is compromised due to inhibition by the autophagy inhibitor 3-methylated (Wei et al., 2017).

With recent studies, autophagy has been found to play a role in HR and PCD. Previous studies have shown that autophagy is a type of PCD that restricts the spread of microbial infection; it contributes to immunity by regulating defense hormone levels and HR (Coll et al., 2014 and Yoshimoto et al., 2009).

Although autophagy has a significant effect in restricting or promoting HR-related cell death, the cross-talk between ROS and autophagy during the HR is still paradoxical. In addition, the synergistic interaction of nitric oxide (NO) and ROS is necessary to initiate cell death mechanisms in plants (Sadhu et al., 2019). The natural immune response in plants usually includes PCD (Hayward et al., 2009). Autophagy-related genes work either by restricting or by promoting PCD when working with HR, but the manifestation of autophagy during the (*Alternaria alternata* toxin) AaT-induced HR response is uncertain (Minina et al., 2016). A study by Sahdu et al. (2019) aimed to reveal unknown correlations between ROS, NO, and autophagy during the HR response. This study observed that the increase in ROS (OH, ROO •, H_2O_2) upon necrotrophic attack activates autophagy as a pro-survival defense strategy. However, prolonged exposure to AaT triggers the Ca^{+2} signaling cascade, increasing oxidative stress and making it easier for NO production to determine cell death.

Furthermore, prevention of NO accumulation was seen as evidence of the emergence of autophagic response. Therefore, these results confirm that NO is a possible regulator of cell survival and cell death during HR, except for ROS. In another study, ROS induced by AaT has been shown to cause macro-autophagy in *Arabidopsis* in the external application of H_2O_2 and methylviology, as it induces autophagy (Xiong et al., 2007). Despite all these studies, there are many unanswered questions about autophagy's exact role in plant defense (Seay et al., 2006).

CONCLUSION

This chapter has focused on some of the biochemical and molecular bases of plant resistance and defense responses to fungal pathogens. However, recently, with the rapid development of multi-level omics, such as transcriptomics, proteomics, metabolomics, and microbiomics, new and significant metabolic pathways in fungal pathogen–plant interactions have been identified. Furthermore, hormone signaling genes, immune receptor genes, pathogen effector genes, and antifungal substances produced by fungi in the plant have been discovered. However, fungus–plant interactions are complex cascade processes, including the fungal infection process, plant defense process, and communications. Therefore, more extensive research is required to fully evaluate interaction-related metabolites and genes, determine their functions, and promote the prevention and control of crop fungal disease.

REFERENCES

AbuQamar, S., Chen, X., Dhawan, R., Bluhm, B., Salmeron, J., Lam, S., Dietrich, R.A., Mengiste, T. 2006. "Expression profiling and mutant analysis reveals complex regulatory networks involved in Arabidopsis response to Botrytis infection." *The Plant Journal*, 48(1), 28–44.

Ahissar, E., Vaadia, E., Ahissar, M., Bergman, H., Arieli, A., Abeles, M. 1992. "Dependence of cortical plasticity on correlated activity of single neurons and on behavioral context." *Science*, 257(5075), 1412–1415.

Alexander, H.M., Mauck, K.E., Whitfield, A.E., Garrett, K.A., Malmstrom, C.M. 2014. "Plant-virus interactions and the agro-ecological interface." *European Journal of Plant Pathology*, 138(3), 529–547

Anding, A.L, Baehrecke, E.H. 2015. "Autophagy in cell life and cell death." *Current Topics in Developmental Biology*, 114, 67–91.

Asselbergh, B., Curvers, K., França, S.C., Audenaert, K., Vuylsteke, M., Van Breusegem, F., Höfte, M. 2007. "Resistance to Botrytis cinerea in sitiens, an abscisic acid-deficient tomato mutant, involves timely production of hydrogen peroxide and cell wall modifications in the epidermis." *Plant Physiology*, 144(4), 1863–1877.

Baenas, N., García-Viguera, C., Moreno, D. 2014. "Elicitation: A tool for enriching the bioactive composition of foods." *Molecules*, 19, 13541–13563.

Bakker, P.A., Raaijmakers, J.M., Bloemberg, G., Höfte, M., Lemanceau, P., Cooke, B.M. eds. 2010. *New perspectives and approaches in plant growth-promoting rhizobacteria research*. New York: Springer Science & Business Media.

Bassham, D.C. 2007. "Plant autophagy more than a starvation response." *Current Opinion in Plant Biology*, 10, 586–593.

Berrocal-Lobo, M., Molina, A., Solano, R. 2002. "Constitutive expression of ETHYLENE-RESPONSE-FACTOR1 in Arabidopsis confers resistance to several necrotrophic fungi." *The Plant Journal*, 29(1), 23–32.

Boba, A., Kostyn, K., Kozak, B., Wojtasik, W., Preisner, M., Prescha, A., Kulma, A. 2020. "Fusarium oxysporum infection activates the plastidial branch of the terpenoid biosynthesis pathway in flax, leading to increased ABA synthesis." *Planta*, 251(2), 1–14.

Bodenhausen, N., Reymond, P. 2007. "Signaling pathways controlling induced resistance to insect herbivores in Arabidopsis." *Molecular Plant-Microbe Interactions*, 20, 1406–1420

Bowling, S.A. et al. 1997. "The cpr5 mutant of Arabidopsis expresses both NPR1-dependent and NPR1-independent resistance." *Plant Cell*, 9, 1573–1584

Boya, P., Reggiori, F., Codogno, P. 2013. "Emerging regulation and functions of autophagy." *Nature Cell Biology*, 15, 1017.

Chern, M.S., Fitzgerald, H.A., Yadav, R.C., Canlas, P.E., Dong, X., Ronald, P.C. 2001. "Evidence for a disease-resistance pathway in rice similar to the NPR1-mediated signaling pathway in Arabidopsis." *The Plant Journal*, 27(2), 101–113.

Clarke, J.D., Volko, S.M., Ledford, H., Ausubel, F.M. & Dong, X. 2000. Roles of salicylic acid, jasmonic acid, and ethylene in cpr-induced resistance in Arabidopsis. *The Plant Cell*, 12(11), 2175–2190.

Coll, N.S., Smidler, A., Puigvert, M., Popa, C., Valls, M., Dangl, J.L. 2014. "The plant metacaspase AtMC1 in pathogen-triggered programmed cell death and aging: Functional linkage with autophagy." *Cell Death and Differentiation*, 21, 1399–1408.

Cooper, R.M., Wood, R.K.S. 1975. "Regulation of synthesis of cell wall degrading enzymes by Veticillium albo-atrum and *Fusarium oxysporum* f. sp. lycopersici." *Physiological Plant Pathology*, 5(2), 135–156.

Cowan, M.M. 1999. "Plant Products as Antimicrobial Agents." *Clinical Microbiology Reviews*, 12, 564–582.

Dagdas, Y.F., Belhaj, K., Maqbool, A. 2016. "An effector of the Irish potato famine pathogen antagonizes a host autophagy cargo receptor." *Life,* 5, 10856.

De Gara, L., de Pinto, M.C., Tommasi, F. 2003. "The antioxidant systems vis-à-vis reactive oxygen species during plant—pathogen interaction." *Plant Physiology and Biochemistry,* 41(10), 863–870.

Dehgahi, R., Subramaniam, S., Zakaria, L., Joniyas, A., Firouzjahi, F.B., Haghnama, K., Razinataj, M. 2015. "Review of research on fungal pathogen attack and plant defense mechanism against pathogen." *International Journal of Agricultural Science Research,* 2(8), 197–208.

Dempsey, D.M.A., Shah, J., Klessig, D.F. 1999. "Salicylic acid and disease resistance in plants." *Critical Reviews in Plant Sciences,* 18(4), 547–575.

De Vleesschauwer, D., Van Buyten, E., Satoh, K., Balidion, J., Mauleon, R., Choi, I.R., Vera-Cruz, C., Kikuchi, S., Höfte, M. 2012. "Brassinosteroids antagonize gibberellin- and salicylate-mediated root immunity in rice." *Plant Physiology,* 158, 1833–1846.

De Vos, M., Van Oosten, V.R., Van Poecke, R.M., Van Pelt, J.A., Pozo, M.J., Mueller, M.J., Buchala, A.J., Métraux, J.P., Van Loon, L.C., Dicke, M., Pieterse, C.M. 2005. "Signal signature and transcriptome changes of Arabidopsis during pathogen and insect attack." *Molecular Plant-Microbe Interactions,* 18, 923–937.

Dowd, C., Wilson, I.W., McFadden, H. 2004. "Gene expression profile changes in cotton root and hypocotyl tissues in response to infection with *Fusarium oxysporum* f.sp. vasinfectum." *Molecular Plant-Microbe Interactions,* 17, 654–667.

Elad, Y., Kirshner, B. 1993. "Survival in the phylloplane of an introduced biocontrol agent (*Trichoderma harzianum*) and populations of the plant pathogen *Botrytis cinerea* as modified by abiotic conditions." *Phytoparasitica,* 21(4), 303.

Ferdes, M. 2018. "Antimicrobial compounds from plants." In *Fighting antimicrobial resistance.* Edited by Budimir, A. Zagreb, Croatia: IAPC-OBP, 273–271.

Freeman, B.C, Beattie, G.A. 2008. "An overview of plant defences against pathogens and herbivores." *The Plant Health Instructor.* doi: 10.1094/PHI-I-2008-0226-01

Gaffney, T., Friedrich, L., Vernooij, B., Negrotto, D., Nye, G., Uknes, S., Ward, E., Kessmann, H., Ryals, J., 1993. "Requirement of salicylic acid for the induction of systemic acquired resistance." *Science,* 261(5122), 754–756.

Geraats, B.P., Bakker, P.A., Van Loon, L.C. 2002. "Ethylene insensitivity impairs resistance to soilborne pathogens in tobacco and *Arabidopsis thaliana.*" *Molecular Plant-Microbe Interactions,* 15(10), 1078–1085.

Gershenzon, J., Croteau, R. 1991. "T'erpenoids." In *Herbivores their interaction with secondary plant metabolites, Vol I: The chemical participants,* 2nd ed. Edited by Rosenthal, G.A. and Berenbaum, M.R. San Diego: Academic Press, 165–219.

Gilbert, G.S. 2002. "Evolutionary ecology of plant diseases in natural ecosystems." *Annual Review of Phytopathology,* 40(1), 13–43.

González-Lamothe, R., El Oirdi, M., Brisson, N., Bouarab, K. 2012. "The conjugated auxin indole 3-acetic acid-aspartic acid promotes plant disease development." *Plant Cell,* 24, 762–777.

Gowthami, L. 2018. "Role of elicitors in plant defense mechanism." *Journal of Pharmacognosy and Phytochemistry,* 7, 2806–2812.

Guil-Guerrero, J.L., Ramos, L., Moreno, C., Zúñiga-Paredes, J.C., Carlosama-Yepez, M., Ruales, P. 2016. "Antimicrobial activity of plant-food by-products: A review focusing on the tropics." *Livestock Science,* 189, 32–49.

Guimaraes, R.L., Stotz, H.U. 2004. "Oxalate production by *Sclerotinia sclerotiorum* deregulates guard cells during infection." *Plant Physiology,* 136, 3703–3711

Hafrén, A., Macia, J.L., Love, A.J., Milner, J.J., Drucker, M., Hofius, D. 2017. "Selective autophagy limits cauliflower mosaic virus infection by NBR1- mediated targeting of viral capsid protein and particles." *Proceedings of the National Academy of Sciences, USA,* 114, E2026–E2035.

Haxim, Y., Ismayil, A., Jia, Q. 2017. "Autophagy functions as an antiviral mechanism against geminiviruses in plants." *ELife,* 6, 23897.

Hayashi, K.I. 2012. "The interaction and integration of auxin signaling components." *Plant Cell Physiology,* 53, 965–975.

Hayward, A.P., Tsao, J., Dinesh-Kumar, S.P. 2009. "Autophagy and plant innate immunity: Defense through degradation." *Seminars in Cell and Developmental Biology,* 20, 1041–1047.

Hegnauer, R. 1988. "Biochemistry, distribution and taxonomic relevance of higher plant alkaloids." *Phytochemistry,* 27(8), 2423–2427.

Hématy, K., Cherk, C., Somerville, S. 2009. "Host—pathogen warfare at the plant cell wall." *Current Opinion in Plant Biology,* 12(4), 406–413.

Kidd, B.N., Kadoo, N.Y., Dombrecht, B., Tekeoglu, M., Gardiner, D.M., Thatcher, L.F., Aitken, E.A., Schenk, P.M., Manners, J.M., Kazan, K. 2011. "Auxin signaling and transport promote susceptibility to the root-infecting fungal pathogen Fusarium oxysporum in Arabidopsis." *Molecular Plant-Microbe Interactions,* 24, 733–748.

Kinkema, M., Fan, W., Dong, X. 2000. "Nuclear localization of NPR1 is required for activation of PR gene expression." *The Plant Cell,* 12(12), 2339–2350.

Klionsky, D.J. 2007. "Autophagy: From phenomenology to molecular understanding in less than a decade." *Nature Reviews Molecular Cell Biology,* 8, 931–937. Excellent synthesis of the key discoveries in the operation of autophagy.

Klionsky, D.J., Cuervo, A.M., Dunn, W.A., Levine, B., van der Klei, I., Seglen, P.O. 2007. "How shall I eat thee?" *Autophagy,* 3(5), 413–416.

Kourtis, N., Tavernarakis, N. 2009. "Autophagy and cell death in model organisms." *Cell Death Differ,* 16, 21–30. A thoughtful account of the potential roles that autophagy plays in programmed cell death in eukaryotes and the interplay with apoptosis.

Leguay, J.J., Jouanneau, J.P. 1987. "Auxin (2,4-dichlorophenoxyacetic acid) starvation and treatment with glucan elicitor isolated from *Phytophthora megasperma* induces similar responses in soybean-cultured cell suspensions." *Developmental Genetics,* 8, 351–364.

Levine, B., Klionsky, D.J. 2004. "Development by self-digestion: Molecular mechanisms and biological functions of autophagy." *Developmental Cell,* 6, 463–477.

Liu, X.H., Gao, H.M., Xu, F. 2012. "Autophagy vitalizes the pathogenicity of pathogenic fungi." *Autophagy,* 8(10), 1415–1425.

Liu, Y., Li, M., Li, T., Chen, Y., Zhang, L., Zhao, G., Zhuang, J., Zhao, W., Gao, L., Xia, T. 2020. "Airborne fungus-induced biosynthesis of anthocyanins in Arabidopsis thaliana via jasmonic acid and salicylic acid signaling." *Plant Science,* 300, 110635.

Loon, L.C., Geraats, B.P., Linthorst, H.J. 2006. "Ethylene as a modulator of disease resistance in plants." *Trends in Plant Science,* 11(4), 184–191.

Mazid, M., Khan, T.A., Mohammad, F. 2011. "Role of secondary metabolites in defense mechanisms of plants." *Biology and Medicine,* 3, 232–249.

McHale, L., Tan, X., Koehl, P., Michelmore, R.W. 2006. "Plant NBS-LRR proteins: Adaptable guards." *Genome Biology,* 7, 212.

Meijer, W.H., van der Klei, I.J., Veenhuis, M., Kiel, J.A. 2007. "ATG genes involved in non-selective autophagy are conserved from yeast to man, but the selective Cvt and pexophagy pathways also require organism-specific genes." *Autophagy,* 3, 604–609.

Melotto, M., Underwood, W., He, S.Y. 2008. "Role of stomata in plant innate immunity and foliar bacterial diseases." *Annual Review of Phytopathology,* 46, 101–122

Meng, X., Zhang, S. 2013. "MAPK cascades in plant disease resistance signaling." *Annual Review of Phytopathology,* 51, 245–266.

Mengiste, T. 2012. "Plant immunity to necrotrophs." *Annual Review of Phytopathology,* 50, 267–294.

Merlot, S., Gosti, F., Guerrier, D., Vavasseur, A., Giraudat, J. 2001. "The ABI1 and ABI2 protein phosphatases 2C act in a negative feedback regulatory loop of the abscisic acid signalling pathway." *The Plant Journal*, 25(3), 295–303.

Michaeli, S., Galili, G., Genschik, P., Fernie, A.R., Avin-Wittenberg, T. 2016. "Autophagy in plants-what's new on the menu?" *Trends in Plant Science*, 21, 134–144.

Minina, E.A., Filonova, L.H., Fukada, K., Savenkov, E.I., Gogvadze, V., Clapham, D., Sanchez-Vera, V., Suarez, M.F., Zhivotovsky, B., Daniel, G. 2016. "Autophagy and metacaspase determine the mode of cell death in plants." *Journal of Cell Biology*, 203, 917–927.

Mizushima, N. 2007. "Autophagy: Process and function." *Genes and Development*, 21, 2861–2873.

Mohr, P.G. Cahill, D.M. 2007. "Suppression by ABA of salicylic acid and lignin accumulation and the expression of multiple genes, in Arabidopsis infected with *Pseudomonas syringae* pv. tomato." *Functional & Integrative Genomics*, 7(3), 181–191

Naseem, M., Dandekar, T. 2012. "The role of auxin-cytokinin antagonism in plant-pathogen interactions." *PLoS Pathogens*, 8(11), 1003026.

Navarro, L., Bari, R., Achard, P., Lison, P., Nemri, A., Harberd, N.P., Jones, J.D.G. 2008. "DELLAs control plant immune responses by modulating the balance of jasmonic acid and salicylic acid signaling." *Current Biology,* 18, 650–655.

Navarro, L., Dunoyer, P., Jay, F., Arnold, B., Dharmasiri, N., Estelle, M., Voinnet, O., Jones, J.D.J. 2006. "A plant miRNA contributes to antibacterial resistance by repressing auxin signaling." *Science*, 312, 436–439.

Nijveldt, R.J., van Nood, E., van Hoorn, D.E., Boelens, P.G., van Norren, K., van Leeuwen, P.A. 2001. "Flavonoids: A review of probable mechanisms of action and potential applications." *The American Journal of Clinical Nutrition*, 74, 418–425.

O'Donnell, P.J., Schmelz, E.A., Moussatche, P., Lund, S.T., Jones, J.B., Klee, H.J. 2003. "Susceptible to intolerance—a range of hormonal actions in a susceptible Arabidopsis pathogen response." *Plant Journal*, 33, 245–257.

Olszewski, N., Sun, T.P., Gubler, F. 2002. "Gibberellin signaling: Biosynthesis, catabolism, and response pathways." *Plant Cell,* 14, S61–S80.

Osbourn, A.E. 1999. "Antimicrobial phytoprotectants and fungal pathogens: A commentary." *Fungal Genetics and Biology*, 26, 163–168.

Peer, W.A., Cheng, Y., Murphy, A.S. 2013. "Evidence of oxidative attenuation of auxin signaling." *Journal of Experimental Botany*, 64, 2629–2639.

Penninckx, I.A., Eggermont, K., Terras, F.R., Thomma, B.P., De Samblanx, G.W., Buchala, A., Broekaert, W.F. 1996. "Pathogen-induced systemic activation of a plant defensin gene in Arabidopsis follows a salicylic acid-independent pathway." *The Plant Cell*, 8(12), 2309–2323.

Pieterse, C.M., Van Pelt, J.A., Ton, J., Parchmann, S., Mueller, M.J., Buchala, A.J., Métraux, J.P., Van Loon, L.C. 2000. "Rhizobacteria-mediated induced systemic resistance (ISR) in Arabidopsis requires sensitivity to jasmonate and ethylene but is not accompanied by an increase in their production." *Physiological and Molecular Plant Pathology*, 57(3), 123–134.

Pieterse, C.M.J., van der Does, D., Zamioudis, C., Leon-Reyes, A., van Wees, S.C.M. 2012. "Hormonal modulation of plant immunity." *Annual Review of Cell and Developmental Biology*, 28, 489–521.

Qin, X., Liu, J.H., Zhao, W.S., Chen, X.J., Guo, Z.J., Peng, Y.L. 2013. "Gibberellin 20-oxidase gene OsGA20ox3 regulates plant stature and disease development in rice." *Molecular Plant-Microbe Interactions,* 26, 227–239.

Ren, W., Zhang, Z., Shao, W., Yang, Y., Zhou, M., Chen, C. 2017. "The autophagy-related gene BcATG1 is involved in fungal development and pathogenesis in Botrytis cinerea." *Molecular Plant Pathology*, 18, 238–248.

Robert-Seilaniantz, A., Navarro, L., Bari, R., Jones, J.D.G. 2007. "Pathological hormone imbalances." *Current Opinion in Plant Biology,* 10, 372–379.

Rock, C.D., Sun, X. 2005. "Crosstalk between ABA and auxin signaling pathways in roots of Arabidopsis thaliana." *(l.) Heynh. Planta,* 222, 98–106.

Sahdu, A., Moriyasu, Y., Acharya, K., Bandyopadhyay, M. 2019. "Nitric oxide and ROS mediate autophagy and regulate Alternaria alternata toxin-induced cell death in tobacco BY-2 cells." *Scientific Reports,* 9, 8973.

Savatin, D.V., Ferrari, S., Sicilia, E., De Lorenzo, G. 2011. "Oligogalacturonide auxin antagonism does not require post-transcriptional gene silencing or stabilization of auxin response repressors in Arabidopsis thaliana." *Plant Physiology,* 157, 1163–1174.

Schweizer, P., Buchala, A., Dudler, R., Métraux, J.P. 1998. "Induced systemic resistance in wounded rice plants." *The Plant Journal,* 14(4), 475–481.

Seay, M., Patel, S., Dinesh-Kumar, S.P. 2006. "Autophagy and plant immunity." *Cell Microbiology,* 8, 899–906.

Serino, L., Reimmann, C., Baur, H., Beyeler, M., Visca, P., Haas, D. 1995. "Structural genes for salicylate biosynthesis from chorismate in Pseudomonas aeruginosa." *Molecular Genetics and Genomics.*

Shah, J., Kachroo, P., Nandi, A., Klessig, D.F. 2001. "A recessive mutation in the Arabidopsis SSI2 gene confers SA-and NPR1-independent expression of PR genes and resistance against bacterial and oomycete pathogens." *The Plant Journal,* 25(5), 563–574.

Shah, J., Tsui, F., Klessig, D.F. 1997. "Characterization of a salicylic acid in sensitive mutant (sai1) of Arabidopsis thaliana, identified in a selective screen utilizing the SA-inducible expression of the tms2 gene." *Molecular Plant-Microbe Interactions,* 10(1), 69–78.

Shan, X., Zhang, Y., Peng, W., Wang, Z., Xie, D. 2009. "Molecular mechanism for jasmonate-induction of anthocyanin accumulation in Arabidopsis." *Journal of Experimental Botany,* 60(13), 3849–3860.

Shin, J., Prabhakaran, V.-S., Kim, K. 2018. "The multi-faceted potential of plant-derived metabolites as antimicrobial agents against multidrug-resistant pathogens." *Microbial Pathogenesis,* 116, 209–214.

Tabata, R., Ikezaki, M., Fujibe, T., Aida, M., Tian, C.E., Ueno, Y., Yamamoto, K.T., Machida, Y., Nakamura, K., Ishiguro, S. 2010. "Arabidopsis auxin response factor6 and 8 regulate jasmonic acid biosynthesis and floral organ development via repression of class 1 KNOX genes." *Plant Cell Physiology,* 51, 164–175.

Talbot, N.J. 2003. "On the trail of a cereal killer: Exploring the biology of *Magnaporthe grisea." Annual Review of Microbiology,* 57, 177–202.

Tanaka, E., Koga, H., Mori, M., Masashi, M. 2011. "Auxin production by the rice blast fungus and its localization in host tissue." *Journal of Phytopathology,* 159, 522–530.

Tatsuki, M., Nakajima, N., Fujii, H., Shimada, T., Nakano, M., Hayashi, K.I., Hayama, H., Yoshioka, H., Nakamura, Y. 2013. "Increased levels of IAA are required for system 2 ethylene synthesis causing fruit softening in peach (*Prunus persica* L. Batsch)." *Journal of Experimental Botany,* 64, 1049–1059.

Tiku, A.R. 2018. "Antimicrobial compounds and their role in plant defense." *Molecular Aspects of Plant-Pathogen Interaction,* 283–307.

Tiryaki, I., Staswick, P.E. 2002. "An Arabidopsis mutant defective in jasmonate response is allelic to the auxin signaling mutant axr1." *Plant Physiology,* 130, 887–894.

Ton, J., Flors, V., Mauch-Mani, B. 2009. "The multifaceted role of ABA in disease resistance." *Trends in Plant Science,* 14(6), 310–317.

Turner, N.C., Graniti, A. 1969, "Fusicoccin: A fungal toxin that opens stomata." *Nature,* 223, 1070–1071

Ueguchi-Tanaka, M., Hirano, K., Hasegawa, Y., Kitano, H., Matsuoka, M. 2008. "Release of the repressive activity of rice DELLA protein SLR1 by gibberellin does not require SLR1 degradation in the gid2 mutant." *Plant Cell,* 20, 2437–2446.

Verberne, M.C., Budi Muljono, A.B., Verpoorte, R. 1999. "Salicylic acid biosynthesis." In *Biochemistry and molecular biology of plant hormones*, vol. 33. Edited by Libbenga, K., Hall, M., Hooykaas, P.J.J. London: Elsevier, 295–312.

Vidhyasekaran, P. 2015. "Plant hormone signaling systems in plant innate immunity, signaling and communication in plants." In *Auxin signaling system in plant innate immunity*. Springer.

Vignutelli, A., Wasternack, C., Apel, K., Bohlmann, H. 1998. "Systemic and local induction of an Arabidopsis thionin gene by wounding and pathogens." *The Plant Journal*, 14(3), 285–295.

Vlot, A.C., Dempsey, D.M.A., Klessig, D.F. 2009. "Salicylic acid, a multifaceted hormone to combat disease." *Annual Review of Phytopathology*, 47, 177–206.

Wang, D., Pajerowska-Mukhtar, K., Culler, A.H., Dong, X. 2007. "Salicylic acid inhibits pathogen growth in plants through repression of the auxin signaling pathway." *Current Biology*, 17, 1784–1790.

Wang, Q., An, B., Hou, X., Guo, Y., Luo, H., He, C. 2018. "Dicer-like proteins regulate the growth, conidiation, and pathogenicity of Colletotrichum gloeosporioides from Hevea brasiliensis." *Frontiers in Microbiology*, 8.

Wang, Y., Noguchi, K., Ono, N., Inoue, S.I., Terashima, I., Kinoshita, T. 2014. "Overexpression of plasma membrane H+-ATPase in guard cells promotes light-induced stomatal opening and enhances plant growth." *Proceedings of the National Academy of Sciences*, 111(1), 533–538.

Wasilewska, A., Vlad, F., Sirichandra, C., Redko, Y., Jammes, F., Valon, C. 2008. "An update on abscisic acid signaling in plants and more." *Molecular Plant*, 1, 198–217.

Wasternack, C., Parthier, B. 1997. "Jasmonate-signalled plant gene expression." *Trends in Plant Science*, 2(8), 302–307.

Wei, Y., Liu, W., Hu, W., Liu, G., Wu, C., Liu, W., Zeng, H., He, C., Shi, H. 2017. "Genome-wide analysis of autophagyrelated genes in banana highlights MaATG8 s in cell death and autophagy in immune response to Fusarium wilt." *Plant Cell Reports*, 36, 1237–1250.

Widmann C, Gibson S, Jarpe MB, Johnson GL. 1999. "Mitogen-activated protein kinase: Conservation of a three-kinase module from yeast to human." *Physiological Reviews*, 79, 143–80

Wild, M., Davière, J.M., Cheminant, S., Regnault, T., Baumberger, N., Heintz, D., Baltz, R., Genschik, P., Achard, P. 2012. "The Arabidopsis DELLA RGA-LIKE3 is a direct target of MYC2 and modulates jasmonate signaling responses." *Plant Cell*, 24, 3307–3319.

Wildermuth, M.C., Dewdney, J., Wu, G., Ausubel, F.M. 2001. "Isochorismate synthase is required to synthesize salicylic acid for plant defence." *Nature*, 414(6863), 562–565.

Wilkinson, S., Davies, W.J. 2010. "Drought, ozone, ABA and ethylene: New insights from cell to plant to community." *Plant, Cell & Environment.*, 33(4), 510–525.

Wilson, R.A., Talbot, N.J. 2009. "Under pressure: Investigating the biology of plant infection by *Magnaporthe oryzae*." *Nature Reviews, Microbiology*, 7, 185–195.

Windsor, D.A. 1998. "Controversies in parasitology, Most of the species on Earth are parasites." *International Journal for Parasitology*, 28(12), 1939–1941.

Xia, Y., Navarre, D., Seebold, K., Kachroo, A., Kachroo, P. 2010. "The glabra1 mutation affects cuticle formation and plant responses to microbes." *Plant Physiology*, 154, 833–846.

Xiong, Y., Contento, A.L., Nguyen, P.Q. Bassham. D.C. 2007. "Degradation of oxidized proteins by autophagy during oxidative stress in Arabidopsis." *Plant Physiology*, 143, 291–299.

Xu, Y.H., Zhu, Y.Y., Zhou, H.C., Li, Q., Sun, Z.X., Liu, Y.G., Lin, H.X., He, Z.H. 2004. "Identification of a 98-kb DNA segment containing the rice Eui gene controlling uppermost internode elongation, and construction of a TAC transgene sublibrary." *Molecular Genetics and Genomics*, 272, 149–155.

Yamaguchi, S. 2008. "Gibberellin metabolism and its regulation." *Annual Review of Plant Biology*, 59, 225–251.

Yang, C.S., Landau, J.M., Huang, M.-T., Newmark, H.L. 2001. "Inhibition of carcinogenesis by dietary polyphenolic compounds." *Annual Review of Nutrition*, 21, 381–406.

Yang, D.L., Li, Q., Deng, Y.W., Lou, Y.G., Wang, M.Y., Zhou, G.X., Zhang, Y.Y., He, Z.H. 2008. "Altered disease development in the eui mutants and Eui overexpressors indicates that gibberellins negatively regulate rice basal disease resistance." *Molecular Plant*, 1, 528–537.

Yang, X., Bassham, D.C. 2015. "New insight into the mechanism and function of autophagy in plant cells." *International Review of Cell and Molecular Biology*, 320, 1–40.

Yang, Y., Shah, J., Klessig, D.F. 1997. "Signal perception and transduction in plant defense responses." *Genes & Development*, 11, 1621–1639.

Yoshimoto, K., Jikumaru, Y., Kamiya, Y., Kusano, M., Consonni, C., Panstruga, R., Ohsumi, Y., Shirasu, K. 2009. "Autophagy negatively regulates cell death by controlling NPR1-dependent salicylic acid signaling during senescence and the innate immune response in Arabidopsis." *The Plant Cell*, 21, 2914–2927.

Zhang, X., Chen, S., Mou, Z. 2010. "Nuclear localization of NPR1 is required for regulation of salicylate tolerance, isochorismate synthase 1 expression and salicylate accumulation in Arabidopsis." *Journal of Plant Physiology*, 167(2), 144–148.

Zhu, X.M., Li, L., Wu, M., Liang, S., Shi, H.B., Liu, X.H., Lin, F.C. 2018. "Current opinions on autophagy in pathogenicity of Fungi." *Virulence*, 481–489.

Zhu, Y., Nomura, T., Xu, Y., Zhang, Y., Peng, Y., Mao, B., Hanada, A., Zhou, H., Wang, R., Li, P., Zhu, X., Mander, L.N., Kamiya, Y., Yamaguchi, S., He, Z. 2006. "ELONGATED UPPERMOST INTERNODE encodes a cytochrome P450 monooxygenase that epoxidizes gibberellins in a novel deactivation reaction in rice." *Plant Cell*, 18, 442–456.

6 Fungal Gene Expression and Interaction With Host Plants

Muhammad Kaleem Sarwar, Anita Payum,
Siddra Ijaz, Qaiser Shakeel and Ghedir Issam

CONTENTS

INTRODUCTION

Unlike animals, plants largely depend on their immune system, triggering a series of biochemical responses to detect the presence of pathogen on/in their system (Chisholm et al., 2006). Genetics and molecular aspects initiate a diverse array of plant–pathogen interactions (Shen et al., 2017). Highly evolved multi-mechanism-based responses are directly or indirectly involved in recognizing a pathogen invasion and triggering an immune response (Nirmala et al., 2007). According to several studies during the last decade, plant resistance in biochemical response/reactions is activated within 5 minutes of initial interaction. Molecular studies indicate that ionfluxes, reactive oxygen species (ROS), protein phosphorylation, and some other pathways of plants and pathogens are activated as early as possible. Many fungal diseases of economically important crops, e.g., tomato *Cladosporium fulvum*, cereal rust fungi, and rice blast, were covered in this overview of recent developments in the research of molecular response of host–fungus interaction at the early infection stage.

Fungi play a vital role in different environments by developing various relationships with plants. These relationships include mainly two types of interactions—mutualistic or pathogenic—with host plants. During the infection period, fungi can efficiently suppress the defense system of host plants by secreting proteins (effectors) that interfere with the physiology and immunity of these plants. Phytopathogens can also disable the host plants' defenses and cause modification in host metabolism,

DOI: 10.1201/9781003162742-6

while entophytic fungi form intercellular contacts with the host plant, usually having a mutualistic interaction. Thus, entophytes can also alter the defenses and metabolism of the host plant. Secretions of enzymes or effectors cause communication between host plants and fungi (Zeilinger et al., 2016).

Effectors disable the defenses of the host plants completely and cause modification in host metabolism; secretions of enzymes or effectors are the cause of communication between host plants and fungi. These signals trigger the immune responses of plants and are often molecularly similar in harmful and beneficial microbiota. This confuses the outcome of host–microbe association. There is a great need to unravel the series of signalling events and metabolical chnages inovoved during host–microbe interactions. Complex response networks have been revealed with the advancement in molecular analyses of the plant immune system. The interaction of host and pathogen molecules trigger the plant immune system and results in complex interaction of their molecules which could be better understood by studying the system biology of the invading plant cell. Many pathogenicity-related genes have been identified that are involved in the degradation of the cell wall, infection structure formation, responses to host plant defenses, toxins production, and signal cascades.

Plant–pathogen interaction has a strong and indispensable association with protein phosphorylation responses, syntaxin-like proteins, calmodulin protein kinases, and mitogen-activated protein kinases (Romeis et al., 2000; Heese et al. 2005; Shen et al., 2017). Kinases and phosphatases are the leading players activating early-stage disease resistance responses (Yang et al., 1997). Activation and phosphorylation regulation are critical aspects of understanding plant–pathogen interactions (Nakagami et al., 2010).

Immunity is a physiological state comprising defensive reactions by which an organism retains its stable cellular and physiological dynamics through rejecting antigenicity (Shen et al., 2017). Specific adaptive and innate immunity are the types of the immune system that trigger different responses in plants (Rajamuthiah and Mylonakis, 2014). Adaptive immunity imparts an immune system that relies on specific recognition and selective invader removal, and it takes a long period for sustained impact. Contrary to this, innate immunity displays an immune response with no need for specialized immune cells (Hoebe et al., 2004); instead, it rely on signaling pathway which make host response rapid against pathgens at the infection site (Sansonetti, 2006). The innate immunity–related cells sense microbe-associated molecular patterns (MAMPs) and pathogen-associated molecular patterns (PAMPs) through pattern recognition receptors (PRRs), including C-type lectin receptors (CLRs), RIG-I-like receptors (RLRs), Toll-like receptors (TLRs), and NOD-like receptors (NLRs) (Takeuchi and Akira, 2010).

Plants have evolved a two-tiered innate immune mechanism against pathogens that comprises PAMP-triggered immunity (PTI) and ETI (Stergiopoulos and De Wit, 2009). ETI is based on resistance genes (R genes) encoding proteins or R proteins by identifying and detecting the pathogen-secreted proteins or effectors. This interaction generates a strong resistance reaction against impeding pathogenesis (Rajamuthiah and Mylonakis, 2014). However, PTI develops a weak reaction, stimulated through PRRs, to identify PAMPs (He et al., 2013).

Besides this interaction, members of plant microbiomes also develop interaction to compete for space and food. Some benefit plant health and inhibit phytopathogens by producing antimicrobial metabolites (de Bruijn and Verhoeven, 2018).

In antagonistic interactions, the pathogen controls gene expression of the host to mislead immune response. Besides this type of cross-species interaction, microbe-controlled cellular modulations exist in other types of cross-species interaction. Some plant beneficial microorganisms secrete metabolites, which impede the phytopathogens' infection in hosts by inhibiting their virulence factors (Chen et al., 2018).

> The cross-species modulation of epigenetic mechanisms is also observed in plant-to-plant interaction. Wheat plants release precursors of histone deacetylase inhibitors into soil intake by neighboring plants, which arrests histone deacetylase activity, resulting in stunted growth (Venturelli et al., 2015). The exchange of small Ribonucelic Acids (RNAs) has documented the cross-species gene expression, whose movement occurs through vesicles to avoid RNase-mediated degradation of moving RNA. This type of exchange has been observed in most plants and pathogens, by which pathogenic fungi suppress the immune system of plants, and pathogens' virulence is repressed by plants.
>
> (Cai et al., 2018)

Can leaves of the plant, for example, be treated with particular short RNAs that target pathogenicity genes as a more long-term strategy of controlling plant pathogens? Or can beneficial microorganisms, naturally occurring, be induced to produce antibiotic metabolites like phenazine, which may be used to combat agricultural pathogens? The mechanisms by which these species, having ability to move RNAs across species, interact and play a role in gene expression to acquire resistance are unclear. More profound knowledge of these systems will open up new avenues for illness research, both basic and applied.

IMMUNE SYSTEM: A FOUNDATION OF FUNGUS–HOST PLANT INTERACTION

Plant innate immune response regulation systems were studied using quantitative phosphoproteomic analysis a decade ago (Nühse et al., 2007). Previous studies on the fungal pathogen–host interaction response suggested that the interactions occurred earlier than previously assumed. According to quantitative phosphoproteomics research, the elicitor activates signaling pathways within minutes of initial interaction in *Arabidopsis* (Benschop et al., 2007). The same research shed light on plant defense signal transduction and early contact responses. Still, the problem is that, after every effort to understand the pathogen–host interaction at each level, the role of early molecular processes during initial host–pathogen interaction is unknown. Familiarity with initial molecular processes is essential for understanding pathways of signals, and ultimately, resistance mechanisms against diseases. Understanding of disease resistance mechanism will lead to genetic improvement of the host plant which can identify the pathogen's resistant or avirulent genes at initial stage of interaction.

The immune system is the cornerstone of the fungus–host relationship. According to some researchers, early elicitors caused the early reaction of plants' tissues or cells. For example, in parsley (*Petroselinum crispum*) cells, a fungal elicitor causes temporary, fast, and sequential phosphorylation of proteins, which activates several pathogen resistance–related genes. After elicitors treatment, phosphorylation of 45 kDa neutral proteins was observed in the microsome, and the cytoplasm, and within one minute phosphorylation of nuclear protein (26 kDa) initiated (Dietrich et al., 1990). The signal transduction mechanism on the plasma membrane and cytoplasm of plant cells was triggered quickly upon initial contact between pathogen and plant (Robatzek, 2014). Many crucial channels were engaged in the early-reaction signal transduction, which can determine later responses of patterns of gene expression at a multi-level. Recent studies have extensively studied the signal components of salicylic acid (SA), ion flux, and other hormones (Robatzek, 2014; Boller and Felix, 2009; Bolouri Moghaddam et al., 2016). The presence of Ca2þ in signal transduction pathways has been linked to the phosphorylation reaction (Dietrich et al., 1990). In elicitor-treated tomato cells, protein phosphorylation events occurred in vivo within minutes. The protein kinase inhibitors K-252a and staurosporine prevented the elicitor's action. Protein kinase inhibitors can also block elicitor-induced early biochemical reactions (Felix et al., 1991). Plant responses to MAMPs had ion flow events in the range of 0.5–2 minutes (Boller, 1995; Nürnberger et al., 2004). Increased inflow of Ca2þ and outflow of K, as well as an efflux of anions, notably nitrate (Wendehenne et al., 2002), are among these alterations. Membrane depolarization is caused by ion fluxes (Mithöfer et al., 2005). MAMPs were shown to trigger an inflow of Ca2þ from the apoplast and a rapid increase in Ca2þ cytoplasm concentrations, perhaps activating calcium-dependent protein kinases (Ludwig et al., 2005).

Plant hormones may have an important function in plant defense against pathogen. In *A. thaliana*, ethylene and SA may promote the expression of defensin-like (DEFL) genes (Penninckx et al., 1996; Penninckx et al., 1998). Depending on the pathogens encountered, the jasmonic acid (JA) and SA defense mechanisms might interact antagonistically or synergistically (Spoel et al., 2003; Beckers and Spoel, 2006; Mur et al., 2006; Van der Does et al., 2013). Plant stomata act as microbial infection barriers. In an SA-dependent way, PAMPs can cause stomatal closure. The virulence factor coronatine (COR) may inhibit PAMP-induced abscisic acid (ABA) signaling in the guard cell from Pst DC3000 [50]. In tobacco, fungal elicitors (a-elicitin and b-elicitin) and SA activate 48 kDa SA-induced protein kinase (SIPK) very quickly (Zhang et al., 1998). In cellular signal transmission, protein phosphorylation, a frequent regulatory method in vivo, plays a crucial function. Proteins are phosphorylated primarily on serine (including threonine), and the enzymes and activities of these two kinds of amino acids are distinct.

RESISTANCE GENE AND AVIRULENCE GENE: GENE FOR THE GENE IN PLANT–PATHOGEN INTERACTION

The early interaction milestone report for barley stem rust was released in 2011 (Nirmala et al., 2011). Barley stem rust (*Puccinia graminis* f. sp. *Tritici*) was a catastrophic barley disease in most parts of North America, until barley cultivals with Rpg1 gene were discovered in 1941. For nearly 70 years, the Rpg1 gene prevented losses of barley production from catastrophic stem rust. Rpg1, a new gene for resistance with receptor kinase homology was found in the short chromosome arm 1(7H) of barley cultivar (Kleinhofs et al., 1993; Brueggeman et al., 2002; Mirlohi et al., 2008). Genetic engineering changed the Rpg1 gene in the greatly vulnerable barley cultivar, Golden Promise, resulting resistance in previously vulnerable barley culti- var (Horvath et al., 2003). Although the Rpg1 gene has a broad resistance range, it does not work against all virulent Puccinia graminis f. sp. tritici (Pgt) strains (Zhang et al., 2006). The gene Rpg1 produces a protein with two domains of tandem kinase: pKI (protein kinase I) and pKII (protein kinase II). Both these domains are essen- tial for stem rust resistance. The pK1 domain is a pseudokinase, while the other is active catalytically. The pK1 domain of the pseudokinase is linked to disease resis- tance, while the pKII domain plays a role in the phosphorylation of protein (Nirmala et al., 2006; Nirmala et al., 2010). The kinase may be found in the cell membrane, endo-membranes, and cytoplasm.

Since the "gene-for-gene" idea was proposed in 1942, several new homologous R and Avr genes have been discovered. Avr9 was the first Avr gene of fungi discovered by molecular cloning in 1991 (van Kan et al., 1991; De Wit et al., 2009). Several kinase genes have been implicated in pathogen–host interactions. The kinase family members of mitogen-activated protein (MAP) are wounding-induced protein kinase (WIPK) and SA-induced protein kinase (SIPK), which are triggered by the tobacco mosaic virus (TMV) in tobacco. The disease-resistance gene N and tyrosine and serine or threonine phosphorylation are required for WIPK activation. Resistance responses and tyrosine phosphorylation were linked to SIPK activation (Zhang and Klessig, 1998a; Zhang et al., 1998; Zhang and Klessig, 1998b).

The Cf-9 gene encodes a protein that protects tomatoes against the *Cladosporium fulvum* fungal disease, and Avr9 is the avirulence gene that corresponds to it. The Cf-9 gene produces extracytoplasmic glycoprotein with 27 membrane-anchored leu- cine-rich replicates (LRRs). It has been found, by analyzing the Cf-9 gene, that the 46 protein kinases 48-kDa, a type of enzyme, and 46-kDa are similar to SIPK and WIPK. Post-translational processes activate both kinases, 48-kDa and 46-kDa, and the activation process contributes to phosphorylation of tyrosine and influx of Ca2 (Van den Ackerveken et al., 1992; Van den Ackerveken et al., 1993; Jones et al., 1994; Blatt et al., 1999; Jones, 1999). Cutinase has several roles, including evoking host-derived signals and attaching fungal spores. Three hours after inoculation of *Curvularia lunata* in maize, ClCUT7 gene (belonging to a gene family of cutinase) expression increased. Cutinase may be involved in fungal–plant interactions (Liu et al., 2016). Within 30 minutes of inoculating elicitor N-acetylchitooligosaccharide in the leaves or roots of the rice plant, an elicitor-responsive gene, EL2, is quickly and instantly expressed. Because the elicitor infected the rice seedling, the hyphal development of the fungus causing rice blast disease was significantly slowed but not stopped. The elicitor induced the disease resistance genes PR-I and PR-X (PBZ1) routinely and locally, although the rice defense response mechanism was unknown (Tanabe et al., 2006).

Another gene related to disease implicated during early contact is Avr9, from tomato leaf mildew. The findings on Avr9 genes were primarily compared with those on fungal rust genes, i.e., RGD-binding gene and VPS9. The Avr9 gene codes for a 63-amino-acid preprotein, which may interact with the *Cladosporium fulvum* fun- gal pathogen and tomato. Within 3–5 minutes, Avr9 increased K outward rectifying by 2.5–3 times and nearly totally repressed K inward rectifying. The K channel responses were irreversible and selective (Jones et al., 1994; Romeis et al., 1999)

EXPRESSION AND INTERACTION OF FUNGAL GENES DURING PLANT DISEASE DEVELOPMENT

Nearly 10% of known fungal species can cause plant diseases. Some of them are obligate in nature, meaning they must stay with the host throughout their lives, while saprophytic fungus may grow on both live and dead plant material. Non-obligate parasites, which rely on the host during some phases to complete their life cycle, fall somewhere in between these two extremes. These lifestyle variations necessitate specialized host adaptations. While saprophytic fungi supposedly need to overcome the host defense system, which they can accomplish effectively by producing specialized poisons and/or secreting lytic enzymes, signals from both parties must carefully control the interaction between biotrophic fungi and plants. Parasitic fungi with an obligatory biotrophic nature, including mildews and rusts, live close to their host plants but do not mediate cell death or defense mechanisms, while fungal pathogens of a necrotrophic nature can produce necrosis before penetrating. Pathogens with a facultative biotrophic nature can survive as saprotrophs, although completion of fungal life cycle on sexual bases, as in *Ustilago maydis*, may necessitate close interaction with the host.

The life cycle of hemibiotrophs, such as *Cladosporium fulvum*, *Colletotrichum* spp. *Phytophthora infestans,* and *Magnaporthe grisea,* may be completed outside the host plant during infection; they produce specialized structures not present during axenic growth. These pathogens have a life cycle that begins with spore germination and ends with hyphal development on the leaf or root surfaces. Appressoria, specialized structures of infection that pierce by the cuticle of the host plant and its epidermal layer by internally creating higher turgor pressure, can be induced by chemical and physical cues. Natural openings such as stomata and wounds are also used as entrance points.

Host metabolism is effectively adjusted to the fungus' requirements to establish the pathogenic haustorium, i.e., the intracellular organ facilitating facultative or obligate parasites (Agrios, 1997). The importance of MAP kinase signaling and Cathelicidin Antimicrobial Peptide (cAMP) during the early communication of plants has been highlighted in numerous processes and is the focus of significant current research (Banuett, 1998; Kronstad, 2000; Lengeler et al., 2000). Plant chemicals that induce fungal growth and development are unknown for the most part. Gene expression analysis concentrates on the expression of genes in response to fungal disease and our understanding of gene functions in disease development and their respective tactics for identifying signals and pathways to induce them. The findings that the expressed genes play a critical role in pathogenic development during fungal interaction with host plants has piqued researchers' curiosity. Subtractive hybridization, differential display, protein purification through reverse genetics, mutational screens, promoter or enhancer trapping methods, and array or Serial Analysis of Gene Expression (SAGE) approaches can all be used to identify plant-induced genes when the sequence of the genome or an adequate amount of expressed sequence tags is accessible.

In *Colletotrichum gloeosporioides*, plant surface characteristics affect early development, due to genes exclusively expressed in this phase. The function of these genes is uncertain; however, it is known that waxes on the host surface preferentially trigger them (Kolattukudy et al., 1995), and CAP20 is required for the functioning of appressorium (Hwang et al., 1995). Conidial contact with a rigid surface is required for the reaction to host wax and ethylene (Flaishman et al., 1995), and it is evident that contact with a rigid surface stimulates calmodulin signaling Ca^{2+} in *C. gloeosporioides* (Kim et al., 1998). Several *C. gloeosporioides* genes (CHIP) expressed due to contact with the rigid surface have been found due to this research (Liu and Kolattukudy, 1998; Kim et al., 2000). CHIP1 enzyme facilitates ubiquitin conjugation. CHIP2 might be a transcription factor, while CHIP3 is a membrane-bound protein. Small proteins released by fungi, i.e., hydrophobins, aggregate at hydrophilic or hydrophobic interfaces and can play a role in adhesion to plants' hydrophobic surfaces (Wessels, 2000). MPG1, an *M. grisea* hydrophobin expressed during appressorium development, might play this role (Beckerman and Ebbole, 1996) and references therein). PTH11 was shown to be an actual pathogenicity gene in the same fungus, and it is essential for effective appressorium production as a reaction to cutin monomers and the hydrophobicity of the contacted surface (DeZwaan et al., 1999). PTH11 is responsible for encoding Pth11p, a new protein that is transmembrane and very faintly apparent in the cell cortex and is relocated to vacuolar membranes during induction of fungal appressorium. Treatment of cAMP for two separate REM1 mutants missing PTH11 restored appressorium development and pathogenicity, indicating that Pth11p is an upregulatory activator of cAMP-mediated signaling that can act in an overlying manner signaling pathway. Pth11p appears to suppress appressorium differentiation on weakly inductive surfaces, directing the fungus to the right entrance point (DeZwaan et al., 1999).

Successful plant penetration may be aided by the creation of infection structures and enzymes capable of destroying its cell wall components. Cutinase genes are upregulated at early stages of contact in different pathosystems (Kolattukudy et al., 1995; Van Kan et al., 1997; Muñoz and Bailey, 1998). In conjunction with particularly produced hydrolytic enzymes, physical and chemical signaling cues define the intricacy of the early infection phase, as shown in these investigations. Due to the fusion of fungal genes, promoters to reporter genes may be utilized to monitor disease development properly and examine the induction involvement of host genotypes. Simultaneously, these genes give the required tools for identifying signals and regulators, subsequently exploring the signaling pathways that lead to induction.

INTRACELLULAR COOPERATION AND PLANT–FUNGUS INTERACTION

In the cooperation process, biological modules pay the price for the advantage of their belonging group (Hamilton, 1963). Labor division—the specialty of many modules (such as cells, tissues, organs, or complete organisms), each committed to particular tasks—is usually linked to the rise of cooperation. Regarding cytoplasmic metabolite exchange between cells (Aktipis et al. 2015; Campbell et al., 2015; West et al., 2015; West and Cooper, 2016), cooperation also entails the interchange of molecules across specialized biological modules, a method known as "resource allocation." As a result of cooperation, a collection of cells can utilize the surrounding resources more effectively. For example, the production of digestive enzymes by specialized cells in some syntrophic microbial communities alters local resource availability, benefiting surrounding cells (H. Koschwanez et al., 2011; Morris et al., 2013). Cooperation is linked to fundamental shifts in life's evolution, such as creating new functions at a greater level of the cellular organization, including multicellularity. Many complex properties in nature are based on it, including virulence in some pathogenic organisms. Several functional and developmental phases, inter-relaying on feeding, were described in the wood decay fungal mycelium (Rayner and Franks, 1987; Rayner, 1991; Fricker et al., 2017) due to fluctuations in the native environment inside the host tissues.

However, since the 1830s, when the cell hypothesis was first proposed, the eukaryotic cell has been seen as a self-contained biological unit (BalušKA et al., 2004). In multicellular animals, mechanisms such as determining the fate of embryonic cells, rhythm of circadian clock, and level of immunology might all be cell autonomous (Randow et al., 2013). Furthermore, malignancies and neurological illnesses are caused by the growth of a single cell type that is harmful to the entire body (Aktipis et al., 2015). The extent to which cooperation plays a role in complicated features like host colonization in viruses with various infection methods is mostly unclear. Genes, cells, and organisms are shaped by natural selection to enhance their evolutionary success at the cost of their rivals. Cooperation is promoted only when specified circumstances are satisfied in this situation (Nowak, 2006). Kin selection, also known as genetic closeness between individuals, is a well-known circumstance that enhances the evolvement chance of cooperative traits (Hamilton, 1963).

Cooperation is likely to win when compliant modules are separated in space and engage with certain neighbors in a preferred manner (Nadell et al., 2010; Rueffler et al., 2012). According to the spatial game theory, the geographical organization of cooperating populations might encourage collaboration (Doebeli and Knowlton, 1998). Studies on synthetic yeast colonies revealed that increasing the colony's size enhances the emergence of local cooperation (Van Dyken et al., 2013). Another element favoring job specialization and collaboration is variation, which can take the mutation forms, phenotypic deviations, or features of the native environment (Barta, 2016). For example, the inoculation of metabolic auxotrophies cumulatively and randomly in yeast populations resulted in the spontaneous formation of metabolically cooperative populations (Campbell et al., 2015).

Fungal infections having a wide range of plant hosts release several complex proteins compared with specialized fungi, implying that these species have a higher metabolic cost of virulence (Badet et al., 2017). These findings show that the choice for forming the labor division is stronger in broad-host-range fungi, which might be linked to fungal lifestyles. The capacity of pathogens for host colonization and causing illness is primarily influenced by the expressed virulence factors (Schulze-Lefert and Panstruga, 2011). The production of virulence components and transport to their sites of action place a metabolic strain on pathogen cells. Many parasite evolution hypotheses rely on communication or exchange between pathogen virulence and spread, or pathogen virulence and growth within the host (Ebert, 1998; Alizon et al., 2009). A rising number of experimental investigations (Thrall and Burdon, 2003; Bruns et al., 2014; Peyraud et al., 2016) have found an exchange between virulence and spread.

As fungi have a compliant nature of virulence, disease control techniques relying on virulence reduction may have unintended consequences. Indeed, less virulent variants can reduce resource efficiency trade-offs and increase the predominance of virulent species in natural communities (Lindsay et al., 2016). Furthermore, the silencing of released virulence factors might be ineffective as it may disturb the natural balnce of pathogen population and replace it with highly hostile pathogenic strains.

REFERENCES

Agrios GN (1997) Plant disease caused by fungi. In: *Plant Pathology* (eds) Agrios GN. Academic Press, London, UK, pp. 245–406.

Aktipis CA, Boddy AM, Jansen G, Hibner U, Hochberg ME, Maley CC, Wilkinson GS (2015) Cancer across the tree of life: Cooperation and cheating in multicellularity. *Philosophical Transactions of the Royal Society B: Biological Sciences* 370(1673):20140219.

Alizon S, Hurford A, Mideo N, Van Baalen M (2009, February) Virulence evolution and the trade-off hypothesis: history, current state of affairs and the future. *Journal of Evolutionary Biology* 22(2):245–259.

Badet T, Peyraud R, Mbengue M, Navaud O, Derbyshire M, Oliver RP, Barbacci A, Raffaele S (2017) Codon optimization underpins generalist parasitism in fungi. *Elife* 6:e22472.

Baluška F, Volkmann D, Barlow PW (2004) Eukaryotic cells and their cell bodies: cell theory revised. *Annals of Botany* 94(1):9–32.

Banuett F (1998) Signalling in the yeasts: an informational cascade with links to the filamentous fungi. *Microbiology and Molecular Biology Reviews* 62(2):249–274.

Barta Z (2016) Individual variation behind the evolution of cooperation. *Philosophical Transactions of the Royal Society B: Biological Sciences* 371(1687):20150087.

Beckerman JL, Ebbole DJ (1996) MPG1, a gene encoding a fungal hydrophobin of *Magnaporthe grisea*, is involved in surface recognition. *Molecular Plant-Microbe Interactions: MPMI* 9(6):450–456.

Beckers GJM, Spoel SH (2006) Fine-tuning plant defence signalling: Salicylate versus jasmonate. *Plant Biology* 8:1–10.

Benschop JJ, Mohammed S, O'Flaherty M, Heck AJR, Slijper M, Menke FLH (2007) Quantitative phosphoproteomics of early elicitor signaling in Arabidopsis. *Molecular & Cellular Proteomics* 6:1198–1214.

Blatt MR, Grabov A, Brearley J, Hammond-Kosack K, Jones JD (1999) K+ channels of Cf-9 transgenic tobacco guard cells as targets for *Cladosporium fulvum* Avr9 elicitor-dependent signal transduction. *The Plant Journal* 19(4):453–462.

Boller T (1995) Chemoperception of microbial signals in plant cells. *Annual Review of Plant Biology* 46(1):189–214.

Boller T, Felix G (2009, June 2) A renaissance of elicitors: perception of microbe-associated molecular patterns and danger signals by pattern-recognition receptors. *Annual Review of Plant Biology* 60:379–406.

Bolouri Moghaddam MR, Vilcinskas A, Rahnamaeian M (2016, April) Cooperative interaction of antimicrobial peptides with the interrelated immune pathways in plants. *Molecular Plant Pathology* 17(3):464–471.

Brueggeman R, Rostoks N, Kudrna D, Kilian A, Han F, Chen J, Druka A, Steffenson B, Kleinhofs A (2002) The barley stem rust-resistance gene Rpg1 is a novel disease-resistance gene with homology to receptor kinases. *Proceedings of the National Academy of Sciences* 99:9328–9333.

Bruns E, Carson ML, May G (2014) The jack of all trades is master of none: a pathogen's ability to infect a greater number of host genotypes comes at a cost of delayed reproduction. *Evolution* 68(9):2453–2466.

Cai Q, Qiao L, Wang M, He B, Lin FM, Palmquist J, Huang SD, Jin H (2018) Plants send small RNAs in extracellular vesicles to fungal pathogen to silence virulence genes. *Science* 360(6393):1126–1129.

Campbell K, Vowinckel J, Mülleder M, Malmsheimer S, Lawrence N, Calvani E, Miller-Fleming L, Alam MT, Christen S, Keller MA, Ralser M (2015) Self-establishing communities enable cooperative metabolite exchange in a eukaryote. *Elife* 4:e09943.

Chen Y, Wang J, Yang N, Wen Z, Sun X, Chai Y, Ma Z (2018) Wheat microbiome bacteria can reduce virulence of a plant pathogenic fungus by altering histone acetylation. *Nature Communications* 9(1):1–4.

Chisholm ST, Coaker G, Day B, Staskawicz BJ (2006) Host-microbe interactions: shaping the evolution of the plant immune response. *Cell* 124:803–814.

de Bruijn I, Verhoeven KJ (2018) Cross-species interference of gene expression. *Nature Communications* 9(1):1–3.

De Wit PJGM, Mehrabi R, Van Den Burg HA, Stergiopoulos I (2009) Fungal effector proteins: past, present and future. *Molecular Plant Pathology* 10:735–747.

DeZwaan TM, Carroll AM, Valent B, Sweigard JA (1999) Magnaporthe grisea pth11p is a novel plasma membrane protein that mediates appressorium differentiation in response to inductive substrate cues. *The Plant Cell* 11(10):2013–2030.

Dietrich A, Mayer JE, Hahlbrock K (1990) Fungal elicitor triggers rapid, transient, and specific protein phosphorylation in parsley cell suspension cultures. *Journal of Biological Chemistry* 265(11):6360–6368.

Doebeli M, Knowlton N (1998) The evolution of interspecific mutualisms. *Proceedings of the National Academy of Sciences USA* 95:8676–8680.

Ebert D (1998) Experimental evolution of parasites. *Science* 282:1432–1435.

Felix G, Grosskopf DG, Regenass M, Boller T (1991) Rapid changes of protein phosphorylation are involved in transduction of the elicitor signal in plant cells. *Proceedings of the National Academy of Sciences* 88(19):8831–8834.

Flaishman MA, Hwang CS, Kolattukudy PE (1995) Involvement of protein phosphorylation in the induction of appressorium formation in *Colletotrichum gloeosporioides* by its host surface wax and ethylene. *Physiological and Molecular Plant Pathology* 47(2):103–117.

Fricker MD, Heaton LL, Jones NS, Boddy L (2017) The mycelium as a network. *The Fungal Kingdom* 1:335–367.

H. Koschwanez JR, Foster KW, Murray A (2011) Sucrose utilization in budding yeast as a model for the origin of undifferentiated multicellularity. *PLoS Biology* 9(8):e1001122.

Hamilton WD (1963) The evolution of altruistic behavior. *The American Naturalist* 97:354–356.

He F, Zhang H, Liu J, Wang Z, Wang G (2013) Recent advances in understanding the innate immune mechanisms and developing new disease resistance breeding strategies against the rice blast fungus Magnaporthe oryzae in rice. *Hereditas* 36:756–765.

Heese A, Ludwig AA, Jones JDG (2005) Rapid phosphorylation of a syntaxin during the Avr9/Cf-9-race-specific signaling pathway. *Plant Physiology* 138:2406–2416.

Hoebe K, Janssen E, Beutler B (2004) The interface between innate and adaptive immunity. *Nature Immunology* 5:971–974.

Horvath H, Rostoks N, Brueggeman R, Steffenson B, von Wettstein D, Kleinhofs A (2003) Genetically engineered stem rust resistance in barley using the Rpg1 gene. resistance gene with homology to receptor kinases. *Proceedings of the National Academy of Sciences* 100:364–369.

Hwang CS, Flaishman MA, Kolattukudy PE (1995) Cloning of a gene expressed during appressorium formation by *Colletotrichum gloeosporioides* and a marked decrease in virulence by disruption of this gene. *The Plant Cell* 7(2):183–193.

Jones DA, Thomas CM, Hammond-Kosack KE, Balint-Kurti PJ, Jones JD (1994) Isolation of the tomato Cf-9 gene for resistance to *Cladosporium fulvum* by transposon tagging. *Science* 266(5186):789–793.

Jones JD (1999) Rapid Avr9-and Cf-9-dependent activation of MAP kinases in tobacco cell cultures and leaves: Convergence of resistance gene, elicitor, wound, and salicylate responses. *The Plant Cell* 11:273–287.

Kim YK, Li D, Kolattukudy PE (1998) Induction of Ca^{2+}-calmodulin signaling by hard-surface contact primes *Colletotrichum gloeosporioides* conidia to germinate and form appressoria. *Journal of Bacteriology* 180(19):5144–5150.

Kim YK, Liu ZM, Li D, Kolattukudy PE (2000) Two novel genes induced by hard-surface contact of *Colletotrichum gloeosporioides* conidia. *Journal of Bacteriology* 182(17):4688–4695.

Kleinhofs A, Kilian A, Maroof MS, Biyashev RM, Hayes P, Chen FQ, Lapitan N, Fenwick A, Blake TK, Kanazin V, Ananiev E (1993) A molecular, isozyme and morphological map of the barley (*Hordeum vulgare*) genome. *Theoretical and Applied Genetics* 86(6):705–712.

Kolattukudy PE, Rogers LM, Li D, Hwang CS, Flaishman MA (1995) Surface signaling in pathogenesis. *Proceedings of the National Academy of Sciences of the United States of America* 92:4080–4087.

Kronstad JW (2000) Triggers and targets of cAMP signalling. *Trends in Microbiology* 8(7):302.

Lengeler KB, Davidson RC, D'souza C, Harashima T, Shen WC, Wang P, Pan X, Waugh M, Heitman J (2000) Signal transduction cascades regulating fungal development and virulence. *Microbiology and Molecular Biology Reviews* 64(4):746–785.

Lindsay RJ, Kershaw MJ, Pawlowska BJ, Talbot NJ, Gudelj I (2016) Harbouring public good mutants within a pathogen population can increase both fitness and virulence. *eLife* 5:e18678.

Liu T, Hou J, Wang Y, Jin Y, Borth W, Zhao F, Liu F, Hu J, Zuo Y (2016) Genome-wide identification, classification and expression analysis in fungal—plant interactions of cutinase gene family and functional analysis of a putative ClCUT7 in *Curvularia lunata*. Mol. *Genetics Genomics* 291:1105–1115.

Liu ZM, Kolattukudy PE (1998) Identification of a gene product induced by hard-surface contact of *Colletotrichum gloeosporioides* conidia as a ubiquitin-conjugating enzyme by yeast complementation. *Journal of Bacteriology* 180(14):3592–3597.

Ludwig AA, Saitoh H, Felix G, Freymark G, Miersch O, Wasternack C, Boller T, Jones JD, Romeis T (2005) Ethylene-mediated cross-talk between calcium-dependent protein kinase and MAPK signaling controls stress responses in plants. *Proceedings of the National Academy of Sciences* 102(30):10736–10741.

Mirlohi A, Brueggeman R, Drader T, Nirmala J, Steffenson BJ, Kleinhofs A (2008) Allele sequencing of the Barley stem rust resistance gene Rpg1 identifies regions relevant to disease resistance. *Phytopathology* 98:910–918.

Mithöfer A, Ebel J, Felle HH (2005) Cation fluxes cause plasma membrane depolarization involved in β-glucan elicitor-signaling in soybean roots. *Molecular Plant-Microbe Interactions* 18(9):983–990.

Morris BE, Henneberger R, Huber H, Moissl-Eichinger C (2013) Microbial syntrophy: Interaction for the common good. *FEMS Microbiology Reviews* 37(3):384–406.

Muñoz CI, Bailey AM (1998) A cutinase-encoding gene from *Phytophthora capsici* isolated by differential-display RT-PCR. *Current Genetics* 33(3):225–230.

Mur LAJ, Kenton P, Atzorn R, Miersch O, Wasternack C (2006) The outcomes of concentration-specific interactions between salicylate and jasmonate signaling include synergy, antagonism, and oxidative stress leading to cell death. *Plant Physiology* 140:249–262.

Nadell CD, Foster KR, Xavier JB (2010) Emergence of spatial structure in cell groups and the evolution of cooperation. *PLoS Computational Biology* 6(3):e1000716.

Nakagami H, Sugiyama N, Mochida K, Daudi A, Yoshida Y, Toyoda T, Tomita M, Ishihama Y, Shirasu K (2010) Large-scale comparative phosphoproteomics identifies conserved phosphorylation sites in plants. *Plant Physiology* 153:1161–1174.

Nirmala J, Brueggeman R, Maier C, Clay C, Rostoks N, Kannangara CG, Von Wettstein D, Steffenson BJ, Kleinhofs A (2006) Subcellular localization and functions of the barley stem rust resistance receptor-like serine/threonine-specific protein kinase Rpg1. *Proceedings of the National Academy of Sciences* 103(19):7518–7523.

Nirmala J, Dahl S, Steffenson BJ, Kannangara CG, von Wettstein D, Chen X, Kleinhofs A (2007) Proteolysis of the barley receptor-like protein kinase RPG1 by a proteasome pathway is correlated with Rpg1-mediated stem rust resistance. *Proceedings of the National Academy of Sciences of the United States of America* 104:10276–10281.

Nirmala J, Drader T, Chen X, Steffenson B, Kleinhofs A (2010) Stem rust spores elicit rapid RPG1 phosphorylation. *Molecular Plant-Microbe Interactions* 23(12):1635–1642.

Nirmala J, Drader T, Lawrence PK, Yin C, Hulbert S, Steber CM, Steffenson BJ, Szabo LJ, Von Wettstein D, Kleinhofs A (2011) Concerted action of two avirulent spore effectors activates Reaction to Puccinia graminis 1 (Rpg1)-mediated cereal stem rust resistance. *Proceedings of the National Academy of Sciences* 108(35):14676–14681.

Nowak MA (2006) Five rules for the evolution of cooperation. *Science* 314:1560–1563.

Nühse TS, Bottrill AR, Jones AME, Peck SC (2007) Quantitative phosphoproteomic analysis of plasma membrane proteins reveals regulatory mechanisms of plant innate immune responses. *The Plant Journal* 51:931–940.

Nürnberger T, Brunner F, Kemmerling B, Piater L (2004) Innate immunity in plants and animals: Striking similarities and obvious differences. *Immunological Reviews* 198(1):249–266.

Penninckx IA, Eggermont K, Terras FR, Thomma BP, De Samblanx GW, Buchala A, Métraux JP, Manners JM, Broekaert WF (1996) Pathogen-induced systemic activation of a plant defensin gene in Arabidopsis follows a salicylic acid-independent pathway. *The Plant Cell* 8(12):2309–2323.

Penninckx IA, Thomma BP, Buchala A, Métraux JP, Broekaert WF (1998) Concomitant activation of jasmonate and ethylene response pathways is required for induction of a plant defensin gene in Arabidopsis. *The Plant Cell* 10(12):2103–2113.

Peyraud R, Cottret L, Marmiesse L, Gouzy J, Genin S (2016) A resource allocation trade-off between virulence and proliferation drives metabolic versatility in the plant pathogen Ralstonia solanacearum. *PLoS Pathogens* 12(10):e1005939.

Rajamuthiah R, Mylonakis E (2014) Effector triggered immunity: activation of innate immunity in metazoans by bacterial effectors. *Virulence* 5:697–702.

Randow F, MacMicking JD, James LC (2013) Cellular self-defense: how cell-autonomous immunity protects against pathogens. *Science* 340(6133):701–706.

Rayner AD (1991) The phytopathological significance of mycelial individualism. *Annual Review of Phytopathology* 29(1):305–323.

Rayner AD, Franks NR (1987) Evolutionary and ecological parallels between ants and fungi. *Trends in Ecology & Evolution* 2(5):127–133.

Robatzek S (2014) Endocytosis: At the crossroads of pattern recognition immune receptors and pathogen effectors. In: *Applied Plant Cell Biology* (eds) Nick P, Opatrny Z. Springer, Berlin, Heidelberg, pp. 273–297.

Romeis T, Piedras P, Jones JDG (2000) Resistance gene-dependent activation of a calcium-dependent protein kinase in the plant defense response. *Plant Cell* 12:803–816.

Romeis T, Piedras P, Zhang S, Klessig DF, Hirt H, Jones JD (1999) Rapid Avr9-and Cf-9-dependent activation of MAP kinases in tobacco cell cultures and leaves: Convergence of resistance gene, elicitor, wound, and salicylate responses. *The Plant Cell* 11:273–287.

Rueffler C, Hermisson J, Wagner GP (2012) Evolution of functional specialization and division of labor. *Proceedings of the National Academy of Sciences* 109(6):E326–E335.

Sansonetti PJ (2006) The innate signaling of dangers and the dangers of innate signaling. *Nature Immunology* 7:1237–1242.

Schulze-Lefert P, Panstruga R (2011) A molecular evolutionary concept connecting non-host resistance, pathogen host range, and pathogen speciation. *Trends in Plant Science* 16(3):117–125.

Shen Y, Liu N, Li C, Wang X, Xu X, Chen W, Xing G, Zheng W (2017) The early response during the interaction of fungal phytopathogen and host plant. *Open Biology* 7:170057.

Spoel SH, Koornneef A, Claessens SM, Korzelius JP, Van Pelt JA, Mueller MJ, Buchala AJ, Métraux JP, Brown R, Kazan K, Van Loon LC (2003) NPR1 modulates cross-talk between salicylate- and jasmonate-dependent defense pathways through a novel function in the cytosol. *The Plant Cell* 15:760–770.

Stergiopoulos I, De Wit PJGM (2009) Fungal effector proteins. *Annual Review of Phytopathology* 47:233–263.

Takeuchi O, Akira S (2010) Pattern recognition receptors and inflammation. *Cell* 140:805–820.

Tanabe S, Okada M, Jikumaru Y, Yamane H, Kaku H, Shibuya N, Minami E (2006) Induction of resistance against rice blast fungus in rice plants treated with a potent elicitor, N-acetylchitooligosaccharide. *Bioscience, Biotechnology, and Biochemistry* 70:1599–1605.

Thrall PH, Burdon JJ (2003) Evolution of virulence in a plant host-pathogen meta-population. *Science* 299:1735–1737.

Van den Ackerveken GF, Van Kan JA, De Wit PJ (1992) Molecular analysis of the avirulence gene avr9 of the fungal tomato pathogen Cladosporium fulvum fully supports the gene-for-gene hypothesis. *The Plant Journal* 2(3):359–366.

Van den Ackerveken GF, Vossen P, De Wit PJ (1993) The AVR9 race-specific elicitor of Cladosporium fulvum is processed by endogenous and plant proteases. *Plant Physiology* 103:91–96.

Van der Does D, Leon-Reyes A, Koornneef A, Van Verk MC, Rodenburg N, Pauwels L, Goossens A, Körbes AP, Memelink J, Ritsema T, Van Wees SC (2013) Salicylic acid suppresses jasmonic acid signaling downstream of SCFCOI1-JAZ by targeting GCC promoter motifs via transcription factor ORA59. *The Plant Cell* 25(2):744–761.

Van Dyken JD, Müller MJ, Mack KM, Desai MM (2013) Spatial population expansion promotes the evolution of cooperation in an experimental prisoner's dilemma. *Current Biology* 23(10):919–923.

van Kan JA, Van den Ackerveken G, De Wit P (1991) Cloning and characterization of cDNA of avirulence gene avr9 of the fungal pathogen *Cladosporium fulvum*, causal agent of tomato leaf mold. *Molecular Plant-Microbe Interaction* 4:52–59.

Van Kan JA, Van't Klooster JW, Wagemakers CA, Dees DC, Van der Vlugt-Bergmans CJ (1997) Cutinase A of Botrytis cinerea is expressed, but not essential, during penetration of gerbera and tomato. *Molecular Plant-Microbe Interactions* 10(1):30–38.

Venturelli S, Belz RG, Kämper A, Berger A, von Horn K, Wegner A, Böcker A, Zabulon G, Langenecker T, Kohlbacher O, Barneche F (2015) Plants release precursors of histone deacetylase inhibitors to suppress growth of competitors. *The Plant Cell* 27(11):3175–3189.

Wendehenne D, Lamotte O, Frachisse JM, Barbier-Brygoo H, Pugin A (2002) Nitrate efflux is an essential component of the cryptogein signaling pathway leading to defense responses and hypersensitive cell death in tobacco. *The Plant Cell* 14(8):1937–1951.

Wessels JG (2000) Hydrophobins, unique fungal proteins. *Mycologist* 14(4):153.

West SA, Cooper GA (2016) Division of labour in microorganisms: An evolutionary perspective. *Nature Reviews Microbiology* 14:716–723.

West SA, Fisher RM, Gardner A, Kiers ET (2015) Major evolutionary transitions in individuality. *Proceedings of the National Academy of Sciences of the United States of America* 112:10112–10119.

Yang Y, Shah J, Klessig DF (1997) Signal perception and transduction in plant defense responses. *Genes & Development* 11:1621–1639.

Zeilinger, S, Gruber, S, Bansal R, Mukherjee, PK (2016) Secondary metabolism in Tricho-derma–chemistry meets genomics. *Fungal Biology Reviews* 30(2):74–90.

Zhang L, Fetch T, Nirmala J, Schmierer D, Brueggeman R, Steffenson B, Kleinhofs A (2006) Rpr1, a gene required for Rpg1-dependent resistance to stem rust in barley. *Theoretical and Applied Genetics* 113:847–855.

Zhang S, Du H, Klessig DF (1998) Activation of the tobacco SIP kinase by both a cell wall-derived carbohydrate elicitor and purified proteinaceous elicitins from *Phytophthora* spp. *The Plant Cell* 10:435–450.

Zhang S, Klessig DF (1998) Resistance gene N-mediated de novo synthesis and activation of a tobacco mitogen-activated protein kinase by tobacco mosaic virus infection. *Proceedings of the National Academy of Sciences USA* 95:7433–7438.

Zhang S, Klessig DF. 1998 The tobacco wounding-activated mitogen-activated protein kinase is encoded by SIPK. *Proceedings of the National Academy of Sciences USA* 95:7225–7230.

7 Biotechnological Interventions for Fungal Strain Improvement

Sidra Anam, Farwa Batool and Nishat Zafar

CONTENTS

INTRODUCTION

Fungi have a pronounced potential in numerous industry applications due to their capacity to secrete a broad range of enzymes, secondary metabolites, proteins, and organic acids (Dufosse *et al.*, 2014; Mäkelä *et al.*, 2016). Among these, enzymes produced by fungi have been used in many commercial applications. For example, the fungus *Aspergillus niger* is extensively used to produce secondary metabolites and extracellular enzymes. Extracellular enzymes include a wide variety of carbo-hydrate-active enzymes (CAZymes) responsible for the degradation of plant biomass and its application in various fields such as biofuels and biochemical detergents, food and feed, and pulp and paper (Bala & Singh, 2019; Champreda *et al.*, 2019; Mäkelä *et al.*, 2014). Similarly, *Penicillium* species have produced many industrially essential

DOI: 10.1201/9781003162742-7

enzymes such as cellulases and xylanases (Saini *et al.*, 2015; Vaishnav *et al.*, 2018). *Penicillium subrubescens* has been reported to proficiently degrade insulin, which can be used as a prebiotics and biogenic agent, a raw material to produce fructose, and a raw material for ethanol production (Chi *et al.*, 2011; Mansouri *et al.*, 2013). Given the extensive application of fungal enzymes in the pharmaceutical industry, increasing the production of these bioactive molecules has become one of the main objectives of the industry over the last decades.

Fungi produce a diverse range of secondary metabolites that are clinically and industrially important. Secondary metabolites are organic products that are not essential for funal growth, but may exhibit antibacterial, antifungal, and antiviral properties. A range of secondary metabolites secreted by filamentous fungi also act as anti-cancer drugs (Bhagwat *et al.*, 2021; Kaur *et al.*, 2020). Thus, the improvement of fungal strains to produce the desired compounds in higher quantities requires modern biotechnology. It has been reported that most of the genes that are involved in the synthesis of secondary metabolites in fungi present themselves as clusters on the chromosomal arms and remain silent or poorly expressed under conventional laboratory conditions (Kjærbølling *et al.*, 2019). Thus, the production of novel secondary metabolites is delayed because of genomic complications and the unavailability of adequate gene-targeting tools (Bhagwat *et al.*, 2021).

For handling these difficulties, many techniques of genetic modifications are being utilized. For example, genome editing is used to alter the genome of the targets through insertion, deletion, rearrangement of nucleotides, and modifications in gene transcription (Barrangou *et al.*, 2007). These variations in the genetic structures are made to engineer new corridors that can help overproduce fungal-derived bioactive molecules (Bhagwat *et al.*, 2021). This chapter summarizes modern approaches/tools that may help to improve the production of bioactive molecules. These techniques include the following:

1) Clustered regularly interspaced short palindromic repeats CRISPR-associated protein 9 (CRISPR)/Cas-editing tools in the production of bioactive molecules
2) Protoplast fusion
3) Genetic engineering for improvement in the efficiency of protein expression and secretion
4) Chemical mutagenesis
5) Genetic transformation

CRISP-Cas9 System

Components of the CRISP-Cas9 System

CRISPR, together with the Cas proteins, avoids many limitations of traditional methods. A microbial adaptive immune system conserves memories of preceding infections by adding short, exogenous genetic material parts, called "spacers," present at CRISPR loci (Sternberg *et al.*, 2016). CRISPR-Cas9 comprises two components: sgRNA and Cas9 protein. First, the sgRNA identifies target sequences in the genomic DNA, and Cas9 endonuclease initiates a site-specific double-strand break into target DNA (Doudna & Charpentier, 2014). The innate cellular repair mechanism repairs double-strand breaks in DNA via homologous recombination or non-homologous end joining. The non-homologous end-joining pathways are vulnerable to errors, introducing frameshifts in open reading frames that lead to the gene knockout on translation.

The targeted mechanism of the CRISPR-Cas9 system allows targeting of metabolic pathways by transmitting the metabolic flux toward desirable biomolecules, thus disregarding those genes bringing undesirable side effects during this process. Furthermore, the CRISPR/Cas9 system is not limited to adding nucleotides at a single locus but uses multiple single-guide RNAs, which allows gene addition at multiple sites, generating multiple knockouts in an individual gene in a single step (Doudna & Charpentier, 2014).

Using a homology-directed repair mechanism, the CRISPR-Cas9 system can induce a double-strand break to generate gene knock-ins within the desired sequence. The insertion of donor DNA at the targeted site will change the coding region of a gene. Lastly, the CRISPR-Cas9 system can be used for DNA-free editing using the protein or RNA component without using a vector. DNA-free gene editing may be a safe option to avoid the unwanted genetic changes introduced by integrating plasmid DNA at the targeted cut sites.

Besides acting as a gene-editing tool, nuclease-dead Cas9 (dCas9), a catalytically deactivated Cas9 variant, may be used to regulate target gene expression by its fusion with a transcriptional activator or repressor (Doudna & Charpentier, 2014). This technique is being applied to develop epigenome in mammal cells by the fusion of dCas9 with different DNA effectors, such as acetylases and histone methylases, that cause site-specific DNA methylation or histone modifications (Hilton *et al.*, 2015). In short, the CRISPR-Cas9 system is becoming a capable method for producing bioactive molecules in microbes.

Use of CRISPR-Cas System

This system proposes excellent opportunities for genetic manipulation of fungi because of its simplicity, and cost-effectiveness. This system has ability to target multiple loci simultaneously offers a great advantage, mainly for use with fungi, for which a restricted set of functional selection markers is available. In this process, the design of the guide RNA and the targeting is flexible, except that a protospacer adjacent motif (PAM) is needed for the process, which could restrict the selection of the site of choice. However, the main criterion is the successful delivery of the CRISPR-Cas system to its respective targets; therefore, the use of this method remains restricted to those organisms for which practical transformation tools are available. The CRISPR-Cas system has been used for genetic modifications of several filamentous fungi, including pathogens and those fungal strains that are important at the industry level (Nødvig *et al.*, 2015). For example, in *Neurospora crassa*, a model species, both sgRNA and Cas9 paradigms were inserted into conidial hyphae using an electric field, coupled with donor plasmids, to switch the clr-2 gene with the β-tubulin promoter. The comparison of the CRISPR-Cas9 system with the usage of a mus-51-deficient strain showed comparable efficiencies in homologous recombination (Matsu-Ura *et al.*, 2015).

In 2015, the CRISPR-Cas9 system was used for *Trichoderma reesei,* applying a two-step method. In this process, a cas9-expressing fungal strain was produced and then transmuted with sgRNAs transcribed *in vitro* (Liu *et al.*, 2015). The ura5 gene was targeted in this study, and all analyzed transformants presented the desired frame-shift mutations. Furthermore, this study has been applied in laboratory to transcribe sgRNAs and has applied to six different species of Aspergillus, including *Aspergillus niger* and *Aspergillus brasiliensis* (Nødvig *et al.*, 2015; Kuivanen *et al.*, 2016). The system was founded on a single plasmid that contains the codon-optimized Cas9 coding genes and a ribozyme-flanked sgRNA along with the activator of the meiotic anaphase (AMA1) sequence involved in support of plasmid replication but confirming that the plasmid is rapidly lost without selection to minimize the assumption out of a destination. Another research group focused on applications of the CRISPR-Cas system in the *Aspergillus niger* strain, and the intentions wthis studyere to generate a strain that can efficiently produce galactaric acid (Sarkari *et al.*, 2017).

Protoplast Fusion

Fungal strains are responsible for producing a wide array of enzymes, and these enzymes have commercial applications in the biopharmaceutical industry (Patil *et al.*, 2015). One of the abundantly occurring polymers is cellulose, a good source of many chemicals and fuels (Kubicek *et al.*, 2009). It has been reported that many microbes secreting cellulose-degrading enzymes drive the degradation of cellulose. *Trichoderma* spp., a non-pathogenic filamentous fungus, is a well-known producer of this enzyme and has been extensively used for over 30 years in the biotechnological industry (Kubicek *et al.*, 2008; Leite *et al.*, 2019). Additionally, *Trichoderma* produces a higher level of extracellular cellulases which are essential enzymes for several industrial applications to be used for bioconversion, pulp and paper, detergents, textiles, and food and feed. The genome of *Trichoderma reesei* comprises over 200 genes that encode flycoside hydrolases (Peterson & Nevalainen, 2012), of which 16 genes have been involved in encoding cellulolytic enzymes, including CBHI, CBHII, EGL3, and BGL1. Given the extensive use of cellulase and the high cost of production, there is a need to develop other strategies/approaches to increase fungal enzyme production. Protoplast fusion, also called "somatic fusion," is a modern approach that can be used to introduce novel genetic characteristics into industrially important fungal strains (Häkkinen *et al.*, 2012).

The process involves the union of protoplasts—isolated from two genetically variant somatic cells—to achieve a hybrid protoplast. The hybrid protoplast contains heteroplasmic cytoplasm along with recombined genomic parts. Fusion of protoplast is comparatively a new multipurpose approach that offers a high frequency of genetic recombination compared with other established genetic manipulation approaches. Protoplast fusion is becoming an important genetic manipulation tool because of its potential to bring genetic changes, and it may be used for strain improvement.

Mechanism of Protoplast Fusion

- The fusion of protoplast is a physical process in which two protoplasts interact and attach themselves to each other either spontaneously or through an artificially induced process.

- When two protoplasts from the same source or different sources come into proximity, that protoplast combination process starts an induction phase whereby changes occur in the membrane's electrostatic potential. Once the fusion is complete, the membrane stabilizes, and surface potential comes to its original state.
- Upon protoplast fusion, there may be a disturbance in the proteins and glycoproteins of the membrane, which increases the fluidity of the membrane and generates a space for the union of lipid molecules, allowing the coalescence of adjacent cell membranes.
- The presence of phosphate groups on the intramembrane develops the negative charge present in protoplasts. Furthermore, the presence of Ca^{2+} ions causes the zeta-potential of the plasma membrane to be reduced. Under reduction conditions, the two protoplasts aggregate.
- The high alkalinity of the solution used in the chemo fusion protoplast process induces the intramembranous production of lysophospholipid, which may be linked with membranous fusion.

Methods of Protoplast Fusion

1. Spontaneous Fusion

During the isolation process, protoplasts often fuse spontaneously because of superficial physical contact among similar parental protoplasts. This phenomenon involves the fusion of protoplasts from neighboring cells via plasmodesmata and the formation of a multinucleate protoplast. The plasmodesmata get enlarged, which eventually allows the entry of organelles into adjoining cells. Spontaneous fusion gives rise to homokaryon.

2. Induced Fusion

Induced fusion involves the fusion of two or more isolated protoplasts via chemical agents. The presence of similar electronegative charges on the surface of the individual protoplasts causes repulsion forces between fused protoplasts. The use of chemical agents decreases the electronegativity of the isolated protoplasts and allows them to fuse. Induced fusion is a valuable technique because of its ability to bring genetic modifications from widely different genotypes and sexually incompatible organisms.

3. Electrofusion

Electrofusion is a comparatively modern approach of protoplast fusion that uses mild electrical fields in protoplast for the induction of fusion. This technique is high speed, easy to use, and simple. It is relatively more efficient than the chemical-induced fusion method. Electrofusion is applicable to those species whose protoplasts exhibit a severe toxic response toward polyethylene glycol in the chemo fusion process. The origin of electrofusion is based on biophysical studies of the cell membrane. The Senda research group was the first to report the protoplast electrofusion when they used two microelectrodes at the ends of pairs adhering Rauwolfia protoplasts and induced protoplast fusion with a 5–12 µAmp DC pulse. In their procedure, fusion yields were restricted to single protoplast pairs. However, another research group in 1981 improved this method for the large-scale fusion of plant protoplasts.

IMPROVEMENT IN THE EFFICIENCY OF EXPRESSION AND SECRETION OF PROTEIN THROUGH GENETIC MODIFICATIONS

Protein has broad applications in various fields of science like biotechnology, medicine, and biochemistry. Due to the high efficiency of post-translational modifications and protein secretion, fungi are well-thought-out cell factories for protein production. An effective strategy for protein production is the optimization of targeting protein codon or the process of transcription (Wang et al., 2020).

In fungi, the pathway involved in protein secretion occurs in three main steps: (1) the transfer of polypeptide molecules from the ribosome to the endoplasmic reticulum (ER), (2) modification in ER, (3) packing as well as translocation of these packed proteins in the form of vesicles to the Golgi apparatus and out of the cell-matrix. It has been reported that first, the transport pathways of co- or post-translational changes are involved in moving polypeptide molecules from the ribosome to ER. For blocking translation, the binding of signal peptide recognition particle (SRP) with signal peptide sequence (SPS) is the first step in the pathway of co-translational transport. After this, SRP leads the ribosome-mRNA-nascent peptide complex for targeting the ER membrane; this is followed by binding of SRP with its receptors. Consequently, SRP is unconstrained from the complex, restarting translation followed by entry of the nascent polypeptide into the lumen of ER via the Sec61p transport complex. Following the pathway of post-translational transport, translation of the newly synthesized proteins occurs in the cytosol, keeping them unfolded through interaction with Hsp70 chaperone and co-chaperones. This complex then targets ER via interacting with the receptors (Sec62p-Sec72p-Sec73p sub-complex) present at the membrane. The entrance of this complex into ER is facilitated through the membrane protein Sec63p and the chaperone-binding immunoglobulin protein (BiP) present in the lumen of ER (Haßdenteufel et al., 2018).

The folding and modification in polypeptides is the second step of the protein preparation pathway, and it occurs in ER. The newly synthesized polypeptides, which are already precisely folded, are moved for the next step of modification, such as glycosylation. This glycosylation step is considerably important for the localization, secretion, and stabilization of proteins. Following this step, the glycosylated protein is safely secreted out of the cell. The unfolded protein response (UPR) senses the unfolded proteins in ER, triggering biosynthesis of folding enzymes and chaperones, and these unfolded proteins are broken down by ER-associated degradation elements (ERAD) (Krasevec et al., 2014; Kuivanen et al., 2016; Xu et al., 2018).

In the third step of protein preparation, the folded protein is transported to Golgi bodies via fusing with the target membrane and then secreted out of the cell. Golgi-derived secretory vesicles are transported to the plasma membrane of the apical site via apical vesicle clusters present in Spitzenkörper, such as in filamentous fungi. The interaction of target vesicles with the target membrane is the most critical step and is facilitated through soluble N-ethylmaleimide-sensitive factor-associated protein receptors elements (SNARE). SNAREs are categorized into two types on the basis of their localization: target membrane SNARE (t-SNARE) and vesicle SNARE (v-SNARE). The complex of three receptors—v-SNARE protein SNC1 and t-SNARE proteins SSO1 and SSO2—is responsible for the fusion of bubbles, such as in filamentous fungi (Valkonen *et al.*, 2007; Haßdenteufel *et al.*, 2018).

CHEMICAL MUTAGENESIS

Generally, any physical or chemical agent involved in genome change (mostly in DNA) is named as mutagen, It has been reported that most mutations are involved in causing cancer and are thus generally named "carcinogens." Mistakes in DNA recombination, repairing and replication, and spontaneous hydrolysis cause spontaneous mutation (Parthiban *et al.*, 2018).

Role of Polyamines in Mutagenesis

Fungi from the phyla Zygomycota and Ascomycota have unique and unusual biochemical pathways. Many products, such as the secondary metabolite, are utilized as vital pharmaceuticals, including antibiotics, statins, and immune-depressing agents. Under different conditions, the individual species can produce more than a hundred secondary metabolites. Improvement in fungal strains helps us obtain a high output of secondary metabolites and a significant decline in useless products. The primary tool for the strain improvement of filamentous fungi is random mutagenesis and screening. Polyamines are involved in random mutagenesis for the improvement of the yield of secondary metabolites. Recently, it has been shown that adding exogenous polyamines may increase the production of such improved strains of filamentous fungi (Zhgun, 2020).

GENETIC TRANSFORMATION

Fungi are considered suitable for genetic transformation as they have an excellent ability to secrete and prepare good metabolites that are used for the benefit of humans. Many techniques have been developed that enable us to modify the target genes so that they may be used to depict the functionality of genes. For genetic transformation, the delivery of exogenous genetic material into the cell is necessary (Li *et al.*, 2017).

Nowadays, the methods that are generally used for genetic modifications are

- Protoplast-mediated transformation (PMT)
- Agrobacterium-mediated transformation (AMT)
- Electroporation transformation
- Biolistic method
- Shockwave-mediated transformation (SWMT)

PMT

This is the most commonly used method for the genetic modification of fungi. Its working is dependent on the proportion of protoplast numbers. The basic principle of PMT is the using of commercially available enzymes. In this method, the cell wall of fungi is removed for the production of protoplast. Following this, some chemicals (like polyethylene glycol) are utilized to successfully fuse foreign genetic material into the protoplast. The cell wall composition of fungi varies from strain to strain. It has been reported that the parts of the spore coat are considerably diverse among the same strains of hyphae. Due to these reasons, the universal transformation method cannot be used universally for various fungal strains. However, to some extent, there are standard methods for protoplast preparation. PTM is being improved to attain excellent efficacy of genetic modification and target the locus of required genes through gene editing. Enzyme treatment is generally used for protoplast preparation via removing the cell wall of fungi. However, physical methods like grinding and supersonic wave shock treatment have been reported. They are rarely used, though, because of low yield and practical inconvenience (Penttilä *et al.*, 1987; Wen-hui *et al.*, 2006).

The basic steps involved in PTM are as follows:

- Preparation of protoplast
- Uptake of exogenous DNA
- Regeneration of protoplasts

AMT

Agrobacterium is a gram-negative bacterium and is generally an inhabitant of soil. *Agrobacterium tumefaciens* (*A. tumefaciens*) causes infection in wounded plants. The tumor-inducing plasmid, also named "Ti plasmid," has a size of more than 200 kb and can be isolated when the infection is at an early stage. In wounded plants, the wound facilitates the entry of *A. tumefaciens* into the plant; it injects its Ti plasmid into the nucleic acid of the wounded plant, causing infection. The injected part of the Ti plasmid is usually named "transfer DNA" or "T-DNA." The T-DNA injects itself into the genomic part of the plant arbitrarily, in a monoclone form. The T-DNA's left and right borders are lined by two defective sequences that contain the genes encoding for the enzymes involved in forming plant hormones responsible for tumor growth. Then, a secondary vector is prepared, in which the gene of interest is inserted between the right and left parts of the Ti plasmid. The resulting modified plasmid is transmuted into *A. tumefaciens*. This positive clone of Agrobacterium is utilized as a carrier to integrate the gene of interest into the genome of fungi. This method was first applied to transform *S. cerevisiae*. Compared with conventional transformation methods, AMT is more efficient and stable. Hygromycin resistance inserted into the plasmid is usually used in the transformation of *Aspergillus awamori*. This genetic transformation method has been utilized for many species of Ascomycetes, including *Monascus purpureus* and Aspergillus (Michielse *et al.*, 2008; Azizi *et al.*, 2013).

The many advantages of AMT are as follows: a variety of recipients for transformation, such as hyphae, spores, and protoplasts; the integration capability of foreign nucleic acid into genetic material resulting in the formation of stable recombinants; and the efficiency of transmutation leading to a large number of recombinants. A secondary vector is also needed in this method, which is a tiresome preparation. For optimizing AMT, many factors, such as the following, should be taken into account (Azizi *et al.*, 2013):

- Type of starting fungal material
- Amount of acetosyringone
- Ratio of Agrobacterium and fungi
- Culturing conditions of fungi and Agrobacterium

Electroporation Transformation

Electroporation is a rapid, modest, and valuable genetic modification method for many filamentous fungi. In this method, a capacitor is used to store electric charges to figure out a high voltage, then the sample is bumped into by voltage impulse, and finally, the foreign DNA is shifted into the cells instantly. Commonly, both exponential and square waves are utilized for transformation. Exponential waves are produced via the charging and discharging of the capacitor, and the value of electric intensity decreases from the highest to the lowest in exponential form. A nonsinusoidal periodic waveform is named a "square wave," which can be depicted as an immeasurable summary of sinusoidal waves, in which the amplitude interchanges at a steady frequency ranging from high to low. In electroporation, various waveforms are utilized to transform fungal species (Weaver, 2003; Li *et al.*, 2017).

In electroporation, exposure of the cell to an electric field changes the structure of the cytomembrane due to a voltage that is tempted between cytomembranes. Due to this electric shock, many small pores are formed in the cytomembrane. This induced permeability of the cell wall is changeable within the verges of the period and voltage. If this permeability is not alterable, the cells will become injured. Therefore, the small pores produced due to electric shock are two types: some can be reversed to their original shape, while others cannot (Weaver, 2003). When an appropriate field intensity is applied, cytomembrane protein and lipid molecules may restore the previous shape.

On the other hand, the electric shock involved in changeable recovery could also cause prolonged recovery, leading to the cell's death. Using electroporation, the foreign nucleic acid can be inserted into the protoplast of plants, fungi, and bacterium and in the animal's cells. This method is effectively utilized for many fungal species. It has been reported that the germinating spores are more docile for genetic alteration (Weaver, 2003; Wu & Geoffrey, 2004). Nowadays, this method has become a reliable way of genetic alteration.

The electroporation method has been expansively utilized for both eukaryotic and prokaryotic cells. Electroporation is a suitable method of interest for research to be used for many new species of fungi. This method is more reliable and cost effective compared with previous transformation methods. However, the exact process of electroporation is not clear yet. The permeability rate of the cytomembrane depends upon a number of factors. Along with this, a proper buffer is required for the optimization of this process.

Factors affecting the efficacy of this method include the following:

- Electric field strength
- Capacitance
- Duration of pulse and frequency rate

Biolistic Transformation

This method is also called "particle bombardment." In this method, the exogenous nucleic acid is adsorbed on tungsten or gold particle surfaces. Using high pressure, the particles are inserted into host cells. Particle bombardment can cause transient as well as stable transformation. Many features affect the capacity of bombardment in patterns of multifarious interfaces. Natural constraints, such as cell type, condition of growth, and density of cell, and instrumental settings, such as size and type f particle, level of pressure and vacuum, and distance from the target site, have considerable significance. This is the most effective transformation method among all gene transmutation processes. It has no limitation for the type of fungal species or the type of host cells. For those fungal species that we cannot culture, or the protoplast preparation is tiresome, this method is excellent. It is an informal and expedient process. However, there are some expenses on consumables and instruments. This method can be considered if others flop. It has been used to successfully transform the Aspergillus species (Sanford *et al.*, 1993; Barcellos *et al.*, 1998).

SWMT

In this method, the basic principle is the change and transfer of energy for the generation of momentary pressure disturbance and twisting force across the cells, which results in the formation of the transient cavitation effect. This technique is utilized in medical fields such as orthopedics and for breaking up stones in the kidney. The permeability in the membrane of cells is changed by SWMT via acoustic cavitation, facilitating the insertion of foreign genetic material into cells. The insertion of foreign nucleic acid into various bacteria, such as *Salmonella typhimurium, Escherichia coli, and Pseudomonas aeruginosa,* is done by this method. For fungi, such as *Fusarium oxysporum, Phanerochaete chrysosporium*, and *Aspergillus niger,* this method was first reported in 2013.

Compared with previous genetic modification methods, SWMT is applicable directly on spores, not on the protoplast. Control of shockwave spores' number, speed, and energy needs to be optimized, although other physical factors are under control. The genetic alteration results are excellent. Compared with the Agrobacterium transformation method, the SWMT method could enhance the transformation efficiency by 5400 fold for *Aspergillus niger* (Ogden *et al.*, 2001; Magaña-Ortíz *et al.*, 2013).

This method has some disadvantages, though, such as the low transformation efficacy of genome to cell due to nucleic acid damage caused by shockwave treatment. However, this efficacy is considerably high for the number of cells being used. To enhance effectiveness, the proportion of DNA to cell should be considered (Loske, 2007; Prakash *et al.*, 2011). It is an easy method for producing a large number of plasmids in the laboratory. However, the resources and instruments for shockwave are somewhat costly as they have primarily been created for application in medical fields. Ultimately, this is the main hurdle for its application in labs.

REFERENCES

Azizi M, Yakhchali B, Ghamarian A, Enayati S, Khodabandeh M and Khalaj V (2013). Cloning and expression of gumboro VP2 antigen in *Aspergillus niger. Avicenna Journal of Medical Biotechnology* 5(1):35.

Bala A and Singh B (2019). Cellulolytic and xylanolytic enzymes of thermophiles for the production of renewable biofuels. *Renewable Energy* 136:1231–1244.

Barcellos FG, Fungaro MHP, Furlaneto MC, Lejeune B, Pizzirani-Kleiner AA and Lúcio de Azevedo J (1998). Genetic analysis of *Aspergillus nidulans* unstable transformants obtained by the biolistic process. *Canadian Journal of Microbiology* 44(12):1137–1141.

Barrangou R, Fremaux C, Deveau H, Richards M, Boyaval P, Moineau S, Romero DA and Horvath P (2007). CRISPR provides acquired resistance against viruses in prokaryotes. *Science* 315(5819):1709–17012.

Bhagwat AC, Patil AM and Saroj SD (2021). CRISPR/Cas 9-based editing in the production of bioactive molecules. *Molecular Biotechnology* 1–7.

Champreda V, MHuantong W, Lekakarn H, Bunterngsook B, Kanokratana P, Zhao XQ, Zhang F, Inoue H, Fujii T and Eurwilaichitr L (2019). Designing cellulolytic enzyme systems for biorefinery: From nature to application. *Journal of Bioscience and Bioengineering* 128(6):637–654.

Chi ZM, Zhang T, Cao TS, Liu XY, Cui W and Zhao CH (2011). Biotechnological potential of inulin for bioprocesses. *Bioresource Technology* 102(6):4295–4303.

Doudna JA and Charpentier E (2014). The new frontier of genome engineering with CRISPR-Cas9. *Science* 346(6213).

Dufosse L, Fouillaud M, Caro Y, Mapari SA and Sutthiwong N (2014). Filamentous fungi are large-scale producers of pigments and colorants for the food industry. *Current Opinion in Biotechnology* 26:56–61.

Haßdenteufel S, Johnson N, Paton AW, Paton JC, High S and Zimmermann R (2018). Chaperone-mediated Sec61 channel gating during ER import of small precursor proteins overcomes Sec61 inhibitor-reinforced energy barrier. *Cell Reports* 23(5):1373–1386.

Häkkinen M, Arvas M, Oja M, Aro N, Penttilä M, SaloheiMo M and Pakula TM (2012). Re-annotation of the CAZy genes of *Trichoderma reesei* and transcription in the presence of lignocellulosic substrates. *Microbial Cell Factories* 11(1):1–26.

Hilton IB, D'ippolito AM, Vockley CM, Thakore PI, Crawford GE, Reddy TE and Gersbach CA (2015). Epigenome editing by a CRISPR-Cas9-based acetyltransferase activates genes from promoters and enhancers. *Nature Biotechnology* 33(5):510–517.

Kaur N, Alok A, Kumar P, Kaur N, Awasthi P, Chaturvedi S, Pandey P, Pandey A, Pandey AK and Tiwari S (2020). CRISPR/Cas9 directed editing of lycopene epsilon-cyclase modulates metabolic flux for β-carotene biosynthesis in banana fruit. *Metabolic Engineering* 59:76–86.

Kjærbølling I, MorTensen UH, Vesth T and Andersen MR (2019). Strategies to establish the link between biosynthetic gene clusters and secondary metabolites. *Fungal Genetics and Biology* 130:107–121.

Krasevec N, Milunović T, Lasnik MA, Lukančič I, Komel R and Porekar VG (2014). Human granulocyte colony stimulating factor (G-CSF) produced in the filamentous fungus Aspergillus niger. *Acta Chimica Slovenica* 61(4):709–717.

Kubicek CP, Komon-Zelazowska M and Druzhinina IS (2008). Fungal genus Hypocrea/Trichoderma: From barcodes to biodiversity. *Journal of Zhejiang University Science B* 9(10):753–763.

Kubicek CP, Mikus M, Schuster A, SchMoll M and Seiboth B (2009). Metabolic engineering strategies for the improvement of cellulase production by *Hypocrea jecorina*. *Biotechnology for Biofuels* 2(1):1–14.

Kuivanen J, Wang YMJ and Richard P (2016). Engineering Aspergillus niger for galactaric acid production: Elimination of galactaric acid catabolism by using RNA sequencing and CRISPR/Cas9. *Microbial Cell Factories* 15(1):1–9.

Leite P, Silva C, Salgado JM and Belo I (2019). Simultaneous production of lignocellulolytic enzymes and extraction of antioxidant compounds by solid-state fermentation of agro-industrial wastes. *Industrial Crops and Products* 137:315–322.

Li D, Tang Y, Lin J and Cai W (2017). Methods for genetic transformation of filamentous fungi. *Microbial Cell Factories* 16(1):1–13.

Liu R, Chen L, Jiang Y, Zhou Z and Zou G (2015). Efficient genome editing in filamentous fungus Trichoderma reesei using the CRISPR/Cas9 system. *Cell Discovery* 1(1):1–11.

Loske AM (2007). *Shock wave physics for urologists*. UNAM, Centro de Física Aplicada y Tecnología Avanzada.

Magaña-Ortíz D, Coconi-LinarEs N, Ortiz-Vazquez E, Fernández F, Loske AM and Gómez-Lim MA (2013). A novel and highly efficient method for genetic transformation of fungi employing shock waves. *Fungal Genetics and Biology* 56:9–16.

Mäkelä MR, Bredeweg EL, Magnuson JK, Baker SE, De Vries RP and Hildén K (2016). Fungal ligninolytic enzymes and their applications. *Microbiology Spectrum* 4(6):16.

Mäkelä MR, Donofrio N and de Vries RP (2014). Plant biomass degradation by fungi. *Fungal Genetics and Biology* 72:2–9.

Mansouri S, HoubRAken J, Samson RA, Frisvad JC, Christensen M, Tuthill DE, Koutaniemi S, HAtakka A and Lankinen P (2013). *Penicillium subrubescens*, a new species efficiently producing inulinase. *Antonie van Leeuwenhoek* 103(6):1343–1357.

Matsu-Ura T, Baek M, Kwon J and Hong C (2015). Efficient gene editing in *Neurospora crassa* with CRISPR technology. *Fungal Biology and Biotechnology* 2(1):1–7.

Michielse CB, Hooykaas PJ, Van Den Hondel CA, and Ram AF (2008). Agrobacterium-mediated transformation of the filamentous fungus *Aspergillus awamori*. *Nature Protocols* 3(10):1671–1678.

Nødvig CS, Nielsen JB, Kogle ME and Mortensen UH (2015). A CRISPR-Cas9 system for genetic engineering of filamentous fungi. *PLoS One* 10(7):e0133085.

Ogden JA, Tóth-Kischkat A and Schultheiss R (2001). Principles of shock wave therapy. *Clinical Orthopaedics and Related Research* 387:8–17.

Parthiban P, Murali RK, ChinniAh C, Ravikumar A and Thagavel K (2018). Genetic Improvement of Fungal Pathogens. *Advances in Plants & Agriculture Research* 8(1):4–9.

Patil N, Patil S, Govindwar S and Jadhav JP (2015). Molecular characterization of intergeneric hybrid between A spergillus oryzae and T richoderma harzianum by protoplast fusion. *Journal of Applied Microbiology* 118(2):390–398.

Penttilä M, Nevalainen H, Rättö M, SalminEn E and Knowles J (1987). A versatile transformation system for the cellulolytic filamentous fungus Trichoderma reesei. *Gene* 61(2):155–164.

Peterson R and Nevalainen H (2012). Trichoderma reesei RUT-C30—thirty years of strain improvement. *Microbiology* 158(1):58–68.

Prakash GD, Anish RV, JaGadeesh G and Chakravortty D (2011). Bacterial transformation using micro-shock waves. *Analytical Biochemistry* 419(2):292–301.

Saini R, Saini JK, Adsul M, Patel AK, MAthur A, Tuli D and Singhania RR (2015). Enhanced cellulase production by *Penicillium oxalicum* for bio-ethanol application. *Bioresource Technology* 188:240–246.

Sanford JC, Smith FD and Russell JA (1993). Optimizing the biolistic process for different biological applications. *Methods in Enzymology* 217:483–509.

Sarkari P, Marx H, Blumhoff ML, Mattanovich D, Sauer M and Steiger MG (2017). An efficient tool for metabolic pathway construction and gene integration for *Aspergillus niger*. *Bioresource Technology* 245:1327–1333.

Sternberg SH, RichtEr H, Charpentier E and Qimron U (2016). Adaptation in CRISPR-Cas systems. *Molecular Cell* 61(6):797–808.

Vaishnav N, Singh A, Adsul M, Dixit P, Sandhu SK, MAthur A, Puri SK and Singhania RR (2018). Penicillium: The next emerging champion for cellulase production. *Bioresource Technology Reports* 2:131–140.

Valkonen M, KalkMan ER, SaloheiMo M, Penttiläö M, Read ND and Duncan RR (2007). Spatially segregated SNARE protein interactions in living fungal cells. *Journal of Biological Chemistry* 282(31):22775–22785.

Wang Q, Zhong C and Xiao H (2020). Genetic engineering of filamentous fungi for efficient protein expression and secretion. *Frontiers in Bioengineering and Biotechnology* 8:293.

Weaver JC (2003). Electroporation of biological membranes from multicellular to nano scales. *IEEE Transactions on Dielectrics and Electrical Insulation* 10(5):754–768.

Wen-hui T, Yan-ping L and Yang X (2006). Factors affect the formation and regeneration of protoplasts of microorganism. *Modern Food Science Technology* 22(3):263–268.

Wu S and Geoffrey JL (2004). High efficiency transformation by electroporation of Pichia pastoris pretreated with lithium acetate and dithiothreitol. *Biotechniques* 36(1):152–154.

Xu Y, Wang YH, Liu TQ, ZHang H, Zhang H and Li J (2018). The GlaA signal peptide substantially increases the expression and secretion of α-galactosidase in *Aspergillus Niger. Biotechnology Letters*, 40(6):949–955.

Zhgun AA (2020). Random mutagenesis of filamentous fungi strains for high-yield production of secondary metabolites: The role of polyamines. In *Genotoxicity and mutagenicity-mechanisms and test methods*. IntechOpen.

8 Molecular Identification and Detection of Phytofungi

Siddra Ijaz, Imran Ul Haq, Nabeeha Aslam Khan and Bukhtawer Nasir

CONTENTS

INTRODUCTION

Fungal phytopathogens are leading biotic factors linked to disease causation in plants (Doehlemann et al., 2017; Fisher et al., 2020), which ultimately has a deleterious impact on agronomic traits and the health of plants species worldwide (Asibi et al., 2019; Marc̆iulynas et al., 2020; Wenneker and Thomma, 2020). They infect plants at any developmental stage, from seedling to seed maturing (Narayanasamy, 2011; Godfray et al., 2016). The rapidly increasing population demand for food security and safety requires sustainable plant disease management (Sarrocco and Vannacci, 2018). Therefore, effective and smart strategies for timely alert and prompt response are critical factors for fighting against pathogenic fungi in plants (Nagrale et al., 2016).

Numerous advancements have been made in the arena of fungal detection and identification. Traditional strategies have made the diagnosis on the basis of visible signs and symptoms of diseases caused by phytofungi (Tör and Woods-Tör, 2017). Conventional approaches, including microscopic and biochemical techniques (Tan et al., 2008; Sharma and Sharma, 2016), possess some downsides in phytopathogens diagnosis, i.e., due to lack of subtle expertise and fungal phenotypic serological plasticity (Luchi et al., 2020). These drawbacks and inefficiencies in the traditional fungal pathogen diagnosis tools led to a paradigm shift toward modern tools aimed at molecular biology. The molecular approaches allow phytofungi diagnosis at preliminary stages of infection, albeit at an infinitesimal inoculum level (Rollins et al., 2016; Johnston-Monje and Mejia, 2020). Besides, molecular biology, in juxtaposition with computation biology, offers a platform for recording nomenclatural and genetic information of phytopathogenic fungi, which promotes the use of molecular tools for the precise and accurate diagnosis of fungi.

Advanced molecular techniques have played an important role in pathogen detection and diagnosis, especially in poly-fungal diseases. These are equally effective for symptomatic and asymptomatic diseases, obligate pathogens, and in-sole and synergistic infections, which were challenging to address using traditional Koch's postulates. Among polymerase chain reaction (PCR)-based assays, qPCR is extensively used to quantify and differentiate plant pathogens, mainly when the sample load is too small. However, these techniques involve sophisticated and expensive instruments that can only be used by a professional with expertise and knowledge of bioinformatics to obtain results without misinterpretations (Moffat et al., 2015; Mancini et al., 2016; Abdullah et al., 2018; Hariharan and Prasannath, 2021).

RESTRICTION FRAGMENT LENGTH POLYMORPHISM (RFLP)

This approach has been used for fungal detection by the restriction digestion of fungal genomic DNA and then separation of DNA fragments of different sizes through gel electrophoresis. In this method, polymorphisms are analyzed on the basis of the genome's cleavage sites of restriction enzymes. This technique has become archaic with the emergence of PCR-based methods for detecting and identifying phytofungi (Martínez-García et al., 2011). However, its modified version, the RFLP–PCR method, will be discussed under PCR-based approaches for phytofungi detection.

FUNGAL DETECTION USING PCR ANALYSIS

The PCR method has brought a huge revolution in molecular biology. It allows the synthesis of multiple copies of a particular genetic region in a genome using random or sequence-specific primers. On the basis of principles and protocols, different types of PCR methods are being used to detect fungal pathogens. These methods include randomly amplified polymorphic DNA (RAPD), simple sequence repeats (SSRs), single nucleotide polymorphism (SNP), PCR–RFLP, amplified fragment length polymorphism (AFLP), reverse transcription PCR (RT–PCR), multiplex PCR, nested PCR, real-time quantitative PCR (qPCR), rolling circle amplification (RCA), and loop-mediated isothermal amplification (LAMP) (Aslam et al., 2017; Cheng et al., 2020).

> *RAPD:* This PCR method is based on arbitrary primers (decamers), which are short stretches of nucleotide sequences. This method relies on the amplification of random regions of the fungal genome and is most suitable for

haploid fungi. It is not suitable for diploids, heterokaryons, and oomycetes (polyploid fungi) (Fourie et al., 2011; McDonald, 1997).

SSRs: SSRs, microsatellites, or short tandem repeats (STRs) are the repetitive non-coding regions in the eukaryotic genome. They are highly polymorphic, co-dominant molecular markers used to explore fungi's genetic diversity and genetic mapping. Their multiallelic nature imparts versatility to them. Sequence-specific primer pairs flanking the microsatellite regions are used in the SSR–PCR approach to characterize the fungi (Capote et al., 2012).

AFLP: This method deals with the genomic DNA restriction digestion followed by the ligation of the adaptors to them. In PCR analysis, primers complementary to their 3′ end adapter are used. The AFLP- based PCR analysis amplifies approximately hundreds of DNA fragments in a single turn to simultaneously detect polymorphisms at different genomic sites of fungi (Groenewald et al., 2006; Liu et al., 2009).

RFLP–PCR: In this method, fungal detection is made by employing a specific set of PCR primers to amplify the target regions. The resultant products are restricted with nucleases to get the restriction profile of fungi (Hyakumachi et al., 2005).

SNPs: Phytofungi exhibit single-to-few base pair differences among each other in particular genetic regions that are being used in the identification and detection of fungi. SNPs are the single loci, biallelic DNA markers used in linkage and association mapping (Brumfield et al., 2003; Narum et al., 2008).

PHYTOFUNGI DETECTION THROUGH MULTIPLEX PCR

Multiplex PCR analysis facilitates the simultaneous detection of multiple fungi. More than one set of target-specific primers is used in a single reaction to detect multiple fungi in this method. Therefore, it is practiced when more than one kind of fungus infects the host plant (Pallás et al., 2018). While investigating a possibly complex disease, it is crucial to have a system that can generate amplicons of all associated pathogens in a single experiment; for this, multiplex PCR is used. It requires several designed primer pairs, all having the same annealing temperature, which results in the simultaneous amplification of multiple DNA fragments. These specific amplified sequences of DNA are then separated using electrophoresis. This system has been successfully used to identify 12 fungal pathogens associated with cranberry fruit rot, with the help of internal transcribed spacer (ITS) region and translation elongation factors 1-alpha (TEF-1 α) (Sint et al., 2012; Zhao et al., 2014; Conti et al., 2019). Quantitative PCR can detect an associated disease pathogen in real time. Fluorescent dyes monitor each amplification step, whose signals are directly related to the number of amplicons produced. It is an ingeniously sensitive technique to detect a single spore of pathogenic fungi in the diseased sample. Various novel pathogens of different plant diseases have been detected and distinguished using this technique, such as *Diaporthe helianthi* and *D. gulyae* in sunflower stem canker and *Verticillium longisporum* in rapeseed.

This technique is fast, sensitive, reliable, and specific (Balodi et al., 2017; Elverson et al., 2020). Bio PCR is highly reliable and effective in diagnosing seed-borne pathogens. Initially, the putative pathogen is grown on artificial growth media to increase its concentration and to avoid other microbial contaminations (Kumar et al., 2020).

PHYTOFUNGI DETECTION THROUGH NESTED PCR

In the nested PCR method, two sets of primer pairs are used in two consecutive rounds of PCR. In the first round, non-specific primers are used; however, the second round of PCR is based on a pair of pathogen-specific primers for amplifying the target region (Porter-Jordan et al., 1990; Bhat and Browne, 2010). In the nested PCR approach, external primers are used to obtain amplification at the generic level, while internal pairs of primers are used for specific characterization of amplicons at the species level. Targeted or multi-pathogen detection is achieved by designing specific or no-specific primers, respectively, succeeded by DNA sequencing in a PCR-based system. Acquired nucleotide sequences undergo basic local alignment search tool analysis so that each isolate can be identified and distinguished (Sikdar et al., 2014; Mirmajlessi et al., 2015). This is done to make PCR amplification more sensitive and specific for identifying fungi. Nested PCR analysis has been successfully used to diagnose the pathogens responsible for the twig blight of pomegranate and *Eucalyptus* dieback. However, these two PCR assays are tedious and may produce false outcomes due to cross-contamination (Bhat and Browne, 2010; Yang et al., 2017).

TRANSCRIPTOMICS-BASED APPROACHES FOR PHYTOFUNGI DETECTION

Transcriptomics-based approaches are commonly practiced for analyzing fungal gene expression during disease prognosis (Yang et al., 2010). The gap in other PCR methods for fungal diagnostics is their failure to differentiate between the living and dead phytofungi in the diseased sample. Therefore, the RT–PCR) approach has tried to bridge this gap by using random primers, oligo (dT) primers, and gene-specific primers for amplifying cDNA from cellular messenger RNA (mRNA) population at a specific stage of a particular disease. Likewise, another highly sensitive, accurate, more reliable method is real-time qPCR, which has become a standard approach for detecting phytofungi (Garrido et al., 2009). In this method, the reaction is monitored through fluorescent signals produced proportionally to the number of target amplicons in a sample. It allows target gene quantification through measured magnitude to a standard curve of known magnitude in a biological sample. Hence, in plant pathology, this attribute possesses significant worth for correlating the fungus quantity in a diseased sample and monitoring pathogenesis and disease progression in an infected plant. The chemistry of the real-time qPCR method is based on the convention of double-stranded DNA-binding dyes (usually SYBR green dye) and fluorescent-labeled probes, including TaqMan probe, Molecular Beacons, and Scorpions probe (Diguta et al., 2010).

ISOTHERMAL AMPLIFICATION METHODS–BASED FUNGAL DETECTION

RCA, an isothermal enzymatic assay, amplifies DNA using DNA polymerase (phi29) and extends primer annealing to circular DNA target using strand displacement activity, consequently releasing ssDNA. Padlock probes are designed to target elongation factors to confirm the pathogenicity of a plant pathogen. Another technique, LAMP, involves two steps: cycling and elongation. It utilizes forward and backward inner primers, loop primers, and outer primers to recognize unique sequences in target DNA. The nucleic acid sequence–-based amplification technique converts the ssRNA to produce cDNA through RT (Gu et al., 2018; Panno et al., 2020).

POST-AMPLIFICATION TECHNIQUES

DNA microarray, or DNA chip, is a collection of thousands of microscopic DNA spots at specific positions used for complementary DNA hybridization, which gives a comparative analysis of transcripts. In contrast, DNA macro arrays are species-specific probes arranged on well plates and fixed on nitrocellulose membranes (Bhatia and Dahiya, 2015).

NUCLEIC ACID PROBE-BASED ASSAYS

In situ hybridization is used to detect the location of a specific sequence in the chromosome. In principle, targeted mRNA is identified by designing a labeled, antisense, single-stranded RNA probe that, after binding, forms complementary base pairs. The binding of the synthetic strand is detectable as it is labeled, usually with radioactive isotopes (^{32}P and ^{35}S), due to their high sensitivity, or otherwise with biotin, tyramide, or bromo-deoxyuridine. This technique was successfully used to distinguish various isolates of rust fungi from their respective host plant tissues (Jensen, 2014; Ellison et al., 2016).

Fluorescent in situ hybridization is a comparatively innovative technology used to detect nucleotide sequences using probes labeled with fluorochromes. Usually, oligonucleotide probes' hybridization with the rRNA of pathogenic cells results in staining visualized through confocal laser microscopy. The fluorescent in situ hybridization technique has been successfully applied to detect the causal organism, *Sclerotium rolfsii*, of southern blight in tomatoes. This technique is highly sensitive and accurate in identifying principal pathogens from mix-species samples (Cui et al., 2016; Milner et al., 2019).

DNA BARCODING AND PHYLOGENETIC ANALYSES FOR PHYTOFUNGI DETECTION

DNA barcoding is a more authenticated molecular approach for fungal identification in a highly precise and more reliable way. It is based on small DNA segments designated for fungal identification and the delimitation of its species boundaries (Geiser,

2004; Hibbett, 2013). Among the most popular taxonomically informative DNA barcode–based markers are the internal transcribed spacer region of ribosomal DNA, tubulin, actin, translation elongation factor 1-alpha, glyceraldehyde 3-phosphate dehydrogenase, and RNA polymerase II second largest subunit. They are usually part of the genome of all fungi lineages. DNA barcoding helps identify fungi at the complex stages of their life cycle, where morphological identification is cumbersome, confusing, and impractical.

Phylogenetic analyses elucidate evolutionary liaison among different populations and different organisms of a population to divulge genetic relatedness and diversity. Different phylogenetic methods generate the phylogram for deciphering the evolutionary relationship among lineages. The phylogram or phylogenetic tree comprises branches, nodes, and roots. External nodes display operational taxonomic units (OTUs); however, hypothetical taxonomic units are the description of internal nodes, which predict the progenitors of OTUs (Pevsner, 2009; Hennell et al., 2012). For phylogram construction, molecular data (nucleotide sequences and amino acid sequences) are required. By using the molecular data, a phylogenetic tree is generated with the assistance of phylogenetic methods. These methods are broadly classified into (1) character state and (2) distance matrix. Unweighted pair group method with arithmetic means, neighbor-joining, Fitch–Margoliash come under the distance matrix clustering algorithm. However, maximum parsimony, maximum likelihood, and Bayesian inference come under character state on the basis of optimal search criteria (Swofford, 2002).

HIGH-THROUGHPUT SEQUENCING AND PHYTOFUNGI DETECTION

Next-generation sequencing uses modern methods, such as pyro-sequencing, parallel signature sequencing, polony sequencing, and oligonucleotide ligation–based sequencing, by involving DNA fragmentation, bioinformatics analysis, and variant interpretation as major steps. Illumina Hi-seq platform, most widely used for RNA-seq, ingeniously explains transcriptome's dynamic nature and standardizes the next-generation sequencing. This platform was used to detect the novel blight-pathogen of the ornamental plants *Calonectria pseudonaviculata*, two isolates that showed only a single nucleotide polymorphism. Another important example is the yellow rust of wheat caused by *Puccinia striiformis* f. sp. *tritici*. Pathogenomics through the RNA-seq of infected wheat leaves revealed that the pathogen population has dramatically shifted in the United Kingdom, possibly due to reemerging or a new set of pathogen lineages. Although novel pathogens are not well defined, these techniques can sequence the whole genome of putative pathogen without specific primers or PCR amplification (Hubbard et al., 2015; Malapi-Wight et al., 2016; Qin, 2019).

REFERENCES

Abdullah AS, Turo C, Moffat CS, Lopez-Ruiz FJ, Gibberd MR, Hamblin J, Zerihun A (2018) Real-time PCR for diagnosing and quantifying co-infection by two globally distributed fungal pathogens of wheat. *Frontiers in Plant Science* 9:1086.

Asibi AE, Chai Q, Coulter JA (2019) Rice Blast: A disease with implications for global food security. *Agronomy* 9(8):451.

Aslam S, Tahir A, Aslam MF, Alam MW, Shedayi AA, Sadia S (2017) Recent advances in molecular techniques for the identification of phytopathogenic fungi—a mini review. *Journal of Plant Interactions* 12(1):493–504.

Balodi R, Bisht S, GhAtak A, Rao KH (2017) Plant disease diagnosis: Technological advancements and challenges. *Indian Phytopathology* 70(3):275–281.

Bhat RG, Browne GT (2010) Specific detection of phytophthora cactorum in diseased strawberry plants using nested polymerase chain reaction. *Plant Pathology* 59:121–129.

Bhatia S, Dahiya R (2015) Concepts and techniques of plant tissue culture science. In: *Modern Applications of Plant Biotechnology in Pharmaceutical Sciences* (eds) Bhatia S, Sharma K, Dahiya R, Bera T. Academic Press, Boston, MA, pp. 121–156.

Brumfield RT, Beerli P, Nickerson DA, Edwards SV (2003) The utility of single nucleotide polymorphisms in inferences of population history. *Trends in Ecology & Evolution* 18(5):249–256.

Capote N, PAstrana AM, Aguado A, Sánchez-Torres P (2012) Molecular tools for detection of plant pathogenic fungi and fungicide resistance. In: *Plant Pathology* (eds) Cumagun CJ. IntechOpen, Rijeka, Croatia, pp. 151–202.

Cheng Y, Tang X, Gao C, Li Z, Chen J, Guo L, Wang T, Xu J (2020) Molecular diagnostics and pathogenesis of fungal pathogens on bast fiber crops. *Pathogens* 9(3):223.

Conti M, Cinget B, Vivancos J, Oudemans P, Bélanger RR (2019) A molecular assay allows the simultaneous detection of 12 fungi causing fruit rot in cranberry. *Plant Disease* 103(11):2843–2850.

Cui C, Shu W, Li P (2016) Fluorescence in situ hybridization: Cell-based genetic diagnostic and research applications. *Frontiers in Cell and Developmental Biology* 4:89.

Diguta CF, RouSseaux S, WeidmaNn S, Bretin N, Vincent B, Guilloux-Benatier M, Alexandre H (2010) Development of a qPCR assay for specific quantification of Botrytis cinerea on grapes. *Fems Microbiology Letters* 313(1):81–7.

Doehlemann G, Ökmen B, Zhu W, Sharon A (2017) Plant pathogenic fungi. *Microbiology Spectrum* 5(1):5–1.

Ellison MA, McMahon MB, Bonde MR, Palmer CL, Luster DG (2016) In situ hybridization for the detection of rust fungi in paraffin embedded plant tissue sections. *Plant Methods* 12:37.

Elverson TR, Kontz BJ, Markell SG, Harveson RM, Mathew FM (2020) Quantitative PCR assays developed for diaporthe helianthi and diaporthe gulyae for phomopsis stem canker diagnosis and germplasm screening in sunflower (*Helianthus annuus*). *Plant Disease* 104(3):793–800.

Fisher MC, Gurr SJ, Cuomo CA, Blehert DS, Jin H, Stukenbrock EH, Stajich JE, Kahmann R, Boone C, Denning DW, Gow NA. (2020) Threats posed by the fungal kingdom to humans. *Wildlife, and Agriculture MBio* 11(3):e449–420.

Fourie G, Steenkamp ET, Ploetz RC, Gordon TR, Viljoen A (2011) Current status of the taxonomic position of Fusarium oxysporum formae specialis cubense within the Fusarium oxysporum complex. *Infection Genetics and Evolution* 11(3):533–542. ISSN 1567-1348.

Garrido C, Carbú M, Fernández-Acero FJ, BooNhAm N, Colyer A, Cantoral JM, Budge G (2009) Development of protocols for detection of colletotrichum acutatum and monitoring of strawberry anthracnose using real-time PCR. *Plant Pathology* 58(1):43–51.

Geiser DM, Jiménez-GaSco M, Kang S, Makalowska I, VeeraraghavaN N, Ward TJ, ZhaNg N, Kuldau GA, O'Donnell K (2004) Fusarium-ID v.1.0: A DNA sequence database for identifying fusarium. *European Journal of Plant Pathology* 110(5–6):473–479.

Godfray HC, Mason-D'Croz D, Robinson S (2016) Food system consequences of a fungal disease epidemic in a major crop. *Philos. Philosophical Transactions of the Royal Society B* 371(1709):20150467.

Groenewald S, Van Den Berg N, Marasas WFO, Viljoen A (2006) The application of high-throughput AFLPs in assessing genetic diversity in *Fusarium oxysporum* f.sp. cubense. *Mycological Research* 110:297–305.

Gu L, Yan W, Liu L, Wang S, Zhang X, Lyu M (2018) Research progress on rolling circle amplification (RCA)-based biomedical sensing. *Pharmaceuticals (Basel Switzerland)* 11(2):35.

Hariharan G, Prasannath K (2021) Recent advances in molecular diagnostics of fungal plant pathogens: A mini review. *Frontiers in Cellular and Infection Microbiology* 10:829.

Hennell JR, D'AgoStino PM, Lee S, Khoo CS, Sucher NJ (2012) Using GenBank® for genomic authentication: A tutorial. In: *Plant DNA Fingerprinting and Barcoding: Methods and Protocols* (eds) Sucher NJ. Humana Press, Springer, New York, pp. 181–200.

Hibbett DS, Taylor JW (2013) Fungal systematics: Is a new age of enlightenment at hand? *Nature Reviews Microbiology* 11(2):129–133.

Hubbard A, Lewis CM, Yoshida K, Ramirez-Gonzalez RH, de Vallavieille Pope C, Thomas J, Kamoun S, Bayles R, Uauy C, Saunders DG (2015) Field pathogenomics reveals the emergence of a diverse wheat yellow rust population. *Genome Biology* 16(1):23.

Hyakumachi M, PriyatMojo A, Kubota M, Fukui H (2005) New anastomosis groups, AG-T and AG-U, of binucleate *Rhizoctonia* spp. Causing root and stem rot of cutflower and miniature roses. *Phytopathology* 95(7):784–792.

Jensen E (2014) Technical review: In situ hybridization. *The Anatomical Record* 297(8):1349–1353.

Johnston-Monje D, Mejia JL (2020) Botanical microbiomes on the cheap: Inexpensive molecular fingerprinting methods to study plant-associated communities of bacteria and fungi. *Applications in Plant Sciences* 8:e11334.

Kumar R, Gupta A, Srivastava S, Devi G, Singh VK, Goswami SK, Gurjar MS, Aggarwal R. (2020) Diagnosis and detection of seed-borne fungal phytopathogens. In: *Seed-Borne Diseases of Agricultural Crops: Detection, Diagnosis & Management* (eds) Kumar R, Gupta A. Springer, Singapore, pp. 107–142.

Liu JH, Gao L, Liu TG, Chen WQ (2009) Development of a sequence-characterized amplified region marker for diagnosis of dwarf bunt of wheat and detection of tilletia controversa Kühn. *Letters in Applied Microbiology* 49(2):235–240.

Luchi N, Ioos R, Santini A (2020) Fast and reliable molecular methods to detect fungal pathogens in woody plants. *Applied Microbiology and Biotechnology* 104:2453–2468.

Malapi-Wight M, Demers JE, Veltri D, Marra RE, Crouch JA (2016) LAMP detection assays for Boxwood blight pathogens: A comparative genomic approach. *Scientific Reports* 6:26140.

Mancini V, Murolo S, Romanazzi G (2016) Diagnostic methods for detecting fungal pathogens on vegetable seeds. *Plant Pathology* 65:691–703.

Marčiulynas A, Marčiulynienė D, Lynikienė J, GedminAs A, Vaičiukynė M, Menkis A (2020) Fungi and oomycetes in the irrigation water of forest nurseries. *Forests* 11(4):459.

Martínez-García LB, Armas C, Miranda JD, Padilla FM, Pugnaire FI (2011) Shrubs influence arbuscular mycorrhizal fungi communities in a semi-arid environment. *Soil Biology & Biochemistry* 43(3):682–689.

McDonald BA (1997) The population genetics of fungi: Tools and techniques. *Phytopathology* 87:448–453.

Milner H, Ji P, Sabula M, Wu T (2019) Quantitative polymerase chain reaction (Q-PCR) and fluorescent in situ hybridization (FISH) detection of soilborne pathogen *Sclerotium rolfsii*. *Applied Soil Ecology* 1362019:86–92.

Mirmajlessi SM, Destefanis M, Gottsberger RA, Mänd M, Loit E (2015) PCR-based specific techniques used for detecting the most important pathogens on strawberry: A systematic review. *Systematic Reviews* 4(1):9.

Moffat CS, See PT, Oliver RP (2015) Leaf yellowing of the wheat cultivar Mace in the absence of yellow spot disease. *Australasian Plant Pathology* 44:161–166.

Nagrale DT, Sharma L, Kumar S, Gawande SP (2016) Recent diagnostics and detection tools: Implications for plant pathogenic alternaria and their disease management. In: *Current Trends in Plant Disease Diagnostics and Management Practices* (eds) Kumar P, Gupta A, Tiwari A, Kamle M. Springer, Switzerland, pp. 111–164.

Narayanasamy P (2011) Diagnosis of fungal diseases of plants. In: *Microbial Plant Pathogens-Detection and Disease Diagnosis* (eds) Narayanasamy P. Springer, Dordrecht, Netherlands, pp. 273–284.

Narum SR, Hatch D, Talbot AJ, Moran P, Powell MS (2008) Iteroparity in complex mating systems of steelhead Oncorhynchus mykiss (Walbaum). *Journal of Fish Biology* 72(1):45–60.

Pallás V, Sánchez-Navarro JA, Delano J (2018) Recent advances on the multiplex molecular detection of plant viruses and viroids. *Frontiers in Microbiology* 9:2087.

Panno S, MAtić S, Tiberini A, Caruso AG, Bella P, Torta L, Stassi R, Davino S (2020). Loop mediated isothermal amplification: Principles and applications in plant virology. *Plants* 9(4):461.

Pevsner J (2009) Access to sequence data and literature information. In: *Bioinformatics and Functional Genomics* (eds) Pevsner J. John Wiley & Sons, Inc., Hoboken, NJ, pp. 12–45.

Porter-Jordan K, Rosenberg EI, Keiser JF, Gross JD, Ross AM, NaSim S, Garrett CT (1990) Nested polymerase chain reaction assay for the detection of cytomegalovirus overcomes false positives caused by contamination with fragmented DNA. *Journal of Medical Virology* 30(4):85–91.

Qin D (2019) Next-generation sequencing and its clinical application. *Cancer Biology & Medicine* 16(1):4–10.

Rollins L, Coats K, Elliott M, Chastagner G (2016) Comparison of five detection and quantification methods for *Phytophthora ramorum* in stream and irrigation water. *Plant Disease* 100:1202–1211.

Sarrocco S, Vannacci G (2018) Preharvest application of beneficial fungi as a strategy to prevent postharvest mycotoxin contamination: A review. *Crop Protection* 110:160–170.

Sharma P, Sharma S (2016). Paradigm shift in plant disease diagnostics: A journey from conventional diagnostics to nano-diagnostics. In: *Current Trends in Plant Disease Diagnostics and Management Practices* (eds) Kumar P, Gupta V, Tiwari A, Kamle M. Springer, Cham, Switzerland, pp. 237–264.

Sikdar P, Okubara P, Mazzola M, Xiao CL (2014) Development of PCR assays for diagnosis and detection of the pathogens *Phacidiopycnis washingtonensis* and sphaeropsis pyriputrescens in apple fruit. *Plant Disease* 98(2):241–246.

Sint D, Raso L, Traugott M (2012) Advances in multiplex PCR: Balancing primer efficiencies and improving detection success. *Methods in Ecology and Evolution* 3(5):898–905.

Swofford DL (2002) *PAUP*: Phylogenetic Analysis Using Parsimony (*and Other Methods)*. Sinauer Associates, Sunderland, MA.

Tan DHS, Sigler L, Gibas CFC, Fong IW (2008) Disseminated fungal infection in a renal trans- plant recipient involving *Macrophomina phaseolina* and *Scytalidium dimidiatum*: Case report and review of taxonomic changes among medically important members of the Botryosphaeriaceae. *Medical Mycology* 46:285–292.

Tör M, Woods-Tör A (2017) Fungal and oomycete diseases. In: *Encyclopedia of Applied Plant Sciences* (eds) Thomas B, Murray BG, Murphy DJ. Academic Press, New York, pp. 77–82.

Wenneker M, Thomma BPHJ (2020) Latent postharvest pathogens of pome fruit and their management: From single measures to a systems intervention approach. *European Journal of Plant Pathology* 156(3):663–681.

Yang F, Jensen JD, Svensson B, Jorgensen HJL, Collinge DB, Finnie C (2010) Analysis of early events in the interaction between fusarium graminearum and the susceptible barley (*Hordeum vulgare*) cultivar Scarlett. *Proteomics* 10(21):3748–3755.

Yang K, Lee I, Nam S (2017) Development of a rapid detection method for Peronospora destructor using loop-mediated isothermal amplification (LAMP). *HortScience* 52(9):S413.

Zhao X, Lin CW, Wang J, Oh DH (2014) Advances in rapid detection methods for foodborne pathogens. *Journal of Microbiology and Biotechnology* 24:297–312.

9 Underpinning the Phylogeny and Taxonomy of Phytofungi Through Computational Biology

Mubashar Raza, Parisa Razaghi, Laith Khalil Tawfeeq Al-Ani and Liliana Aguilar-Marcelino

CONTENTS

INTRODUCTION

Fungi present an enormous diversity of taxa and numerically are among the most abundant eukaryotes on the earth (Morton 2021). Fungi enjoy much popularity in agriculture, biotechnological, and pharmaceutical applications. The total number of estimated fungal species in the world is 2.2–3.8 million, out of which 120,000 have currently been accepted and 3% named so far (Hawksworth and Lücking 2017).

DOI: 10.1201/9781003162742-9

In the last decade, many taxa including novel phyla, classes, orders, families, and genera, have been established by molecular phylogeny (Wijayawardene et al. 2020). The increasing number of infectious diseases caused by microorganisms, especially fungi, is regarded as a global threat to food security (Hyde et al. 2018). Over the past decades, many virulent infections, mainly caused by fungi and fungi-like organisms, have resulted in severe die-offs (Fisher et al. 2012), most of which were previously undescribed. Hence, accurate identification and description of fungal species can help prevent the disasters caused by fungal pathogens.

New techniques and recent advancements in whole fungal genomes have become available to ease genetic manipulation, resulting in new taxa and an increasing number of fungi available for industrial purposes. Previously, organisms were classified only on the basis of a comparison of morphological characteristics as comparative anatomy was the foundation of morphological systematics (Hillis 1987), and advanced molecular techniques were not present to support the traditional taxonomy. Classification of organisms, including micro and living, is a problem in computer science and biology. Over time, DNA barcoding has significantly gained taxonomists' attention to find an effective and efficient species identification method (Seifert 2009). In DNA-based techniques, selecting diagnostic markers in coding and noncoding DNA regions of the nuclear and mitochondrial genome holds great significance. White et al. (1990) employed three main components of the fungal ribosomal operon—internal transcribed spacer (ITS), smaller-subunit (SSU), and large subunit (LSU) rRNA—as markers to identify the fungal species. To date, these gene regions are robust enough to be used in fungal molecular systematics.

Consequently, DNA sequence as a straightforward and powerful identification technique, is based on polymerase chain reaction (PCR). The use of PCR-based methods has revolutionized research in biology, as this strategy has reinforced the expansion of molecular application in fungal identification (Gherbawy and Voigt 2010). No matter the source of isolate or how its presumptive novelty is determined, the first step is to find its phylogenetic placement and relationship with closely related or previously described species. This chapter is an overview of all the steps toward taxonomic placement determination of fungal isolates on the basis of their sequence data.

METHODS OF FUNGI IDENTIFICATION

1. **Morphological identification:** Fungi grow and develop a wide range of morphological characteristics shared with other fungi-like organisms. For example, oomycetes such as downy milder and powdery mildew are obligate parasites. Most fungi are filamentous, and morphologies of micro- and macro-fungi are diverse. For example, we can recognize the variety of toadstools and mushrooms. Considering the current estimation of fungal diversity, micro-fungi are more diverse than macro-fungi, and due to overlapping characteristics, morphology-based identification of fungi remains a daunting task. Mycologists have used phenotypic characteristics

(morphology) for fungal identification, including conidiogenous cells resulting from mitosis (asexual reproduction) or meiosis (sexual reproduction). Mycologists still adopt this approach to understand the evolution of morphological characteristics. However, it helps classify fungi only at a certain level, such as order or family, and may not be helpful at the species level (Panasenko 1967; Lutzoni et al. 2004). Morphology-based identification may be misleading due to cryptic speciation, hybridization, and convergent evolution (Worrall 1999; Stenlid 2002; Kohn 2005; Giraud et al. 2008; Foltz et al. 2013; Hughes et al. 2013; Lücking et al. 2014), and this practice is no longer acceptable according to latest International Code of Nomenclature for algae, fungi, and plants (Hawksworth 2012; Hibbett and Taylor 2013).

2. **Molecular identification:** Fungi identification based on phenotypic characteristics alone can be challenging, and the use of molecular data began when nuclear ribosomal primers were introduced in 1990s (White et al. 1990). Many molecular techniques or approaches have been evaluated for fungi identification, but PCR amplification is the most promising followed by sequencing. The molecular data generated from fungal DNA sequences using primers has led to the new era of fungal identification (Bruns et al. 1991; Mitchell and Zuccaro 2006). Target gene regions for sequence-based approaches show sufficient conservation among fungi, while being variable enough to allow durable discrimination between closely related species. It has been observed between the three most common ribosomal gene regions that SSU possesses the lowest variation among taxa and ITS exhibits the highest variation. While SSU can be used for the higher taxonomic levels, such as phylum, class, order, and family, ITS is useful for species-level identification, and it has been chosen as a universal barcode for identifying fungi (Schoch et al. 2012). At the same time, for some fungal groups, such as *Aspergillus, Bipolaris, Cladosporium Curvularia, Fusarium, Phoma, Penicillium,* and *Trichoderma,* ITS is insufficient at the species level because of low level of resolution in ITS in the genera (Raza et al. 2019). So, protein-coding genes must be utilized for species fungal identification instead of ITS due to the presence of intron regions (Tkacz et al. 2004; Schoch et al. 2009).

Modern concept of fungal taxonomy: This concept includes genealogical approaches such as single- or multi-gene phylogenies, phylo-genomics, and genealogical concordance (Lücking et al. 2020). A combination of morphological and molecular studies developed a trustworthy natural classification structure that reveals the true phylogenetic relationship. Taxonomists prefer to apply phylogenetic studies combined with morphological characteristics for classification purposes because the species concept is more reliable for identifying fungi, especially micro-fungi when their morphological characteristics are similar or overlap through phylogenetic analysis. This approach has helped reconstruct novel lineages by comparing phenotypic characteristics with molecular phylogenetic analysis (Liu et al.

2017a; Phookamsak et al. 2019; Zhang et al. 2021). Recently, "divergence times" have been proposed and applied to certain types of fungi as the comprehensive, standardized criterion for comparing phylogeneticists and morphology (Zhao et al. 2016; Liu et al. 2017b).

STEPS IN PHYLOGENETIC ANALYSIS

The critical steps in phylogenetic tree construction are as follows:

1. Sequence acquirement
2. Multi-sequence alignment and removal of unreliable sites
3. Evolutionary model selection for maximum likelihood and Bayesian analysis
4. Execution of maximum likelihood or Bayesian analysis for clade support
5. Tree representation with bootstrap values

Sequence quality assurance and contiguous sequence: Fungal DNA sequence (forward and reverse) is generated by the sequencing approach and visualized using different types of software. The quality of chromatograms can be checked by the peak's sharpness using Chromas software, supported by the chromatogram file format, and some other related software, including BioEdit, Chromatogram explorer, and 4Peaks. The visualization of chromatograms is adjusted with tracing the vertical and horizontal scale bars shown in Fig. 1. Both forward and reverse sequences reconstruct the original DNA sequence known as "contiguous or contig" sequence, which is usually larger in nucleotides (bp) length of target gene region. A forward sequence fragment and its complementary fragment (reverse sequence) are ideal for generating a consensus DNA sequence (Conig) (Fig. 2). SeqMan software could be used for contiguous sequence assembly, and it starts by matching forward with reverse fragments, resulting in a consensus sequence. It must be checked whether the consensus sequence is gapped or has some ambiguous nucleotide symbols, such as D, H, K, M, N, R, Y, which may be compared and replaced with forward and reverse fragment peaks' sharpness (Fig. 2, B). The consensus sequence is to be saved in the Fasta format notepad (Fig. 2).

BLAST search liability: The basic local alignment search tool (BLAST) finds similarity between sequences. It is preferable to do a BLAST search after sequencing and consensus assembly, and it is more significant to verify the congruence of the sequence resulting from fungal DNA and compare it with other markers genes. If differences are detected, that could be due to the hybrid status of some species. BLAST guidelines are available at www.ncbi.nlm.nih.gov/blast/BLAST_guide.pdf. BLAST search steps using default parameters are as follows:

1. Go to the National Center for Biotechnology Information (NCBI) web portal and click on nucleotide blast (Fig. 3, A).

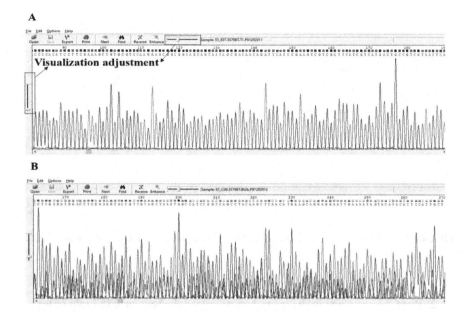

FIGURE 9.1   Chromatograms visualized in Chromas. A. Quality-assured chromatogram. B. Unqualified chromatogram.

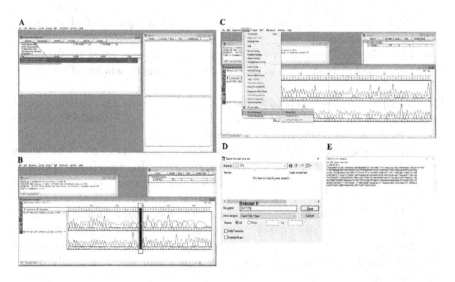

FIGURE 9.2   Consensus sequence assembly. A. Forward and reverse fragment assembly. B. Contig quality check. C–D. Save contig in a single file. E. View in notepad.

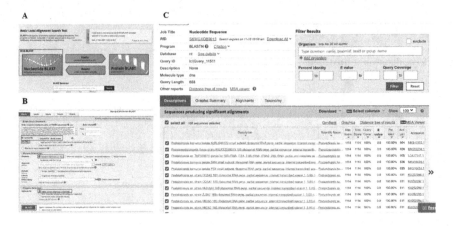

FIGURE 9.3 BLAST search liability. A. NCBI web portal. B. Blast query box. C. Blast results.

2. Paste the sequence in the query box or upload the sequence in Fasta or text file format (Fig. 3, B).
3. Enter a job title.
4. Choose Search Set as "Standard databases (nr etc.)" and change database to "Nucleotide collection (nr/nt)."
5. Optimize the program selection for "Highly similar sequences (megablast)." BLAST guidelines suggest selecting the "megablast" option to find out what has been sequenced.
6. Click "BLAST," and users may choose to tick the box "Show results in a new window" if they want to see the result in a separate window (Fig. 3, C).

ITS is a preliminary method used for fungi identification, and it is capable of differentiating species in most genera. Other gene regions, such as LSU and SSU could be sequenced depending on the liability of ITS blast search.

Reference sequences and outgroup selection: Sequences should be chosen among those fungal type species or authentic sequences from published literatures to obtain a more accurate phylogenetic analysis. The nucleotide sequences of types can be downloaded directly from the primary DNA sequence databases, i.e., NCBI, EMBL-EBI, and DDJB, but the user should be cautious before using this data in any analysis. Checking the validity of sequences and comparing any information (GenBank accession numbers, culture codes, species' names, depositors' names) against the authentic publication where the type strain was firstly published is recommended.

A phylogenetic tree can be made as a rooted or an unrooted tree. In the unrooted tree, only relationships between the taxa are shown, without any direction to the evolutionary changes. In contrast, a rooted tree has a direction to indicate the evolutionary time and shows the last common ancestor of ingroups (Kinene et al. 2016).

Among the different rooting methods, such as midpoint rooting, molecular clock rooting, and outgroup rooting, the last mentioned technique is widely used for rooting a tree (Kinene et al. 2016). The choice of one or more taxa that are divergent from the rest of the taxa (ingroup) is fundamentally based on prior knowledge of the group of interest or may become apparent during the sequence alignment. Wrong selection of outgroup(s) influences the phylogeny reconstruction, such as different tree topologies (De Dreu et al. 2016); therefore, to ensure the robustness of the outgroup rooting method, it is suggested that more than one outgroup be added to the analysis.

An appropriate outgroup is the actual sister group or sequence(s) that are neither phylogenetically close nor phylogenetically distant from ingroups. In the case of trivial phylogenetics, such as species delimitation and resolving species in a set of sequences at the genus level, a proper outgroup is chosen from sister groups rather than less closely related taxa. However, at higher taxonomic levels, outgroup taxa would be distantly related to the ingroup but not that much distant, causing long branch attraction due to either a large divergence time and/or an increased rate of evolution (Tarrío et al. 2000).

Sequence alignment: Sequence information has become a fundamental tool of systematic evolutionary research, and all analyses of sequence data are fundamentally based upon alignment. The basis of alignment is the arrangement of sequence data to identify regions of similarity. The main object of alignment is to create the most efficient statement of initial homology, as different sequence alignment methods affect phylogenetic topology. The resulting phylogeny is highly dependent on the mode of alignment, which includes pairwise alignment and multiple alignments, rather than on the method of phylogenetic reconstruction (Ogden and Rosenberg 2006). The pairwise alignment method is used between two sequences at a time when searching a database for sequences with high similarity to a query, while multiple alignments are used for more than two sequences at a time, and the focus is placed on aligning all of the sequences in a given query set to identify conserved sequence regions across a group of sequences.

A sequence alignment is produced by a multiple sequence program named "Multiple Alignment using Fast Fourier Transform (MAFFT)." All sequences can be inserted manually into the input box (Fig. 4, A); the default options generally remain unchanged, resulting in a file in fasta format (Fig. 4, B). Sequences that have an incorrect direction relative to the other sequences appear as blue lines (Fig. 4, B), and a user can easily check and convert them into input data to make sure all sequences are in one direction, i.e., forward or reverse direction. Aligned sequences of nucleotides can be opened in Bioedit, and they are found in the form of columns of bases in the data matrix so that identical or similar characters are aligned in successive columns, and gaps are inserted between sequences where sequences differ in length. Visual alignment helps remove the improperly aligned against the precise measurements of default settings. It is suggested that the conserved domain be checked first, and then the intron regions to find the best-matching sequences. Furthermore, removing two ends of misaligned sequences maximizes the similarity of sequences (Fig. 5).

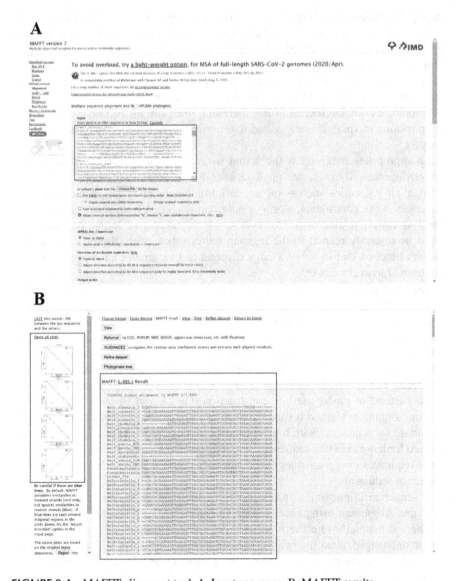

FIGURE 9.4   MAFFT alignment tool. A. Input sequence. B. MAFFT results.

CONSTRUCTION OF PHYLOGENETIC TREES

Indeed, the construction of phylogenetic trees is possible using software such as Mesquite, MEGA, MrBayes, PAUP, RAxML, Tree Gen, PAML, TREECON, and PHYLIP. These phylogeny program packages are used to build phylogenetic trees, but the question is, which one is the best for molecular taxonomy? The best software for the analytic methods is selected mainly on the basis of the principal purposes of evolutionary research and how to better understand phylogenetic relationships

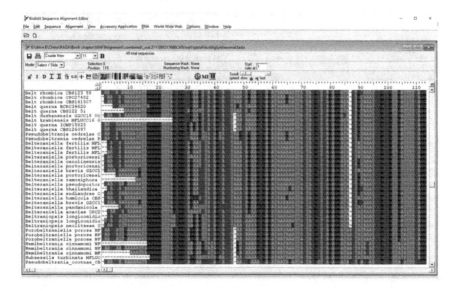

FIGURE 9.5   Aligned sequences.

between organisms. For example, molecular evolutionary genetics analysis can be used for phylogenetic construction, while Mesquite software converts the fasta file into different formats supported by MrBayes, PAUP, and RAxML. Phylogenetic trees are constructed by following these four special steps.

FIRST STEP: ACQUISITION OF HOMOLOGOUS DNA OR PROTEIN SEQUENCES

Use your favorite browser to collect the sequences for some interesting samples of fungi; for example, open a browser on Chrome, then pull up the NCBI website (a website containing sequence data of organisms) or go to NCBI by alignment software. This website is free for users and contains genetic sequence databases, such as GenBank. It also allows users to perform an alignment in special programs to see whether the sequences are homologous or not.

Collecting the sequences from NCBI to build the phylogenetic tree requires certain steps to be followed. First, open the website (www.ncbi.nlm.nih.gov/) (Figure 9.6) and press "Blast" on the page. The page "Web Blast," "https://blast.ncbi.nlm.nih.gov/Blast.cgi" appears; if the inserted sequence is a protein, choose the "**Blastp**" tab, and if it is a nucleotide sequence, click the "**Blastn**" tab (Figure 9.6). As an assumption, when you have a nucleotide sequence, you click the "Web Blast," and it shows this page "https://blast.ncbi.nlm.nih.gov/Blast.cgi?PROGRAM=blastn&PAGE_TYPE=BlastSearch&LINK_LOC=blasthome" by placing a set of five tabs in the front of the page, near the top (blastn, blastp, blastx, tblastn, and tblastx) (Figure 9.6).

In the large box, paste your sequence of interest (**Enter accession number(s), gi(s), or FASTA sequence(s)**) and also select the text box "Database": (Standard databases (nr etc.) :) with the box of the list: (**Nucleotide collection (nr/nt)**) as

FIGURE 9.6 Acquisition of homologous DNA or protein sequences.

default. The next step is to select from the three useful choices as follows to determine the types of blast search:

◉ Highly similar sequences (megablast)

○ More dissimilar sequences (discontiguous megablast)

○ Somewhat similar sequences (blastn)

As a default selection, in **Blastn**, click the button: "**BLAST**." The first choice (Highly similar sequences (megablast)) is more suitable for very similar sequences among databases. Other options could be used depending on your research and to find short matches between sequence data that are not very close to a query.

Upon clicking the option "**BLAST**," the BLAST search results page appears in the same or a separate window (Figure 9.6). The description of results appears on the panel of the page. Then, select those sequences that will be helpful for you in research and save them in a file as text. From the results page, select one or more suitable species and create the file for one more sequence, as follows:

1. To create a file of one sequence: Click on the species, a new page appears (Figure 9.6); Click on the button "**Send to:**" and find several choices. Choose **Complete Record: File: FASTA** (see Figure 9.6); then create file and save in a special folder with a special name.
2. To create a file of more sequences: From the results page, select one or more species and click on the button "**Download**" and save as ". **fasta" (Aligned Sequences)**" (see Figure 9.6). Save in a special folder with a unique name.

At this time, the first step is completed, and the file of sequences is ready for alignment in programs such as MEGA11, Clustal X, or others.

Second Step: Align the Collection of Sequences

The alignment for the collected sequences begins with one sequence (either isolated strain sequence or downloaded from NCBI), opening in MEGA11 (Figure 9.7). From the assignment bar at the top, click on (**Edit = Insert sequence from file**) to show

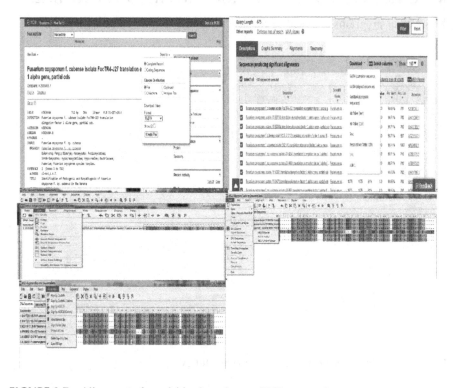

FIGURE 9.7 Alignment of acquisitive homologous DNA or protein sequences

all sequences collected from NCBI (Figure 9.7). Select all sequences before selecting alignment (**Alignment = Align by MUSCLE**) (Figure 9.7). Arrangement of aligned sequences is shown as "TTTTT," "GGGGG," "CCCCC," etc. The aligned file can be saved by clicking the icon (**Data = Export Alignment = FASTA Format**) or save the file as (**MTSX File (.mtsx)**. The file is ready for further processing.

THIRD STEP: EVALUATE A TREE OF THE ALIGNED SEQUENCES

Return to the main page of MEGA11, click on the icon (**Analysis = phylogeny = Construct/Test neighbor-joining tree**) (Figure 9.8). All fields can be left at their default values. To generate the tree, click on Compute.

FOURTH STEP: EXHIBIT THE TREE CLEARLY TO CONVEY THE RELEVANT INFORMATION

The drawing of a phylogenetic tree appears in a new window after the third step is completed, as shown in Figure 9.8. In the new window of the phylogenetic tree, there are several options on the top and left side that can help control the phylogenetic tree's drawing, such as in Circle or Radiation or Traditional (including three types: Rectangular or Straight or Curved) (Figure 9.8). Save the tree in a folder as "**MEGA Tree Sessions**."

MORE PROGRAMS FOR CONSTRUCTING PHYLOGENETIC TREES

Many other programs are used to construct the phylogenetic trees, as follows:

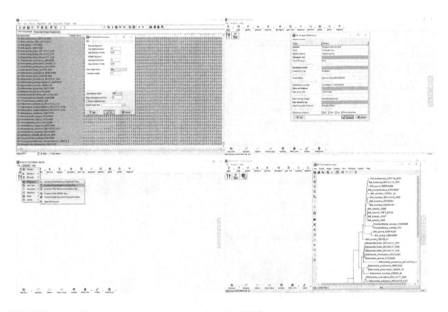

FIGURE 9.8   Phylogenetic tree construction by MEGA

MESQUITE

Mesquite is helpful for students and researchers to analyze sequence data of fungal isolates and build the phylogenetic tree. It can be downloaded from the following link:

"https://github.com/MesquiteProject/MesquiteCore/releases/download/v3.70-build-940/Mesquite.3.70-Windows.zip".

Start from "**File = Open file**" and choose text document and translate to "**FASTA**" (Figure 9.9). Then, merge all sequences by clicking on "**Characters = Merge Matrices from file**" from the Menu bar and select taxa (Figure 9.9). Make alignment by clicking the icon "**Text = Align Multiple Sequences = Align Muscle.**" The alignment process takes some time. Data will then be ready to construct a phylogenetic tree. For building a phylogenetic tree, click "**Taxa&Trees = Tree Inference = Tree Search = Mesquite Heuristic Search (Add & Rearrange)**" (Figure 9.9). A new window opens and shows "**Criteria For Tree Search.**" Click icon "OK" directly by choosing default options, and a phylogenetic tree results in a new window (Figure 9.9).

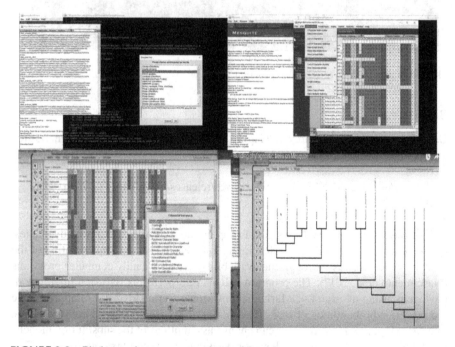

FIGURE 9.9 Phylogenetic tree construction by Mesquite

FIGURE 9.10 A view of Bayesian program

BAYESIAN

Bayesian inference of phylogeny (Figure 9.10) is used to build a phylogenetic tree and can be downloaded from the following website:

"https://github.com/NBISweden/MrBayes/releases/download/v3.2.7/MrBayes-3.2.7-WIN.zip".

The Bayesian phylogenetic method is popular now with regard to two advantages: (1) the power in data analysis by developing models and (2) the implementing of the models, which are easy through the availability of user-friendly computer programs (Nascimento et al. 2017). The program MRBAYES performs Bayesian inference of phylogeny using a variant of Markov chain Monte Carlo (MCMC) to approximate the posterior probabilities of trees (Green 1995). MRBAYES has a command-line interface (Figure 9.11) as the user can change assumptions of the substitution models, the details of the MC of the analysis, or delete or restore the taxa or character. The command line for MrBayes can be updated on the basis of model results obtained from JmodelTest software (Figure 9.12 A, B).

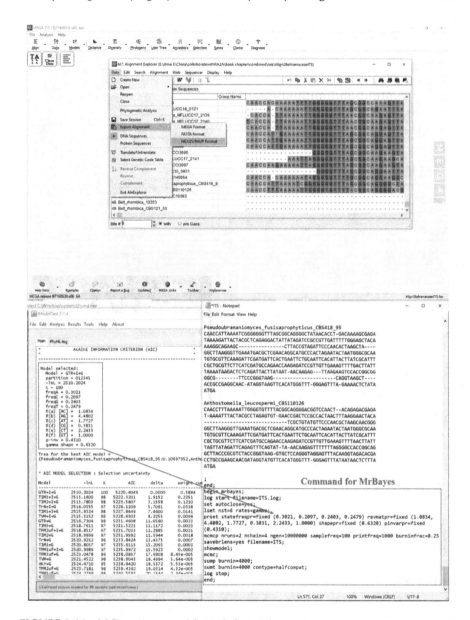

FIGURE 9.11 MrBayes command for a phylogenetic tree

PAUP

PAUP is another phylogenetic software used by many researchers. This program is helpful to construct the phylogenetic tree into several methods and models, comprising maximum parsimony, maximum likelihood, distance-based methods like neighbor joining, and SVDquartets (the multispecies coalescent models with statistically consistent methods). All versions of PAUP* require data and commands

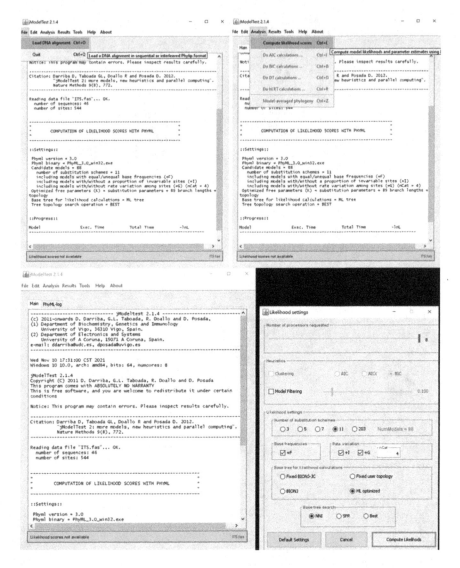

FIGURE 9.12   (A) JmodelTest

to be present in NEXUS format, and it is run by entering the command as shown in Figure 9.13. You can download PAUP software from this website: https://paup. software.informer.com/.

RAxML

RAxML is the phylogenetic software used to evaluate the maximum probability of phylogenies depending on sequences of DNA and proteins (Stamatakis 2014).

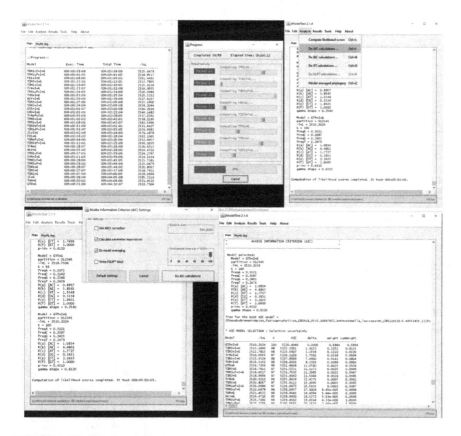

FIGURE 9.12 (B): JmodelTest

Many options are helpful in the construction of a more accurate phylogenetic tree, as follows:

1. Calculating the internode Certainty, and the tree Certainty
2. Calculating the Robinson-Foulds distances between an immense phylogeny and much smaller, more accurate reference phylogenies that contain a rigid subset of the taxa in the immense tree. That potential is utilized to evaluate the fineness of immense trees, which cannot be investigated by the eye automatically.

For building the phylogenetic tree, start by selecting "BestTree.tree_out.text_out file" and open it on the page of the program. This file opens while containing many sequences and also opens as a new window = "please choose an interpreter for this file." You must select a command from this page = "Phylip (DNA/RNA)" and click on this command. The new window opens with all sequences arranged with

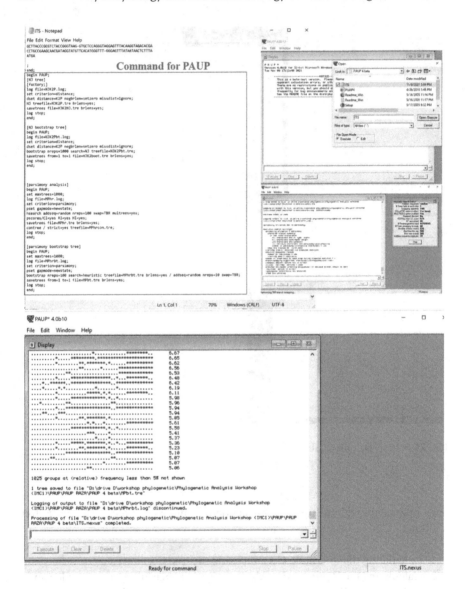

FIGURE 9.13 A view of PAUP program

window = "Export Phylip Options." Select "current system default"; then, click the button "Export" (Figure 9.14). Finally, you will make an alignment file in Phylip format and select the "Bootstrap.raxml_trres" command. The phylogenetic tree opens in the next stage with values for the branch of the tree and not length for the branch.

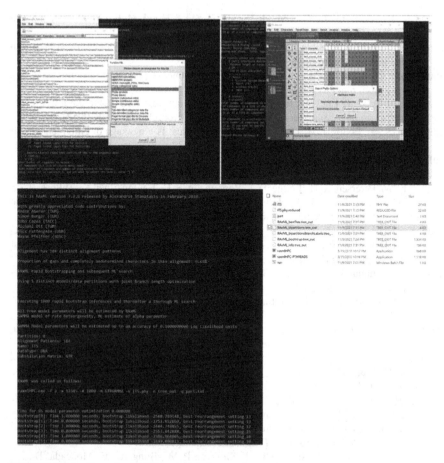

FIGURE 9.14 Steps for building the phylogenetic tree in RAXML

REFERENCES

Bruns, T. D., White, T. J., & Taylor, J. W. (1991). Fungal molecular systematics. *Annual Review of Ecology and Systematics, 22*(1), 525–564.

De Dreu, C. K. W., Gross, J., Méder, Z., Giffin, M., et al. 2016—In-group defense, out-group aggression, and coordination failures in intergroup conflict. *Proceedings of the National Academy of Sciences of the United States of America, 113*, 10524–10529.

Fisher, M. C., Henk, D. A., Briggs, C. J., Brownstein, J. S., Madoff, L. C., McCraw, S. L., & Gurr, S. J. (2012). Emerging fungal threats to animal, plant and ecosystem health. *Nature, 484*(7393), 186–194. doi:10.1038/nature10947

Foltz, M. J., Perez, K. E., & Volk, T. J. (2013). Molecular phylogeny and morphology reveal three new species of Cantharellus within 20 m of one another in western Wisconsin, USA. *Mycologia, 105*(2), 447–461.

Gherbawy, Y., & Voigt, I. K. (2010). *Molecular Identification of Fungi* (pp. 1–501). Springer. doi:10.1007/978-3-642-05042-8

Giraud, T., Refrégier, G., Le Gac, M., de Vienne, D. M., & Hood, M. E. (2008). Speciation in fungi. *Fungal Genetics and Biology, 45*(6), 791–802.

Green, P, J. (1995). Reversible jump Markov chain Monte Carlo computation and Bayesian model determination. *Biometrika, 82,* 711–732.

Hawksworth, D. L. (2012). Managing and coping with names of pleomorphic fungi in a period of transition. *Ima Fungus, 3*(1), 15–24.

Hawksworth, D. L., & Lücking, R. (2017). Fungal diversity revisited: 2.2 to 3.8 million species. *Microbiology Spectrum, 5*(4). doi:10.1128/microbiolspec.FUNK-0052-2016

Hibbett, D. S., & Taylor, J. W. (2013). Fungal systematics: Is a new age of enlightenment at hand? *Nature Reviews Microbiology, 11*(2), 129–133.

Hillis, D M. (1987). Molecular versus morphological approaches to systematics. *Annual Review of Ecology, Evolution, and Systematics, 18*(1), 23–42. Web.

Hughes, K. W., Petersen, R. H., Lodge, D. J., Bergemann, S. E., Baumgartner, K., Tulloss, R. E., ... Cifuentes, J. (2013). Evolutionary consequences of putative intra-and interspecific hybridization in agaric fungi. *Mycologia, 105*(6), 1577–1594.

Hyde, K. D., Al-Hatmi, A., Andersen, B., Boekhout, T., Buzina, W., Dawson, J. T. L., Eastwood, D. C., Gareth Jones, E. B., de Hoog, S., Kang, Y., et al. 2018. The world's ten most feared fungi. *Fungal Divers, 93,* 161–194.

Kinene, T., Wainaina, J., Maina, S., & Boykin, L. M. (2016). Rooting trees, methods for. *Encyclopedia of Evolutionary Biology,* 489–493.

Kohn, L. M. (2005). Mechanisms of fungal speciation. *Annual Review of Phytopathology, 43,* 279–308.

Liu, F., Hou, L., Raza, M., & Cai, L. (2017a). Pestalotiopsis and allied genera from Camellia, with description of 11 new species from China. *Scientific Reports, 7*(1), 1–19.

Liu, J. K., Hyde, K. D., Jeewon, R., Phillips, A. J., Maharachchikumbura, S. S., Ryberg, M., Liu, Z. Y., & Zhao, Q. (2017b). Ranking higher taxa using divergence times: A case study in Dothideomycetes. *Fungal Divers, 84,* 75–99.

Lücking, R., Aime, M. C., Robbertse, B., Miller, A. N., Ariyawansa, H. A., Aoki, T., Cardinali, G., Crous, P. W., Druzhinina, I. S., Geiser, D. M., & Hawksworth, D. L. (2020). Unambiguous identification of fungi: Where do we stand and how accurate and precise is fungal DNA barcoding? *IMA Fungus, 11*(1), 1–32.

Lücking, R., Dal-Forno, M., Sikaroodi, M., Gillevet, P. M., Bungartz, F., Moncada, B., ... Lawrey, J. D. (2014). A single macrolichen constitutes hundreds of unrecognized species. *Proceedings of the National Academy of Sciences, 111*(30), 11091–11096.

Lutzoni, F., Kauff, F., Cox, C. J., McLaughlin, D., Celio, G., Dentinger, B., ... Vilgalys, R. (2004). Assembling the fungal tree of life: Progress, classification, and evolution of subcellular traits. *American Journal of Botany, 91*(10), 1446–1480.

Mitchell, J. I., & Zuccaro, A. (2006). Sequences, the environment and fungi. *Mycologist, 20*(2), 62–74.

Morton J. B. (2021). *Chapter 6 — Fungi in principles and applications of soil microbiology* (3rd ed., pp. 149–170). Morgantown, WV: West Virginia University. doi.org/10.1016/B978-0-12-820202-9.00006-X

Nascimento, F. F., Reis, M. D., & Yang, Z. (2017). A biologist's guide to Bayesian phylogenetic analysis. *Nature Ecology & Evolution, 1*(10), 1446–1454. https://doi.org/10.1038/s41559-017-0280-x.

Ogden, T. H., & Rosenberg, M. S. (2006). Multiple sequence alignment accuracy and phylogenetic inference. *Systematic Biology, 55*(2), 314–328.

Panasenko, V. T. (1967). Ecology of microfungi. *The Botanical Review, 33*(3), 189–215.

Phookamsak, R., Hyde, K. D., Jeewon, R., Bhat, D. J., Jones, E. B. G., Maharachchikumbura, S. S. N., Raspé, O., Karunarathna, S. C., Wanasinghe, D. N., HongSanan, S., Doilo, M. M., Tennakoon, D. S., Machado, A. R., Firmino, A. L., Ghosh, A., KArunArathna, A., Mešić A., Dutta, A. K., ThongBai, B., Devadatha, B., Norphanphoun, C., Senwanna, C., Wei, D. P., Pem, D., Ackah, F. K., Wang, G. N., Jiang, H. B., Madrid, H., Lee, H. B., Goonasekara,

I. D., Manawasinghe, I. S., Kušan, I., Cano, J., Gené, J., Li, J. F., Das, K., Acharya, K., Raj, K. N. A., Latha, K. P. D., Chethana, K. W. T., He, M. Q., Dueñas, M., Jadan, M., Martín, M. P., Samarakoon, M. C., Dayarathne, M. C., Raza, M., Park, M. S., Telleria, M. T., Chaiwa, N. N., Matočec, N., de Silva, N. I., Pereira, O. L., Singh, P. N., Manimohan, P., Uniyal, P., Shang, Q. J., Bhatt, R. P., Perera, R. H., Alvarenga, R. L. M., Nogal-Prata, S., Singh, S. K., Vadthanarat, S., Oh, S. Y., Huang, S. K., Rana, S., Konta, S., Paloi, S., Jayasiri, S. C, Jeon, S. J., Mehmood, T., Gibertoni, T. B., Nguyen, T. T. T., Singh, U., Thiyagaraja, V., Sarma, V. V., Dong, W., Yu, X. D., Lu, Y. Z., Lim, Y. W., Chen, Y., Tkalčec, Z., Zhang, Z. F., Luo, Z. L., Aranagama, D. A., Thambugala, K. M., Tibpromma, S., CamporEsi, E., Bulgakov, T. S., Dissanayake, A. J., Senanayake, I. C., Dai, D. Q, Tang, L. Z., Khan, S., ZHang, H., Promputtha, I., Cai, L., Chomnunti, P., Zhao, R. L., Lumyong, S., Boonmee, S., Wen, T. C., Mortimer, P. E., & Xu, J. C. (2019). Fungal diversity notes 929–1035: Taxonomic and phylogenetic contributions on genera and species of fungi. *Fungal Divers, 95*, 1–273.

Raza, M., Zhang, Z. F., Hyde, K. D., Diao, Y. Z., & Cai, L. (2019). Culturable plant pathogenic fungi associated with sugarcane in southern China. *Fungal Diversity, 99*(1), 1–104.

Schoch, C. L., Seifert, K. A., Huhndorf, S., Robert, V., Spouge, J. L., Levesque, C. A., & Chen, W. 571 Fungal BarCoding C, Fungal Barcoding Consortium Author L. (2012). Nuclear 572 ribosomal internal transcribed spacer (ITS) region as a universal DNA barcode marker for 573 Fungi. *Proceedings of the National Academy of Sciences of the United States of America, 109*, 6241–6246.

Schoch, C. L., Sung, G. H., López-Giráldez, F., Townsend, J. P., Miadlikowska, J., Hofstetter, V., Robbertse, B., Matheny, P. B., Kauff, F., Wang, Z., & Gueidan, C. (2009). The Ascomycota tree of life: A phylum-wide phylogeny clarifies the origin and evolution of fundamental reproductive and ecological traits. *Systematic Biology, 58*(2), 224–239.

Seifert K. A. (2009). Progress towards DNA barcoding of fungi. *Molecular Ecology Resources, 9*(1), 83–89. doi:10.1111/j.1755-0998.2009.02635. x

Stamatakis A. (2014). RAxML version 8: A tool for phylogenetic analysis and post-analysis of large phylogenies. *Bioinformatics, 1;30*(9), 1312–1313. doi:10.1093/bioinformatics/btu033. Epub 2014 Jan 21. PMID: 24451623; PMCID: PMC3998144

Stenlid, J. (2002). Pathogenic fungal species hybrids infecting plants. *Microbes and Infection, 4*(13), 1353–1359.

Tarrío, R., Rodríguez-Trelles, F., & Ayala, F. J. (2000). Tree rooting with outgroups when they differ in their nucleotide composition from the ingroup: The *Drosophila saltans* and *willistoni* groups, a case study. *Molecular Phylogenetics and Evolution, 16*(3), 344–349.

Tkacz, J. S., & Lange, L. (Eds.). (2004). *Advances in fungal biotechnology for industry, agriculture, and medicine.* New York; Boston, Dordrecht, London, Moscow: Kluwer Academic/Plenum Publishers.

White, T. J., Bruns, S. T., Lee, S. J. W. T., & Taylor, J. (1990). Amplification and direct sequencing of fungal ribosomal RNA genes for phylogenetics. In M. A. Innis, D. H. Gelfand, J. J. Sninsky, & T. J. White (Eds.), *PCR protocols: A guide to methods and applications* (pp. 315–322). San Diego, CA: Academic Press.

Wijayawardene, N. N., Hyde, K. D., Al-Ani, L. K. T., Tedersoo, L., et al. (2020). Outline of Fungi and fungi-like taxa. *Mycosphere, 11*(1), 1060–1456. doi:10.5943/mycosphere/11/1/8

Worrall, J. J. (Ed.). (1999). *Structure and dynamics of fungal populations* (Vol. 25). Springer, Dordrecht: Kluwer Academic Publishers.

Zhang, Z. F., Zhou, S. Y., Eurwilaichitr, L., Ingsriswang, S., Raza, M., Chen, Q., . . . Cai, L. (2021). Culturable mycobiota from Karst caves in China II, with descriptions of 33 new species. *Fungal Diversity, 106*(1), 29–136.

Zhao, R. L., Zhou, J. L., Chen, J., Margaritescu, S., Sánchez-Ramírez, S., Hyde, K.D., Callac, P., Parra, L. A., Li, G. J., & Moncalvo, J. M. (2016). Towards standardizing taxonomic ranks using divergence times—a case study for reconstruction of the *Agaricus* taxonomic system. *Fungal Diversity, 78*(1), 239–292.

10 Application of System Biology in Plant–Fungus Interaction

*Qaiser Shakeel, Rabia Tahir Bajwa, Guoqing Li,
Yang Long, Mingde Wu, Jing Zhang
and Ifrah Rashid*

CONTENTS

INTRODUCTION

Because of their nutritional diversity and multiple interactions with host plants, fungi are important in ecological systems and contemporary agriculture. Fungi are significant organic substances and recycling centers; their interaction with below-ground and above-ground plant parts has negative or positive impacts. The relationships between host plants and their related fungus are complicated, leading to

DOI: 10.1201/9781003162742-10

various outcomes. Because many fungi may mix various lifestyles, including pathogenic, saprophytic, or symbiotic, their limits are frequently unclear yet (Grigoriev 2013). Plants can establish an effective defense system and are resistant to most of the diseases in the ecology; symbiosis and innocuous interactions predominate; however, in some cases, parasitic interactions are exceptional, i.e., plants defence system cannot restrict them. The plant genotype influences plants enzymatic secretions, which act as crucial cues for fungal enlistment into the rhizosphere of plants. The result of an association can be determined by plant sensors and the signaling pathways of defense-related proteins that connect with certain fungus-derived compounds. The creation of unique tests for examining the associations has resulted from improved microscopy methods to describe microorganisms, even to the molecular scale. A change may shift plant–pathogen relationships from resistant to vulnerable in the pathogenic fungi or the hosting receptor's genomes (Giraldo and Valent 2013; Stracke *et al.* 2002). Zamioudis and Pieterse (2012) reported that beneficial microorganisms had developed techniques to inhibit or disguise the host plant's defense system, enabling them to invade their hosts endophytically or epiphytically. Surprisingly, both pathogenic and helpful fungi create necessary connections with plants through colonization processes that are comparable in many respects, such as feeding mechanism growth (Corradi and Bonfante 2012). Such interactions are diametrically opposite, since, in one circumstance (symbiosis), the plant is compensated, while in the other (parasitism), the plant must suffer.

From an evolutionary viewpoint, both sorts of interactions (symbiosis and parasitism) can be aspected: as in parasitism (pathogenic microbes) case, the fungus element has diversified to be an effective parasite, while in case of symbiosis, fungus has emerged as a plant companion to act as an effective symbiont. To appreciate the complex signal connection between host plants and fungi, we must first interpret the activities of both plant and microbial signals, their receptors, and their involvement in initiating plant immunity. Each of these characteristics contributes to the effective fungal colonization of host plants and their resistance capacities. Selective use of beneficial microbes' mixtures have been a sustainable method to overcome biotic and abiotic aggravation situations and assure output consistency in future sustainable agricultural scenarios. Furthermore, most helpful microorganisms have similar pathogenic cousins, and it is still unknown how plant immunity distinguishes between pathogenic and beneficial microorganisms to resist the invasion of the pathogens and enable colonization by the beneficial microorganisms. From an evolutionary point of view, even the first eukaryotes were most likely inhabited by diverse prokaryotes, and eukaryotic immune systems evolved to discriminate between dangerous and benign microorganisms. As a result, a complex interplay between microbes and their hosts is expected, impacting all aspects of eukaryote life. Studying the host–microorganism interactions will thus need both traditional and systems biology "genomics" and quantitative modeling techniques. This chapter illustrates the ongoing advancement in our knowledge of the interaction of the plant with fungi based on novel methodologies and current findings. Because of the decrease in sequencing costs, which has enabled an in-depth analysis of the variety and evolution of host-related bacteria, the world of microorganisms has attracted much attention in recent years.

EVOLUTION AND DIVERSIFICATION OF PLANT–FUNGUS INTERACTION

Plant–fungal interactions are as old as the advent of higher plants, including vascular plants. Although plant colonization of territory is thought to have been aided by fungal associates, such interactions stretch back to about 400 million to 460 million years ago, when vascular plants first appeared (Field *et al.* 2012).

Fungus consumes carbohydrates supplied by the plant, resulting in a long-term link between the properties (Buscot *et al.* 2000). Endophytes and mycorrhizae are two examples of symbiotic plant–fungal relationships. In 1885, AB Frank proposed the Greek term "mycorrhizal," which means "fungus roots," and mycorrhiza is regarded as the symbiotic relationship between the plant roots and a fungus (Trappe 2005). On the other hand, endophytes develop symptomlessly within the living cells of leaves, stems, or roots till the death of the host. Here, the fungus may act as mildly harmful (Brundrett 2004). Because endophytic fungi are numerous, most terrestrial plants are expected to harbor one or several such fungi, improving resistance against pests, pathogens, and stress (Strobel and Daisy 2003).

Mycorrhizae are distinguished from endophytic interactions mainly by nutritional transport at their base and coordinated plant–fungal growth (Brundrett 2004). Mycorrhizal fungi are hosted by most terrestrial plant species for adequate nutrient absorption, with over 6,000 fungal species in the phylum Ascomycota, Basidiomycota, and Glomeromycota recognized as mycorrhizae fungi (Bonfante and Anca 2009). The beneficial effects of the fungal symbiotic association with plant roots (enhanced plants' nutritional condition and improved resistance to abiotic and biotic stress conditions) certainly allowed host plants to shift from an intertidal zone, where nutritional sources are readily accessible, to terrestrial environments, in which evaporation regions develop swiftly after absorption of components through plant roots (Corradi and Bonfante 2012). Mycorrhizas are distributed into two types—ectomycorrhizas and endomycorrhizas—on the basis of whether the fungal associates' hyphae are extracellular, surrounding lateral roots of host plants or invading between the plant root system, or intracellular, t penetrating the plant root system (Bonfante and Anca 2009). Approximately 80% of the total plants interact with the endomycorrhizal fungus belonging to the phylum Glomeromycota; most are obligate biotrophic mycorrhizal symbiotic organisms (Karandashov *et al.* 2004). Such fungi are called "arbuscular mycorrhizae" because they generate extremely branching haustoria-like intracellular structures known as "arbuscules" (Buscot 2015). Glomeromycota has been connected with plants over evolution and has survived without modification in morphological aspects for about 400 million years. Other mycorrhizal fungi have polyphylogenetical lines, indicating linear or divergent evolution (Cairney 2000; Wang and Qiu 2006; Parniske 2008). New research supports the concept that the ectomycorrhizal fungus developed polyphylogenetically from several species of saprophytes. A unified evolutionary tree was developed by Kohler *et al.* (2015) on the basis of molecular analysis; it includes 49 fungal taxa with symbiotic or saprophytic habitats, indicating that the ectomycorrhizal fungus most certainly emerged from numerous lines less than 200 million years ago.

Furthermore, an examination of 16 genetic families linked with cell wall degradation of the host plants in primitive white-rot wood-decaying fungus and ectomycorrhizal lines revealed that all symbiotic microbes in such families exhibit significant
genetic loss. In particular, enzymes involved in lignin degradation were lost in ectomycorrhizal fungi, but endomycorrhizal ericoid and orchid fungi retained a diverse
range of cell wall-degrading enzymes. Evolutionary genetic loss and the concurrent
generation of genes that mainly lead to the development of symbiotic association may
be followed by equivalent genome multiplication in host plants, by which future
genome sequencing can eventually provide deeper insights (Kohler et al. 2015).

Several fungal species further progressed by disrupting the delicate balance of
mutual gain to act as plant pathogens categorized as biotrophic, hemibiotrophic, and
necrotrophic microbes, each having a unique form of interaction with its host plants
(Gardiner et al. 2013). Pathogens gain access into the plant system through injuries
and leaf stomata, although plant cell wall-degrading enzymes secreted by the fungus
and specialized structures of infection can also assist pathogens to gain entry into
the plant. Necrotrophic pathogens having a broad host range inflict quick disruption
of significant tissues. Plant cells are damaged through a combination of cell wall-degrading enzymes, reactive oxygen species (ROS), and toxins, and pathogen virulence
has been associated with toxin formation (Wang et al. 2014). These activities reach
membrane degradation in the host plant and the discharge of several nutrients, followed by host colonization and decomposition (Wolpert et al. 2002).

Conversely, toxins are not produced by biotrophic pathogens; instead, they release
effectors to suppress host immunity (Perfect and Green 2001). Biotrophs cannot survive on non-living tissues and complete their whole life cycle in living host tissues,
resulting in symptoms appearing on the host surface after a particular period, in
response to infection. These pathogenic fungi are host specific; they use specialized
hyphae to interact with the host plant at the interfacial region, at which point both the
associates dynamically secrete biomolecules.

Ectoparasitic powdery mildews form highly specialized infectious structures
on the plant cuticle, including primary and appressorium germ tubes, allowing the
pathogen to penetrate the host cell wall through a confluence of cell wall-degrading
enzymes' mechanical forces (Pryce-Jones et al. 1999). The cell membrane is recessed
after penetration of the host cell wall to enclose the haustorium (a newly created
nutritive cell).

As a result, a strong metabolic relationship between the biotrophic pathogen
and the plant is developed as well, as the pathogen seeks to disrupt the host immune
system to support the fungal activities necessary for feeding and development. The
living pattern of hemibiotrophic pathogens is an intermediary between the biotrophic
and necrotrophic patterns, and they grow as biotrophs before shifting to the necrotrophic living pattern (Struck 2006; Gardiner et al. 2013). In hemibiotrophs, the pathogen completes half of its life cycle as biotrophs and the other half as necrotrophs, and
physiological and molecular variables influence such a shift. Certain hemibiotrophs
need lengthy colonization durations of biotrophic pathogens to develop an infection,
but others require just a few hours, and the shift to necrotrophy is quick (Kabbage et
al. 2015). One proposed reason for this disparity is the requirement of enough time
for biotrophy to evade host immunity and prevent programmed cell death via effector

secretion. Hence, disease development occurs instead of the buildup of host defenses, indicating that more significant stress from plant defenses can induce the shift from biotrophy to necrotrophy. However, it is possible that after reaching the defense limit, the pathogen recognizes no edge anymore and switches to the necrotrophy mechanism as a more feasible infection approach.

Consequently, the host cell mortality induced by fungus may occur during the period when plant defenses are increasing. The frequency and duration of the biotrophic phase are probably not entirely dependent on how well the fungal pathogen handles host obstacles. On the other hand, necrotrophy can be related to the fungus necessity for enhanced nutrient uptake. (Kabbage *et al.* 2015).

Biotrophy is an ancient life mode for fungal infections, while necrotrophy is considered a comparatively new evolutionary accomplishment (Pieterse *et al.* 2009). Hemibiotrophs may represent the switch between both dietary methods in this scenario (Horbach *et al.* 2011). The immunity of host plants against necrotrophs differs on the basis of fungus species and might be antagonistic, similar, or dissimilar with biotrophi–c immunity responses. Generally, necrotrophs are regarded as the instinctive strength of pathogens, with little physical contact with their hosts because of poorly formed infection-linked morphogenesis and a plethora of biochemical substances that overload the host plant. In many situations, the approach of necrotrophic fungal infection is less sophisticated compared with that of obligatory biotrophs. Plant–pathogen interactions are prone to concurrent or coevolution, in which diseases must devise novel ways to effectively invade their hosts, while plants must devise new detection systems, along with stronger defense systems, to fight against fungal invasions. Fungi use specific toolkits for morphological and biochemical processes to expand their interaction with host plants that have developed phylogenetically and deductively in order to involve sophisticated elements that exploit and govern host routes. Primary branches of development include species with various host spans and species with a variety of trophic methods of life (Horbach *et al.* 2011).

APPROACHES OF SYSTEM BIOLOGY TO MOLECULAR PLANT–FUNGAL INTERACTION

Studies on genetic and molecular processes in plant immunity systems in the disease scenario have altered our overall understanding of plant–fungal interactions. Therefore, application of identification and information-processing mechanisms to differentiate between beneficial and pathogenic microbes will be a crucial challenge for system biology in plants in the approaching decades.

MOLECULAR PLANT

Interactions between symbionts and plants reveal processes underlying the beneficial versus antagonist difference. Upon first encounter with the plants, Arbuscular mycorrhizal fungi and rhizobia activate temporary suppressed defense-like mechanisms (Libault *et al.* 2010). It has been proposed that Nod and Myc factor signaling play a role in this suppression. A few of such receptors seem to regulate pathogen identification as well. Chitin elicitor receptor kinase 1 (OsCERK1) is a lysin motif receptor like kinase (LysM-RLK) required to form mycorrhizal symbiosis and resistance to rice blast disease, implying that it functions as a "molecular signal" between symbiotic and defensive mechanisms (Zhang *et al.* 2015).

However, the mechanistic mechanism behind this double activity is unclear; it is hypothesized that selectivity is derived through relationships with other LysM-RLKs (Gourion *et al.* 2015). Some other similar instances of dual functioning show that this might be a more extensively employed approach. The Nod factor receptor for Medicago truncatula is NFP, which facilitates detection and defense against the *Colletotrichum trifolii* fungus and the oomycetes, including Aphanomyces euteiches *Phytophthora palmivora* (Gough and Jacquet 2013; Rey *et al.* 2013, 2015). Extensive research on instance PRRs and LysM-RLK suggests that physical relationships between receptors and coreceptors are critical for signal selectivity and signal assimilation. Naturally, plant roots come into contact with a profusion of MAMPs and a stew of diverse signaling molecules. So, it is plausible, if not likely, that a customized reaction is built to particular microbial populations detected through an interacting receptors' network by compositional and quantitative detection of the varied signaling chemicals. As a result, methods of PRR signaling based on globally integrated platforms will be necessary. A study conducted on a transcriptomic interactome is a crucial step toward gaining complete knowledge of this critical plant sensing system. The physical cell surface contact network created by 225 LRRRKs (CSILRR) in *Arabidopsis thaliana* was characterized by employing biochemical pull-down assays. CSILRR demonstrated a high interconnectedness of all LRR-RKs, which grouped in numerous modules with unknown biological significance. Significantly, the researchers demonstrated that, like direct connections, indirect network connections alter the downstream signaling output, and the entire network contributes to the immune system of plants' well-balanced activities. Understanding the plant immune system will require the characterization of the integrated information procedure following the network system of LRR-RK (Smakowska-Luzan *et al.* 2018).

GENETICS AND BIOCHEMICAL FEATURES OF PLANT–FUNGUS INTERACTIONS

EFFECTOR PROTEIN AND CELL WALL-DEGRADING ENZYMES

The cell wall of plants comprises cellulose, hemicelluloses, lignin, and pectin. As a result, the fungal mechanism of the lignocellulose-degrading enzyme consists primarily of laccases and peroxidases for hydrolases of glycoside and degradation of lignin, including cellulases for polysaccharides cellulose degradation, hemicellulases for hemicellulose degradation, and pectinases for pectin degradation (Kubicek 2013). The latest genomic investigation for 103 fungi indicated that plant pathogenic fungi had a higher quantity of carbohydrate-active enzymes, (including PL1 pectate lyases and carbohydrate esterases) than saprophytic fungi as later saprophytes can consume only dead lignocellulose components and cannot colonize living plant hosts (Zhao *et al.* 2013). As a result, *Magnaporthe oryzae* and *Fusarium graminearum* (the hemibiotrophic pathogens) regulated genes-encoding cell wall-degrading enzymes during plant host infection, while the *Botrytis cinerea* (the necrotrophic pathogen) showed an association between pathogen virulence and several cell wall-degrading enzymes (such as xylanases and pectinases (Brito *et al.* 2006; Fernandez-Acero *et al.* 2010; Kawahara *et al.* 2012; Zhao *et al.* 2013). Most biotrophs feeding on living plant host cells for nourishment have fewer cell wall-degrading enzymes encoded in plant genomes, and the activities of cellobiohydrates and endoglucanase of glycoside hydrolase family 6 are absent, contrary to that of fungal hemibiotrophs and necrotrophs (Zhao *et al.* 2013). Furthermore, most of the phytopathogenic fungus proteins are tiny effectors that cannot encode catalytic activity. Such effectors aid pathogens in establishing themselves in the plant by derestricting the host immune system and encouraging plant colonization (Rovenich *et al.* 2014).

Fungal apoplastic effectors can either be discharged into the host's extracellular medium or persist in the cytoplasm and aggregate in the biotrophic interface complex, a membrane-rich structure of plants linked with invading hyphae of the fungus (Giraldo *et al.* 2013). Protease inhibitors that attack plant proteases, proteins that shield the cell walls of fungus from plant chitinases or recognition by the host plant, and tiny compounds that reduce the level of ROS are all examples of apoplastic effectors. Cytoplasmic effectors are detected via resistance proteins of plants, resulting in a hypersensitive response, a response characterized by quick cell death in the immediate infected site to prevent the fungal growth and multiplication (Giraldo and Valent 2013). *Ustilago maydis* genome indicates the synthesis of 550 secreted proteins, and most of them are virulence effectors that increased during colonization of the host plant (Djamei and Kahmann 2012). According to recent research, *U. maydis* is capable of sensing and adapting to the plant and secreting various individual effector combinations like a 1st 'core' set effectors for repressing plant defensive system during the phase of pathogenic penetration, followed by a 2nd set, i.e., cell-type and organ-specific effectors to infect various tissues of the host plant (Skibbe *et al.* 2010; Djamei and Kahmann 2012). The genome of *U. maydis* organizes most of the effector-encoding genes in clusters, and the studies revealed the cluster 19A as the most significant effector gene cluster in which 23 genes are variably triggered upon colonization of various plant parts. In maize, tumor growth was eliminated when the whole cluster 19A was removed, but strains removed for specific effector genes just exhibited a slight drop in pathogen virulence (Brefort *et al.* 2014). Although the proteins are the most recognized effectors, some effectors that depend on metabolites are also known. Such metabolic effectors involve host-specific toxins secreted by *Alternaria*, *Cochliobolus*, and a few *Pyrenophora* species, fuminosin mycotoxins by *Fusarium verticillioides*, and avirulence conferring enzyme 1 (Ace1) and pyrichalasin H by *M. oryzae* (Tsuge *et al.* 2013). Some small non-coding RNAs were found to silence the immunity genes of tomato and Arabidopsis by taking over the RNA interfering machinery, as in the case of *B. cinerea* (Weiberg *et al.* 2013).

BIOCHEMICAL DEFENSES OF HOST PLANT UPON FUNGAL INVASION

Plant immune responses at pathogen infection typically lead to a hypersensitive response, localized aggregation of phytoalexins, and an increase in numerous enzymatic activities (such as -1, 3-glucanase, chitinase, lipoxygenase, catalase, and peroxidase) (Lebeda *et al.* 2001). During the hypersensitive response, the death of cells is assumed to be dependent on the balanced synthesis of nitric oxide and ROS, active signal molecules in resistance to disease, and interaction with the necrotrophic pathogen (Delledonne *et al.* 2001; Sarkar *et al.* 2014). Such defense responses seek to isolate the invasive fungal pathogen at a site lacking an adequate supply of nutrients essential for life, to prevent pathogen transmission (Bolwell *et al.* 2002). The fast, temporary generation of massive levels of ROS, known as "oxidative burst," activates a significant amount of PR proteins.

PR-4 proteins either indicate DNAs or/and RNase, chitinase, and antifungal activities or lack chitinase activities but exhibit RNA, DNA properties (Bai *et al.* 2013). The transcriptional control of encoding genes based on PR protein is equally complex, with poor intrinsic or imperceptible activity under natural physiological circumstances and activation in response to injury, pathogen assault, or environmental factors. The buildup of PR proteins at the infection site is frequently linked with systemic acquired resistance, an extensive, broad-spectrum entire plant immunity that shields distant, intact host plant tissues against consecutive pathogenic fungal invasion (Durrant and Dong 2004; Sabater-Jara *et al.* 2010). In induced resistance mechanisms, metabolic pathways involving jasmonic acid, salicylic acid, and ethylene operate in conjunction (Sticher *et al.* 1997). Such plant hormones act as signal transduction, causing transcription factors to be synthesized in the host cell. The jasmonic acid pathway promotes the production of osmotin, defensin, phytoalexins, proteinase inhibitors, and proline-rich glycoproteins (Sticher *et al.* 1997; Wasternack 1997). Jasmonic acid and salicylic acid are both implicated in the modulation of primary resistance for many infections. According to thorough research, the hormonal signal may differ depending on the pathogen type, with salicylic acid especially controlling the defense system against fungal biotrophs, and jasmonic acid and ethylene mainly controlling defense responses to fungal necrotrophs (Mengiste 2012). On the other hand, beneficial fungi, i.e., symbiotic fungi, can produce jasmonic acid and ethylene-mediated induced systemic resistance, improving defense mechanisms despite directly activating the defense response (Pieterse *et al.* 2009).

THE GENE FOR GENE ASSOCIATION

Harold Henry Flor discovered the gene for gene association while investigating the genetics of rust pathogen (Melampsora lini) and flaxseed (Flor 1942). A one-to-one correlation between the pathogenic avirulence (avr) gene (leading to the affirmation of effector proteins set) and the cognate resistance gene (leading to the interpretation of resistance proteins) is envisaged in this framework to stimulate a signaling pathways' cascade that influences plants' race-specific resistance (Flor 1971). The avr9 and Cf9 genes have been reported for the *Cladiosporum fulvum*–tomato interaction, where the fungal race-specific avr9 gene triggers a hypersensitive response in tomato plants bearing the complementary resistance gene Cf9. Fungal races that are pathogenic on tomato genotypes (Cf9), on the other hand, do not generate the effector because the avr9 gene is absent in them. It has been established that avr9 is an actual gene that follows the hypothesis of the "gene for gene" association when inserted into a *C. fulvum* race and is aggressive to tomato genotype Cf9. Following that, a variety of avr and Cf genes were identified, and findings revealed that every effector had a specific purpose, including binding and modifying host proteins or playing a subservient role in hiding the fungal pathogen (Wulff *et al.* 2001).

PLANT RECEPTORS ORGANIZING DEFENSE VIA SIGNALING IN RESPONSE TO INTERACTION BETWEEN PLANT AND FUNGI

A plant should discriminate between beneficial and pathogenic fungi on several levels to confirm optimal cellular responses. The detection of pathogen-associated molecular patterns (PAMPs) or microbial-associated molecular patterns (MAMPs) via plant tissue surface-resident receptor proteins known as "pattern recognition receptors" is a primary stage in microbial detection (Dodds and Rathjen 2010). The resultant PAMP-triggered immunity provides safety against non-adapted fungal pathogens while also providing limited essential immunity against host-adapted microorganisms. The difference between PAMPs or MAMPs and effectors is not apparent in some cases. Both cell membrane local receptors and cytoplasmic resistance proteins similar to Nod-like receptors can mediate recognition (Thomma *et al.* 2011; Bohm *et al.* 2014).

THE INSIGHT ON PLANT-FUNGUS INTERACTION
VIA PATTERN RECOGNITION RECEPTORS

Plants recognize PAMPs or MAMPs via tiny epitopes, which serve as substrates for cell membrane resident receptors. These PR receptors are highly particular and sensitive, allowing plant tissues to detect a particular pattern of chemicals at subnanomolar levels (Boller and He 2009). Plant PR receptors include receptor-like proteins and receptor kinases. Receptor-like proteins lack the intracellular signaling region, while the receptor kinase has an extracellular region, a membrane-spanning domain, and an intracellular tyrosine or serine/threonine kinase region (Han *et al.* 2014). The FLAGELLIN SENSING 2 (FLS2) PR receptor is the best described of the more than 600 receptor-like kinases and more than 50 receptor-like proteins in the genome of Arabidopsis thaliana. Arabidopsis and rice detect chitin in fungus via the lysine motif Receptor-like kinase chitin elicitor receptor kinase-1 (RLK-CERK1), which triggers the CERK1 oligomerization required for the activation of a signaling pathway (Miya *et al.* 2007; Wan *et al.* 2008). Ligand-induced signaling via PR receptors, a particularly receptor-like protein lacking an intracellular signaling region, can necessitate additional collaborators' involvement to trigger corresponding immunological responses. In Arabidopsis, BRI1-associated kinase-1 forms ligand-induced heteromers with numerous receptor kinases and is one of the important regulatory bacterial proteins of FLS2-mediated signaling. BRI1 kinase-1 is also required for resistance against fungal infection by obligatory biotrophic and hemibiotrophic pathogens, and includes Ve1 inducing resistance in tomatoes against Verticillium wilt and possibly regulating Eix1 in reaction to ethylene-induced xylanase (Han *et al.* 2014). The FLS2–BRI–associated kinase-1 immunological receptor structure includes the receptor-like cytoplasmic kinase BRI1, actively phosphorylated by BRI-associated kinase-1 associated kinase. BIK1 is phosphorylated; it is distinguished from the receptor complex, activating subsequent signaling pathways and plant immunity.

Furthermore, BIK1 is required for resistance in Arabidopsis against necrotrophic diseases and is activated during *B. cinerea* invasion (Wang *et al.* 2014). BIK1 and BAK1 could also interact with other PR receptors, including CERK1, to influence responses of PAMP, but understanding on how diverse PR receptors can focus on such primary promoters is limited (Wang *et al.* 2014).

Symbiosis Pathway

The host plant recognizes the arbuscular mycorrhizae fungus during the formation of mycorrhizae through the general symbiosis pathway, which is also conserved by the symbiosis of Rhizobium-legume. Symbiosis receptor kinase has been identified to physically interface with many proteins, including the E3 ubiquitin ligase SEVEN IN ABSENTIA4 (SINA4) and the mitogen-activated protein kinase kinase (MAPKK) that control signaling of symbiosis by inversely regulating symbiosis receptor kinase concentration at the cell membrane. It has also been revealed that symbiosis receptor kinase, along with associates like SINA4, is found in membrane microspheres that operate as signaling substrates (Tax and Kemmerling 2012). Latest investigations using rice deletion mutants of CERK1 demonstrated that CERK1 performs dual functions, i.e., defense and symbiosis, since mutants were deficient for chitin-triggered defense responses against pathogenic bacteria and fungi and the symbiosis of arbuscular mycorrhizae as well (Miyata *et al.* 2014). CERK1 has been related to the attributes of unornamented chitin tetrasaccharides and pentasaccharides, and arbuscular mycorrhizal fungi symbiosis signals that cause Ca^{2+} spiking. In rice plants, the function of CERK1 as a molecular switch that triggers either symbiotic or defense responses depending on the invading microorganism implies a close evolutionary link between both mechanisms and proofs distinct receptor associates, which allows CERK1 to identify diverse ligands (Zhang *et al.* 2015). Aside from cell membrane–localized receptors, proteins found in the nuclear membrane and endoplasmic reticulum, such as Ca^{2+} networks and a calcium ATPase engaged in Ca^{2+} spiking, are required for symbiosis signaling. The nuclear resident calcium and calmodulin-dependent protein kinase—in which the transcription factor is subsequently phosphorylated, targeting to activate the symbiotic genes' appearance—is thought to decode the Ca^{2+} signal (Singh and Parniske 2012). Rapidly raised concentration of intracellular Ca^{2+} is also detected during disease occurrence and contact with the root-colonizing biocontrol agent *T. atroviride* fungus, suggesting that Ca^{2+} signaling is a typical route for plants to communicate with their fungal associates (Navazio *et al.* 2007).

Escape From Phyto-defensive Mechanism

Microorganisms correlated with plants may interact with plant signaling mechanisms to avoid recognition and plant defense by releasing effectors physically binding with and decreasing the kinase functioning of PR receptors or BRI1 associated kinase-1. This has been demonstrated for both the AvrPtoB and AvrPro effectors from the bacteria *Pseudomonas syringae* (Boller and He 2009), while fungi modify receptor role after using LysM effectors to connect soluble particles of chitin and prevent these from being detected through plant chitin receptors, as revealed for fungal pathogens such as *M. oryzae* and *C. fulvum*. Symbiotic microorganisms also use effectors. An effector is undertaken from the arbuscular mycorrhizal fungus *G. intraradices* through the plant host and operates by modifying the activation of nuclear defense–related genes (Bapaume and Reinhardt 2012).

SIGNALS AND PATHWAYS IN PLANT–FUNGUS INTERACTION

The detection of suitable hosts is one of the most important processes in plant–fungus interaction, and it frequently occurs before actual contact between the associates. Fungi detect and respond to physical and chemical stimuli by differentiating, migrating to suitable sites of infection, or/and forming invasion-associated structures (Bonfante and Genre 2010). Recent information on the signal detection and signaling pathways included in the interaction of plant and fungus is highlighted in the following section, emphasizing the fungal interaction associate. The following section reflects that several interaction-related genes from host plants, compared with fungus, have been studied so far and that the current understanding of pathogenic, compared with symbiotic and saprotrophic, interactions is more developed.

CHEMICAL DISCHARGE FROM ROOT SECRETIONS

Plant roots emit several chemicals that aid in plant–fungal interaction; these are emitted into roots surroundings, known as the "rhizosphere." Ions, amino acids, sugars, free oxygen, phenolics, organic acids, and other secondary metabolites are examples of low molecular weight compounds, while proteins and mucilage (polysaccharides) are examples of high molecular weight root secretions (Bais *et al.* 2006). As a carbon-rich habitat, the rhizosphere draws both helpful and harmful bacteria; however, volatiles produced by the roots of host plants operate as belowground defense chemicals with antiherbivore and antimicrobial activitities (Baetz and Martinoia 2014). Root secretions may be generated both chronically (as phytoanticipins) and in reaction to external stimuli such as pathogen infection (as phytoalexins) (Baetz and Martinoia 2014). Secretion of phenylpropanoids, plant phenolics exhibiting antifungal property, is selectively promoted in barley at infection through the *F. graminearum* soil-borne pathogen (Boddu *et al.* 2006). Likewise, when *Fusarium culmorum* and *Cochliobolus sativus* attack roots of the barley plant, the formation of several toxic antimicrobials, mostly terpenes, assists in activating the host plant's induced systemic resistance (Fiers *et al.* 2013).

Fungal infections also produce volatile chemicals, such as sesquiterpene-derived trichotecene toxins produced by *F. culmorum*, which are powerful regulators of protein biosynthesis and suppress activation of plant defense response genes prior to physical contact of the host with the pathogen (Fiers *et al.* 2013).

Root exudates are significant signaling molecules in plant–symbiotic microbial interaction in their contribution to chemical agents between host plants and pathogens. The sensing of root secretions via the pre-symbiotic mycelium of fungi initiates root colonization by arbuscular mycorrhizal fungi. The responsible molecules have been recognized as strigolactones, carotenoid-derived hormones of the plant found in secretions of plant roots from multiple taxa, and thus may be considered generic necessary signaling components for the formation of arbuscular mycorrhizal symbiosis (Bonfante and Genre 2010). Strigolactones serve as hyphal branching mediators in arbuscular mycorrhizal fungi, encouraging root invasion (Akiyama *et al.* 2005), but they are required during the pre-symbiotic phase and not during intracellular fungal growth (Koltai 2014). The strigolactone derivative GR24, suggesting that strigolactones are specialized signals for arbuscular mycorrhizal fungi (Steinkellner *et al.* 2007). Root exudates are high in carbohydrates, including disaccharide sucrose, which acts as a signaling molecule in various functions, including stimulating plant immunological responses. During the plant–microorganism interaction, sucrose breakdown by the host plant tissue provides a source of carbon for beneficial microorganisms (Koch 2004). However, most mycorrhizal fungi depend on the supplied monosaccharides due to the absence of succrolytic enzymes in them; the sucrose-degrading enzyme invertase is produced after infection in some fungal plant diseases (Voegele *et al.* 2006). Likewise, Metarhizium robertsii and *T. virens*, the rhizosphere-competent biocontrol fungi, break down sucrose using invertase, and in *T. virens*, the activity of invertase is critical for regulation of root colonization (Vargas *et al.* 2011; Liao *et al.* 2013). The pre-symbiotic stage of mycorrhiza development between host plant and arbuscular mycorrhizal fungi exemplifies the concept of precontact association between plant roots and fungi, including fungal- and plant-derived signals. The mechanism of diffusible arbuscular mycorrhizal fungal signaling molecules, known as "Myc" factors, stimulates biochemical reactions in plant roots that are essential for initial mycorrhization. These is a combination of sulphated and non-sulphated diffusible lipochitooligosaccharides that, like the Nod factors recognized from rhizobia, may drive arbuscular mycorrhizal production and branching of host root (Maillet *et al.* 2011).

OXYLIPINS

The involvement of oxylipins, a secondary metabolite class of oxygenated lipid, in cross-bridge signaling has recently received much attention. Endogenous oxylipin-mediated signals, and those generated by the associating partner, affect both the host plant and fungus (Borrego and Kolomiets 2012). Because oxylipins originating from plants and fungi are physically similar, it is not unexpected that they might partially replace each other. Fungal oxylipins, for example, can regulate plant lipid metabolism and change host defense responses by imitating endogenous signaling molecules in diseased plant parts (Tsitsigiannis and Keller 2007; Brodhagen *et al.* 2008). Contrarily, plant-derived oxylipins (such as jasmonates) have direct impacts on fungal reproduction and synthesis of secondary metabolites (Burow *et al.* 1997) or through impacting the viability of overwintering complexes of fungi (Calvo *et al.* 1999). By activating jasmonic acid–responsive genes, oxylipins are involved in disease development and pathogenicity (Thatcher *et al.* 2009). However, jasmonic acid–induced defense mechanisms are frequently engaged in resistance to necrotrophic infections, e.g., jasmonate signaling in *A. thaliana* via the jasmonic acid perception protein coronatine insensitive 1 has been known to be involved against vulnerability to wilt disease caused by *F. oxysporum*. This research implies that fungi can use defense-independent components of the jasmonic acid signaling pathway to induce infection (Thatcher *et al.* 2009). G protein-coupled receptors (GPCRs) have extensively been thought to be involved in oxylipin detection. The activation of GPCR is typically related to cAMP signaling, and in *Aspergillus nidulans*, plant oxylipins were shown to promote a spike in cAMP that was absent upon knockdown of the gprD GPCR-encoding gene. Endogenous oxylipins influence a transformation in *Aspergillus flavus*, the soil-borne pathogen impacting spore and sclerotia formation, and the mycotoxin aflatoxin synthesis activities that have been identified to be controlled by the *A. flavus* GprC and GprD. The studies suggested that GprC and GprD might be crucial for plant–fungal interactions based on the concept that endogenous oxylipins are probably comparable to exogenous oxylipins derived from the host plant (Affeldt *et al.* 2012).

REACTIVE OXYGEN SPECIES

During the plant–fungus interaction, oxylipin-interceded signaling is intimately linked to ROS-stimulated cell signaling. ROS operate as signaling molecules, influencing the defense expression of genes through redox regulation of transcriptional regulators or interaction with certain other signaling factors, including phosphorylation pathways. Furthermore, non-enzymatic oxygenation of lipid byproducts, such as oxylipins, might be caused by ROS activity. Because fungal invasion creates superoxide, ROS generation is not restricted to the plant (Reverberi *et al.* 2012) ROS synthesis is mediated by fungus NADPH (nicotinamide adenine dinucleotide phosphate) oxidase enzymes and ROS gathering at the plant–fungus contact serving as a signal to initiate pathogen attack and defense reactions. During plant invasion, the activation of Nox1 and Nox2 NADPH oxidases linked with fungal appressorium development causes transient oxidative stress in *M. grisea*. Fungal pathogens must resist the plant's protective oxidative burst by using the scavenging enzymes of ROS and altering this species' buildup (Egan *et al.* 2007). Defense suppressor 1 is the pathogenicity factor in *M. oryzae* required to collect extracellular ROS within plant host cells and regulate counterdefense responses. Yes-associated protein is a basic leucine zipper transcription factor that is employed by *U. maydis* as a redox detector, preventing the buildup of hydrogen peroxide generated by plant NADPH oxidases and allowing the fungi to resist initial host defenses (Chi *et al.* 2009). According to new research, secondary metabolites may help the antioxidant defense in response to high levels of ROS against fungi. Various transcription factors linked to the SAPK (stress-activated protein kinase) or MAPK signal pathway have been discovered to regulate the expression of the target genes, such as those involved in secondary and antioxidant metabolic activities, coordinating metabolic activities with response to cellular stress (Hong *et al.* 2013). ROS are also engaged in plant–symbiont relationships, including perennial ryegrass link with the endophyte *Epichloe festucae*. ROS generated by the *E. Festucae* NoxA NADPH oxidase is important for controlling hyphal development inside the plant and sustaining the symbiotic connection.

Breakdown of NADPH oxidase system elements, such as NoxR, NoxA, and RacA (rho-associated C3 botulinum toxin precursor), reverses the symbiotic correlation fungal mutations exhibiting unlimited development in plants (Tanaka *et al.* 2006). The plant interaction with AM fungus results in enhanced levels of ROS, and in mycorrhizal roots, changes occur in the structure of antioxidative enzymes. Catalase in AM has been proposed as a viable strategy for preventing the activation of defense response genes by degrading ROS-implicated signaling pathways in the host plant (Garcia-Garrido and Ocampo 2002). Likewise, ROS play a role in the mutualistic relationship of ECM fungus with plants. While significant root colonization by the fungi cannot promote cell death at disease sites, oxidative blasts in the host plant are generated during the initial encounter with the fungal interaction in the Spanish chestnut (*Castanea sativa*) with dyebell fungus (*Pisolithus tinctorius*). The fungus avoided the conventional plant defensive reaction by stimulating catalase regularly during mycorrhizal growth (Baptista *et al.* 2007).

CONCLUSION

Plant–fungal interactions are experiencing a renaissance due to the latest in-depth discoveries into the phylogenetic links between host plant and fungus and the establishment of novel tools for exploring them. For an extensive period, scientists have examined such interactions using very reductive methodologies. On the other hand, plant–fungal interactions are significantly more complicated than previously assumed, including not just specific fungal associates but an entire plant associated with fungi. Tremendous progress has been made to understand the fungal functions as key plant interacting partners; however, much must be discovered, particularly in relationships between biotrophic fungal diseases and obligate fungi. We have recently joined the genomics and super tenacity age and are ready to use the accompanying genomes, metagenomics, proteomics, transcriptomics, and sophisticated technologies to explore new paths for investigating plant–fungal interactions.

Fungi–plant interactions are now characterized as parasitic, mycorrhizal, or endophytic, with many fungal relationships performing substantial plant growth and maintenance activities. It remains challenging to understand how the fungal associate changes its living pattern to integrate with a host plant, particularly the transformation of endophytic fungi into parasites. A thorough study on fungus biodiversity and the fungal impact on the biology of host plants will not just increase scientific understanding of ecological processes, but will also be critical for managing plant diseases and maximizing the opportunities for beneficial fungi.

REFERENCES

Affeldt KJ, Brodhagen M and Keller NP (2012). *Aspergillus* oxylipin signaling and quorum sensing pathways depend on g protein-coupled receptors. *Toxins* 4(9):695–717.

Akiyama K, Matsuzaki K and Hayashi H (2005) Plant sesquiterpenes induce hyphal branching in arbuscular mycorrhizal fungi. *Nature* 435(7043):824–827.

Baetz U and Martinoia E (2014) Root exudates: The hidden part of plant defence. *Trends in Plant Science* 19(2):90–98.

Bai S, Dong C, Li B and Dai H (2013) A PR-4 gene identified from *Malusdomestica* is involved in the defence responses against *Botryosphaeria dothidea*. *Plant Physiology and Bioch*emistry 62:23–32.

Bais HP, Weir TL, Perry LG, Gilroy S and Vivanco JM (2006) The role of root exudates in rhizosphere interactions with plants and other organisms. *Annual Review of Plant Biology* 57:233–266.

Bapaume L and Reinhardt D (2012) How membranes shape symbioses: Signalling and transport in nodulation and arbuscular mycorrhiza. *Frontiers Plant Science* 3:1–29.

Baptista P, Martins A, Pais MS, Tavares RM and Lino-Neto T (2007) Involvement of reactive oxygen species during early stages of ectomycorrhiza establishment between *Castanea sativa* and *Pisolithus tinctorius*. *Mycorrhiza* 17(3):185–193.

Boddu J, Cho S, Kruger WM and Muehlbauer GJ (2006) Transcriptome analysis of the barley-*Fusarium graminearum* interaction. *Molecular Plant-Microbe Interactions* 19(4):407–417.

Bohm H, ALbert I, FAn L, Reinhard A and Nürnberger T (2014) Immune receptor complexes at the plant cell surface. *Current Opinion in Plant Biology* 20:47–54.

Boller T and He SY (2009) Innate immunity in plants: An arms race between pattern recognition receptors in plants and effectors in microbial pathogens. *Science* 324:742–744.

Bolwell GP, Bindschedler LV, Blee KA, Butt VS, Davies DR, Gardner SL and Minibayeva F (2002) The apoplastic oxidative burst in response to biotic stress in plants: A three-component system. *Journal of Experimental Botany* 53(372):1367–1376.

Bonfante P and Anca IA (2009) Plants, mycorrhizal fungi, and bacteria: A network of interactions. *Annual Review of Microbiology* 63:363–383.

Bonfante P and Genre A (2010) Mechanisms underlying beneficial plant-fungus interactions in mycorrhizal symbiosis. *Nature Communications* 1:48.

Borrego EJ and Kolomiets MV (2012) Lipid-mediated signaling between fungi and plants. In: Witzany V (ed.) *Biocommunication of fungi*. The Netherlands: Springer, pp. 249–260.

Brefort T, Tanaka S, Neidig N, Doehlemann G, Vincon V and Kahmann R (2014) Characterization of the largest effector gene cluster of *Ustilago maydi*. *PLOS Pathogens* 10(7): e1003866.

Brito N, Espino JJ and González C (2006) The endo-beta-1,4-xylanase xyn11A is required for virulence in *Botrytis cinerea*. *Molecular Plant-Microbe Interactions* 19(1):25–32.

Brodhagen M, Tsitsigiannis DI, Hornung E, Goebel C, Feussner I and Keller NP (2008) Reciprocal oxylipin-mediated cross talk in the Aspergillus-seed pathosystem. *Molecular Microbiol*ogy 67(2):378–391.

Brundrett M (2004) Diversity and classification of mycorrhizal associations. *Biolological Review* 79:473–495.

Burow G, Nesbitt T, Dunlap J and Keller NP J (1997) Seed Lipoxygenase products modulate *Aspergillus* mycotoxin biosynthesis. *Molecular Plant- Microbe Interactions* 10(3):380–387.

Buscot F (2015) Implication of evolution and diversity in arbuscular and ectomycorrhizal symbioses. *Journal of Plant Physiology* 172:55–61.

Buscot F, Munch JC, Charcosset JY, Gardes M, Nehls U and Hampp R (2000) Recent advances in exploring physiology and biodiversity of ectomycorrhizas highlight the functioning of these symbioses in ecosystem. *FEMS Microbiology Reviews* 601–614.

Cairney JWG (2000) Evolution of mycorrhiza systems. *Naturwissenschaften* 87(1):467–475.

Calvo AM, Hinze LL, Gardner HW and Keller NP (1999) Sporogenic effect of polyunsaturated fatty acids on development of *Aspergillus* spp. *Applied and Environmental Microbiology* 65(8):3668–3673.

Chi MH, Park SY, Kim S and Lee YH (2009) A novel pathogenicity gene is required in the rice blast fungus to suppress the basal defences of the host. *PLoS Pathogens* 5(4):e1000401.

Corradi N and Bonfante P (2012) The Arbuscular mycorrhizal symbiosis: Origin and evolution of a beneficial plant infection. *PLoS Pathogens* 8(4):8–10.

Delledonne M, Zeier J, MArocco A and Lamb C (2001) Signal interactions between nitric oxide and reactive oxygen intermediates in the plant hypersensitive disease resistance response. *Proceedings of the Natural Academy of Science* 98(23):13454–13459.

Djamei A and Kahmann R (2012) Ustilago maydis: Dissecting the molecular interface between pathogen and plant. *PLoS Pathogens* 8(11):e1002955.

Dodds P and Rathjen J (2010) Plant immunity: Towards an integrated view of plant—pathogen interactions. *Nature Reviews Genetics* 11(8):539–548.

Durrant WE and Dong X (2004) Systemic acquired resistance. *Annual Reviews Phytopathology* 42:185–209.

Egan MJ, Wang ZY, Jones MA Smirnoff and Talbot NJ (2007) Generation generation of reactive oxygen species by fungal NADPH oxidases is required for rice blast disease. *Proceedings of the Natural Academy Sciences* 104(28):11772–11777.

Fernandez-Acero FJ, Colby T, Harzen A, Carbú M, Wieneke U, Cantoral JM and Schmidt J. (2010) 2-DE proteomic approach to the *Botrytis cinerea* secretome induced with different carbon sources and plant-based elicitors. *Proteomics* 10(12):2270–2280.

Field KJ, Cameron DD, Leake JR, Tille S, Bidartondo MI and Beerling DJ (2012) Contrasting arbuscular mycorrhizal responses of vascular and non-vascular plants to a simulated Palaeozoic CO_2 decline. *Nature Communications* 3(1):835.

Fiers M, Lognay G, Fauconnier ML and Jijakli MH (2013) Volatile compound-mediated interactions between barley and pathogenic fungi in the soil. *PLoS One* 8(6):e66805.

Flor HH (1942) Inheritance of pathogenicity in *Melampsora lini*. *Phytopathology* 32:653–669.

Flor HH (1971) Current status of the gene-for-gene concept. *Annual Review of Phytopathology* 9(1):275–296.

Garcia-Garrido JM and Ocampo JA (2002) Regulation of the plant defence response in arbuscular mycorrhizal symbiosis. *Journal of Experimental Botany* 53(373):1377–1386.

Gardiner DM, Kazan K and Manners JM (2013) Cross-kingdom gene transfer facilitates the evolution of virulence in fungal pathogens. *Plant Sciences* 210:151–158.

Giraldo MC, Dagdas YF, Gupta YK, Mentlak TA, Yi M, Martinez-Rocha AL, SaitoH H, TeRauchi R, Talbot NJ and Valent (2013) Two distinct secretion systems facilitate tissue invasion by the rice blast fungus *Magnaporthe oryzae*. *Nature Communications* 4(1):1996.

Giraldo MC and Valent B (2013) Filamentous plant pathogen effectors in action. *Nature Reviews Microbiology* 11(11):800–814.

Gough C and Jacquet C (2013) Nod factor perception protein carries weight in biotic interactions. *Trends in Plant Science* 18(10):566–574.

Gourion B, Berrabah F, Ratet P and Stacey G (2015) Rhizobiumlegume symbioses: the crucial role of plant immunity. *Trends in Plant Science* 20(3):186–194.

Grigoriev I (2013) Fungal genomics for energy and environment. In: Horwitz B, Mukherjee P, Mukherjee M (eds.) *Genomics of Soil- and Plant-Associated Fungi: Soil Biology* (Vol. 36). Berlin, Heidelberg: Springer, pp. 11–27.

Han Z, Sun Y and Chai J (2014) Structural insight into the activation of plant receptor kinases. *Current Opinion in Plant Biology* 20:55–63.

Hong SY, Roze LV and Linz JE (2013) Oxidative stress-related transcription factors in the regulation of secondary metabolism. *Toxins* 5(4):683–702.

Horbach R, Navarro-Quesada AR and Knogge W (2011) When and how to kill a plant cell: Infection strategies of plant pathogenic fungi. *Journal of Plant Physiology* 168(1):51–62.

Kabbage M, Yarden O and Dickman MB (2015) Pathogenic attributes of *Sclerotinia sclerotiorum*: Switching from a biotrophic to necrotrophic lifestyle. *Plant Science* 233:53–60.

Karandashov V, Nagy R, Wegmuller S, Amrhein N and Bucher M (2004) Evolutionary conservation of a phosphate transporter in the arbuscular mycorrhizal symbiosis. *Proceedings of the National Academy of Sciences* 101(16):6285–6290.

Kawahara Y, Oono Y, Kanamori H, MaTsumoTo T, Itoh T and Minami E (2012) Simultaneous RNA-seq analysis of a mixed transcriptome of rice and blast fungus interaction. *PLoS One* 7(11):e49423.

Koch K (2004) Sucrose metabolism: Regulatory mechanisms and pivotal roles in sugar sensing and plant development. *Current Opinion in Plant Biology* 7(3):235–246.

Kohler A, Kuo A, Nagy LG, Morin E, Barry KW, Buscot F, CanBäCk B, Choi C, Cichocki N, Clum A and Colpaert J (2015) Convergent losses of decay mechanisms and rapid turnover of symbiosis genes in mycorrhizal mutualists. *Nature Genetics* 47(4):410–415.

Koltai H (2014) Implications of non-specific strigolactone signaling in the rhizosphere. *Plant Science* 225:9–14.

Kubicek CP (2013) *Fungi and Lignocellulosic Biomass*. New York: Wiley.

Lebeda A, Luhova L, Sedlarova M and Jančová D (2001) The role of enzymes in plant—fungal pathogens interactions. *Journal of Plant Diseases and Protection* 108:89–111.

Liao X, Fang W and Lin L (2013) *Metarhizium robertsii* produces an extracellular invertase (MrINV) that plays a pivotal role in rhizospheric interactions and root colonization. *PloS One* 8(10):e78118.

Libault M, Farmer A, Brechenmacher L, Drnevich J, Langley RJ, Bilgin DD, Radwan O, Neece DJ, Clough SJ, May, GD and Stacey G (2010) Complete transcriptome of the soybean root hair cell, a single-cell model, and its alteration in response to *Bradyrhizobium japonicum* infection. *Plant Physiology* 152(2):541–552.

Maillet F, POinsot V, André O, Puech-Pagès V, Haouy A, Gueunier M, Cromer L, GirauDet D, Formey D, Niebel A and Martinez EA (2011) Fungal lipochitooligosaccharide symbiotic signals in arbuscular mycorrhiza. *Nature* 469(7328):58–63.

Mengiste T (2012) Plant immunity to necrotrophs. *Annual Reviews of Phytopathology* 50:267–294.

Miya A, Alber T P, ShinYa T, DesaKi Y, Ichimura K, Shirasu K, Narusaka Y, Kawakami N, Kaku H and Shibuya N (2007). CERK1, a LysM receptor kinase, is essential for chitin elicitor signaling in Arabidopsis. *Proceedings of the Natural Academy Sciences* 104(49):19 613–618.

Miyata K, Kozaki T, Kouzai Y, Ozawa K, Ishii K, Asamizu E, Okabe Y, Umehara Y, MiyAmoto A, Kobae Y and Akiyami K (2014). The bifunctional plant receptor, OsCERK1, regulates both chitintriggered immunity and arbuscular mycorrhizal symbiosis in rice. *Plant Cell Physiology* 55(11):1864–1872.

Navazio L, Baldan B, Moscatiello R, Zuppini A, Woo SL, Mariani P and Lorito M (2007) Calcium-mediated perception and defence responses activated in plant cells by metabolite mixtures secreted by the biocontrol fungus *Trichoderma atroviride*. *BMC Plant Biology* 7(1):41.

Parniske M (2008) Arbuscular mycorrhiza: The mother of plant root endosymbioses. *Nature Reviews Microbiology* 6(10):763–775.

Perfect SE and, Green JR (2001) Infection *structures* of biotrophic and hemibiotrophic fungal plant pathogens. *Molecular Plant Pathology* 2(2):101–108.

Pieterse CMJ, Leon-Reyes A and Van der Ent S (2009) Networking by small-molecule hormones in plant immunity. *Nature Chemical Biology* 5(5):308–316.

Pryce-Jones E, Carver TIM and Gurr SJ (1999) The roles of cellulase enzymes and mechanical force in host penetration by *Erysiphe graminis* f.sp. *hordei*. *Physiological and Moecular Plant Pathology* 55(3):175–182.

Reverberi M, Fabbri AA and Fanelli C (2012) Oxidative stress and oxylipins in plant—fungus interaction. In: Guenther W (eds.) *Biocommunication of Fungi*. The Netherlands: Springer, pp. 273–290.

Rey T, Chatterjee A, ButtayM, Toulotte J and Schornack S (2015) *Medicago truncatula* symbiosis mutants affected in the interaction with a biotrophic root pathogen. *New Phytologist* 206:497–500.

Rey T, Nars A, BonhoMme M, Bottin A, Huguet S, Balzergue S, Jardinaud MF, Bono JJ, Cullimore J, Dumas B and Gough C (2013) NFP, a LysM protein controlling Nod factor perception, also intervenes in *Medicago truncatula* resistance to pathogens. *New Phytologist* 198(3):875–886.

Rovenich H, Boshoven Thomma BPHJ JC (2014) Filamentous pathogen effector functions: Of pathogens, hosts and microbiomes. *Current Opinion Plant Biol*ogy 20:96–103.

Sabater-Jara AB, Almagro L, Belchí-Navarro S, Ferrer MÁ, BARceló AR and Pedreño MÁ (2010) Induction of sesquiterpenes, phytoesterols and extracellular pathogenesis-related proteins in elicited cell cultures of *Capsicum annuum*. *Journal of Plant Physiol*ogy 167(15):12731281.

Sarkar TS, Biswas P, Ghosh SK and Ghosh S (2014) Nitric oxide production by necrotrophic pathogen *Macrophomina phaseolina* and the host plant in charcoal rot disease of jute: Complexity of the interplay between necrotroph—host plant interactions. *PLoS One* 9(9):e107348.

Singh S and Parniske M (2012) Activation of calcium- and calmodulin-dependent protein kinase (CCaMK), the central regulator of plant root endosymbiosis. *Current Opinion in Plant Biology* 15(4):444–453.

Skibbe D, Doehlemann G, Fernandes J and Walbot V (2010) Maize tumors caused by Ustilago maydis require organ-specific genes in host and pathogen. *Science* 328(5974):89–92.

Smakowska-Luzan E, Mott GA, Parys K, StegMann M, Howton TC, Layeghifard M, Neuhold J, Lehner A, Kong J, Grunwald K and Weinberger N (2018) An extracellular network of Arabidopsis leucine-rich repeat receptor kinases. *Nature* 553(7688):342–346.

Steinkellner S, Lendzemo V, Langer I, Schweiger P, Khaosaad T, Toussaint JP and Vierheilig H (2007) Flavonoids and strigolactones in root exudates as signals in symbiotic and pathogenic plant—fungus interactions. *Molecules* 12(7):1290–1306.

Sticher L, Mauch-Mani B and Metraux JP (1997) Systemic acquired resistance. *Annual Review of Phytopathol*ogy 35(1):235–270.

Stracke S, KiStner K, Yoshida S, MuLder L, SaTo S, KaNeko T, Tabata S, Sandal N, Stougaard J, SzczyglowsKi K and Parniske M (2002) A plant receptor-like kinase required for both bacterial and fungal symbiosis. *Nature* 417(6892):959–962.

Strobel G and Daisy B (2003) Bioprospecting for microbial endophytes and their natural products. *Microbiology and Molecular Biology Reviews* 67(4):491–502.

Struck C (2006) Infection strategies of plant parasitic fungi. In: Cooke BM, Jones DG, Kaye B (eds.) *The Epidemiology of Plant Diseases*. The Netherlands: Springer, pp. 117–137.

Tanaka A, Christensen MJ, Takemoto D, Park P and Scott B (2006) Reactive oxygen species play a role in regulating a fungus-perennial ryegrass mutualistic interaction. *The Plant Cell* 18(4):1052–10166.

Tax F and Kemmerling B (2012) *Receptor-Like Kinases in Plants (eds): From Development to Defense* (Vol. 13). New York: Springer Science & Business Media.

Thatcher LF, Manners JM and Kazan K (2009) *Fusarium oxysporum* hijacks COI1-mediated jasmonate signaling to promote disease development in Arabidopsis. *The Plant Journal* 58(6):927–939.

Thomma BP, Nurnberger T and Joosten MH (2011) Of PAMPs and effectors: The blurred PTI-ETI dichotomy. *The Plant Cell* 23(1):4–15.

Trappe JM (2005) AB Frank and mycorrhizae: The challenge to evolutionary and ecologic theory. *Mycorrhiza* 15(4):277–281.

Tsitsigiannis DI and Keller NP (2007) Oxylipins as developmental and host-fungal communication signals. *Trends in Microbiol*ogy 15(3):109–118.

Tsuge T, Harimoto Y, AKimitsu K, Ohtani K, KodaMa M, Akagi Y, Egusa M, YaMamoto M and Otani H (2013) Host-selective toxins produced by the plant pathogenic fungus *Alternaria alternata*. *FEMS Microbiology Reviews* 37(1):44–66.

Vargas WA, Crutcher FK and Kenerley CM (2011) Functional characterization of a plant-like sucrose transporter from the beneficial fungus *Trichoderma virens*, regulation of the symbiotic association with plants by sucrose metabolism inside the fungal cells. *New Phytol*ogist 189(3):777–789.

Voegele RT, Wirsel S, Moll U, Lechner M and Mendgen K (2006) Cloning and characterization of a novel invertase from the obligate biotroph Uromyces fabae and analysis of expression patterns of host and pathogen invertases in the course of infection. *Molecular Plant- Microbe Interactions* 19(6):625–634.

Wan J, Zhang XC, Neece D, Ramonell KM, Clough S, Kim SY, Stacey MG and Stacey G (2008) A LysM receptor like kinase plays a critical role in chitin signaling and fungal resistance in Arabidopsis. *The Plant Cell* 20(2):471–481.

Wang B and Qiu YL (2006) Phylogenetic distribution and evolution of mycorrhizas in land plants. *Mycorrhiza* 16(5):299–363.

Wang X, Jiang N, Liu J, Wende Liu and Guo-Liang Wang (2014) The role of effectors and host immunity in plant—necrotrophic fungal interactions. *Virulence* 5(7):722–732.

Wasternack C (1997) Jasmonates signal plant gene expression. *Trends in Plant Science* 2:302–307.

Weiberg A, Wang M, Lin FM, ZHao H, Zhang Z, KaloshIan I, Huang HD and Jin H (2013) Fungal small RNAs suppress plant immunity by hijacking host RNA interference pathways. *Science* 342(6154):118–123.

Wolpert TJ, Dunkle LD and Ciuffetti LM (2002) Host-selective toxins and avirulence determinants: What's in a name? *Annual Review of Phytopathology* 40(1):251–285.

Wulff BB, ThoMas CM, Smoker M, Grant M and Jones J (2001) Domain swapping and gene shuffling identify sequences required for induction of an Avr-dependent hypersensitive response by the tomato Cf-4 and Cf-9 proteins. *The Plant Cell* 13(2):255–2272.

Zamioudis C and Pieterse CM (2012) Modulation of host immunity by beneficial microbes. *Molecular Plant-Microbe Interactions* 25(2):139–150.

Zhang X, Dong W, Sun J, Feng, F, Deng Y, He Z, Oldroy GE and Wang E (2015) The receptor kinase CERK1 has dual functions in symbiosis and immunity signalling. *The Plant Journal* 81(2):258–267.

Zhao Z, Liu H, Wang C and Xu JR (2013) Comparative analysis of fungal genomes reveals different plant cell wall degrading capacity in fungi. *BMC Genomics* 14(1):27.

11 Mycorrhizal Symbiosis: *Molecular Mechanism, Application, and Prospects*

Imran Ul Haq, Zakria Faizi, Shehla Riaz,
Younus Raza and Zaianb Malik

CONTENTS

DOI: 10.1201/9781003162742-11

INTRODUCTION

The mycorrhizal association is the most primitive and widespread fungal association with plants. The word "myco" means "fungus," and "rhizha" means "roots."
Mycorrhiza can be defined as the mutualistic relationship between the roots of plants
and soil fungi. In this relationship, fungus, in the plant's active growth period, penetrates the secondary roots and changes the root's anatomy; the modified root is
finally called "mycorrhizae" (Hacskaylo, 1972; Janerette, 1991). Mycorrhizal structures can exist in many shapes, depending on the properties of both partners and the
type of plant–fungal combination (Harley, 1984). The fungus attacks the epidermis
and cortex of the root during infection, but it does not disturb the vascular tissues and
meristematic tissues (Janerette, 1991).

B. Frank Eames in 1885, while describing the altered structures of forest roots,
coined the term "Mycorrhiz" (Frank, 2005; Finlay, 2008). Many categories of fungal symbiosis with plants are recognized on the basis of the morphology of structures formed by Mycorrhiza and plant and fungal species involved. In the symbiosis
between fungi and higher plants, fungi being achlorophyllous, cannot manufacture
their food, so they have a well-adapted mechanism to scavenge food from their hosts.
In a mycorrhizal association with the host, fungi get food from their partner, and in
return, provide many benefits to the plant, i.e., they increase the host plant's capacity for nutrient absorption (Dalpé and Monreal, 2004; Alizadeh, 2011). In addition,
the fungus produces and secretes some growth hormones, enables better absorption of water, protects against plant pathogens (Alizadeh, 2011), increases nitrogen
fixation, increases drought resistance, increases the rate of photosynthesis, and
lowers the concentration of toxic elements such as mercury, arsenic, and cadmium
(Ramanankierana et al., 2007; Alizadeh, 2011).

CLASSES OF MYCORRHIZA

J L Harley in 1961, divided symbiotic mycorrhizal association into ectotheraphic and
endotheraphic. Hence, since 1961, these two terms have been used. Recently, based
on the type of relationship between mutualists and the mode of fungal penetration,
mycorrhiza was classified into three categories: Ectomycorrhiza, Endomycorrhiza,
and Ectendomycorrhiza (Smith and Read, 2010; Alizadeh, 2011).

ECTOMYCORRHIZA

This type of fungal association is formed by a class of higher fungi, i.e., basidiomycetes
and ascomycetes (Alexopoulos and Mims, 1979; Janerette, 1991). The ectomycorrhizal association is found in many plant species: spruce, pine, oak, walnut, chestnut,
spruce, hemlock, eucalyptus, willow, and birch. Ectomycorrhizal fungi form a specific ectomycorrhizal pattern. After successful establishment of a connection with a
compatible host, the fungus forms specific hair-like projections, known as "hyphae,"
that surround the root, and hence, a dense sheath is formed, which is called "mantle."
This mantle physically detaches the roots from their surroundings, and the fungus
releases enzymes that enable it to penetrate the root and the cortex. The hyphae are

confined to spaces between root cells and form a hartig network. This hartig network plays a vital role in the material exchange between the fungus and the plant. After progression of the relationship, the fungus releases certain growth-regulating chemicals that enhance the development of roots (Hacskaylo, 1972; Janerette, 1991).

ENDOMYCORRHIZA

Endomycorrhizal association is more abundant than the ectomycorrhizal association, and it occurs in >90 % of plants. The endomycorrhizal association is primarily found in non-woody plants such as strawberries, legumes, tomatoes, onions, wheat, corn, and grasses. It is also found in some woody plants like orange and apples. Hyphae from germinating spores penetrate the root in the mycorrhizal association. The classification of endomycorrhizal fungi is based on spore morphology and structures bearing spores (Janerette, 1991).

Fungi belonging to zygomycetes, like Glomales and Endogonales, usually form this type of association. During the symbiotic relationships, spores or hyphae of the fungi get stuck to the receptive root; the fungi secrete certain enzymes that dissolve the cell wall. This cell wall dissolution allows the hyphae to enter the roots (Hacskaylo, 1972; Janerette, 1991). Hyphae penetration is usually confined to the epidermis, but sometimes it goes into the cortical cells. Roots look normal when they are examined externally; hence, microscopy is needed to confirm the symbiotic association. When roots are examined through a microscope, threadlike hyphae invading the inner cells can be seen as hyphal strands or swellings (vesicles) and coils (arbuscules).

Due to the presence of vesicles and arbuscules, this association is known as vesicular–arbuscular mycorrhizae. The function of arbuscules is to transfer nutrients, while vesicles have a storage function. These arbuscules and vesicles collectively increase the area between both partners (Harley, 1969; Janerette, 1991). In some endomycorrhizal associations, vesicles are not found, such as among fungi belonging to scutellospora and gigaspora (Duponnois and Plenchette, 2003; Johansson et al., 2004).

ECTENDOMYCORRHIZA

A third type of association is also found in plants, known as the "ectendomycorrhizal association." Plants found in nurseries with ectomycorrhizal associations also form this symbiotic association. The ectendomycorrhizal association has characteristics of both types of associations. A thin mantle can be seen in this association, which cannot be easily detected. Inside root cells, a hartig network can be seen with hyphae. According to some scientists, this symbiosis is an intermediate state between ectomycorrhizal and endomycorrhizal symbiosis (Melin, 1948). This type of fungal association is rarely seen. In this type of association, fungal hyphae penetrate into intercellular spaces, form arbuscules inside and vesicles outside the cells, and form spores or spore-bearing structures (Assigbetse et al., 2005; Bedini et al., 2008).

BIOLOGY OF MYCORRHIZA

Mycorrhizal symbiosis is present everywhere. In this symbiotic relationship, hyphae of fungi suck nutrients from the soil and provide them to the plant they are associated with. Fungi involved in this symbiosis may belong to any one of four phyla—Ascomycota, Basidiomycota, Zygomycota, and Chytridiomycota—and a wide range of host plants from mosses to ferns, liverworts, and seeded plants. The exact mycorrhizal status of many plants association is still unknown, but examination of 6,507 species revealed that 82% plants form mycorrhizal association. Symbiotic relationships are categorized as endomycorrhizae, ectodomycorrhizae, and ectendomycorrhiza. The mycorrhizal association also includes monotropoid, arbutoid, ericoid, and orchid forms, in addition to the three major types of associations (Alizadeh, 2011).

MORPHOLOGY OF MYCORRHIZAL FUNGI

Two types of mycelium can be seen in mycorrhizal association: external mycelium and internal mycelium. The external mycelium flourishes and multiplies inside the soil. It penetrates tiny pores of soil and makes the food available for the plant where food is not easily available availability is difficult (Ramanankierana et al., 2007; Smith and Read, 2010). The internal mycelium grows within or around the parenchymatous cells of the host. It forms various branches inside root cells. These branches are collectively called "arbuscules." These arbuscules provide a bridge for nutrients between the symbionts. After inoculation of roots, arbuscules are formed within two to four days and have a short life span of about 4–15days. After this, they are destroyed and engulfed by host plants. Vesicles, another structure, are formed inside the roots of the host; they are spherical and a site of fat storage and do not have a significant role in nutrient uptake. They last during the whole life of fungi (Wang and Qiu, 2006; Smith and Read, 2010; Alizadeh, 2011).

IMPORTANCE OF MYCORRHIZA

Mycorrhizal symbiosis has many benefits for the plant, such as the following:

- Increases nutrient and water uptake
- Enhances root life and improves its health
- Enables tolerance to drought, high pH, extreme temperatures, and transplant shock in plants
- Reduces the uptake of toxic heavy metals

Furthermore, fungi transfer metabolites from plant to plant and promote plant growth by producing plant growth promoters. Fungi, in return, are benefited with food and shelter (Harley, 1969; Maronek et al., 1981).

The application of mycorrhizal technology in the field may have a significant impact on agriculture globally on account of the following:

- Aid the growth of crops in poor soils
- Enable the truck-farmer to produce more vigorous plants having a mycorrhizal association

- Increase the size of the plant in a short duration, which ultimately increases the biomass and plant production
- Reduce the production cost and fertilizer contamination in the environment by minimizing the requirement for fertilizers (Janerette, 1991)

MOLECULAR MECHANISM INVOLVED IN MYCORRHIZAL SYMBIOSIS

When the genomes of *Laccaria bicolor* and truffle *Tuber melanosporum* were compared with the genome of saprobic and pathogenic fungi, findings highlighted mycorrhiza's biology and the critical factors that regulate the development of mycorrhiza. It was revealed that a small number of enzymes, i.e., pectin lyases and pectinases, target the components of the cell wall (Martin *et al.*, 2009). After the fungus development inside plant tissues, genes coding these enzymes are activated during different steps of the symbiosis. *L. bicolor* secretes many enzymes that degrade the substrate, e.g., glycosyl hydrolases, which hydrolyze polysaccharides of bacteria and many proteases, glucanases, and chitinases that degrade organic matter in the litter (Martin and Nehls, 2009). Despite the increasing number of studies on identifying fungal genes, it is not yet possible to distinguish between ectommycorrizal fungi and arbuscular mycorrhizal fungi. Intriguingly, *STE12-like*, a gene from *G. intraradices*, restores the infectivity of the non-pathogenic mutant of *Colletotrichum lindemuthianum* (Tollot et al., 2009). Furthermore, the gene responsible for the virulence of *Magnaportheoryzae*, which codes Era-like GTPase, complements Gin-N protein from *G. intraradices*4 (Selosse and Roy, 2009). These findings imply that pathogenic and symbiotic fungi share certain common molecular mechanisms to invade plant tissues, regardless of their diverse trophic habits and distant phylogenetic associations.

ROOT COLONIZATION: It is an essential step in a mycorrhizal relationship. In the case of AM fungi, germinating hyphae from spores get food by catalyzing lipids for a few days (Smith and Read, 2010). During this period, they search for a host in the soil, but in the absence of any host, they again become spores by retracting their cytoplasm.

Roots of plants release some compounds, known as "strigolactones," responsible for the perception of the host by the fungal mycelium (Akiyama *et al.*, 2005; Besserer *et al.*, 2006). These strigolactones enhance the interaction of mycorrhizal fungi with the host and have some hormonal role in reducing the lateral branches (Umehara *et al.*, 2008).

Bioactive molecules of AM fungi were 3 kDa, having a chitin backbone and being partially lipophilic (Navazio *et al.*, 2007; Bucher *et al.*, 2009). These diffusible signals, called the "Myc factor" (Kosuta *et al.*, 2003), can be recognized by plants, i.e., membrane-steroid-binding protein, a protein necessary for mycorrhizal association (Kuhn *et al.*, 2010). These factors are the main constituent of the common symbiotic pathway (SYM pathway), which transduces signals and enables the plant to associate with mycorrhizal fungi and other nitrogen-fixing bacteria (Bonfante and Genre, 2010). Plant response to these Myc factors varies from molecular to organ level.

ROLE OF CALCIUM ION FOR NODULATION

The SYM pathway is primarily involved in controlling pre-contact responses, early steps in the mycorrhizal establishment, and root colonization. Any mutation in genes that controls this pathway inhibits fungal colonization in layers of epidermal and sub-epidermal cells. Calcium ion is the main factor involved in the signaling pathway. The rapid induction of calcium ions by Nod factors (Mitra et al., 2004) and fungal exudates (Kosuta et al., 2003), and recurrent oscillation of calcium ions in the nucleus of root hairs are observed due to rhizobial signals (Mitra et al., 2004). The contiguity of branched hyphae has been reported to cause Ca^{2+} oscillations around the nuclear cytoplasm of M. truncatula. Cytoplasmic calcium sensors sense these cytoplasmic oscillations, called "Cameleon," present in root hairs of M. truncatula (Kosuta et al., 2008). The pattern of these calcium ion oscillations varies compared with those that are induced due to nitrogen-fixing bacteria (Kosuta et al., 2008; Hazledine et al., 2009).

PHYSICAL CONTACT BETWEENSYMBIONTS

After successful chemical contact between mycorrhizal fungi and the plant, roots and fungal hyphae proliferate and split in a small volume of the rhizospheric zone. When the tip of hyphae touches the surface of roots, the pre-symbiotic stage of mycorrhizal interaction progresses toward physical contact between the symbionts. In the case of AM fungi, their hyphae spread several centimeters long, form a straight or curved shape, swell and flatten after some time on the epidermal cell wall, and develop a structure known as "hyphodium" (Genre et al., 2005). These hyphodia stick to the root epidermis and penetrate the plant's outer layer. The exact molecular mechanism that mediates the linear apical growth into swollen and branched form is unknown (Bonfante, 2001).

Conversely, hyphodium development is initiated due to a pause in fungal growth that lasts for 4–6 h before the development of penetration hyphae (Genre et al., 2005). During this period, cells and tissues of plants get prepared for colonization. A gene coding for the protein necessary for SYM pathway–dependent expressions, ENOD11, has been reported from hyphodium (Chabaud et al., 2002). Many other genes involved in remodeling the cell wall and defense response are activated during the development of hyphodia the AM-specific structures needed for successful penetration (Bonfante, 2001; Genre et al., 2005).

These AMF-specific structures, also called the PPA (pre-penetration apparatus), are formed by the reorganization of the cytoplasm of epidermal cells needed for successful penetration. Contacted epidermis cells assemble and form an interface compartment for fungi to reside. At contact site, cytoplasm aggregats and forms a thick column that provides a route for hyphae across cell (Genre et al., 2005). All elements involved in this pathway are aggregated in the PPA: endoplasmic reticulum, secretory vesicles, and Golgi bodies (Genre et al., 2008). But the main factor involved is the nucleus that controls the movement across the contact site (Genre et al., 2005).

After the completion of the PPA, the fungus starts to regrow its hyphal tip penetrates the epidermal cells and goes along the track of the PPA. Meanwhile, PPA secretory vesicles are fused to form an invagination in the plant plasma membrane, and the perifungal membrane is assembled. The development of the perifungal membrane determines the symbiotic interface (Bonfante and Genre, 2010).

Arbusculated structures, considered sites of nutrient exchange, are produced within root cortical cells when intracellular hyphae get branched. The mechanism of development of arbuscules is not much known. Some genes involved in the development of arbuscules have recently been identified by reverse genetics (Paszkowski, 2006).

THE PROCESS OF FUNGAL ACCOMMODATION

The process of fungal accommodation into arbuscules requires substantial modification of cortical cells. Thin branches of arbuscules are covered by a perifungal membrane, called "periarbuscular membrane." These arbusculated cells are located where the interface gets more complex (Bonfante, 2001). The assembly of the periarbuscular interface requires the proliferation of Golgi apparatus, Golgi bodies, trans- Golgi network, and other secretory vesicles (Genre *et al.*, 2008; Pumplin and Harrison, 2009). Plastids are multiplied around the fungus, interconnected by stomules (Strack and Fester, 2006). Chromatin gets condensed, while the nucleolus and the nucleus gain size, and in turn, the transcription rate increases (Genre *et al.*, 2008). When PPA and arbuscules occupy sufficient cell volume, the vacuole is reorganized and reduces in size (Pumplin and Harrison, 2009). Arbuscules have an estimated life of 4–5 days. After completion of a cycle, cytoplasm retracts and walls of fungi are disintegrated in fine branches; this disintegration leads toward the development of a compact mass of hyphal cell wall. In the end, the fungus disappears, and the cells of the host regain their previous state (Bonfante, 2001).

MYCORRHIZAL FUNGI AND PLANT DISEASE MANAGEMENT

ENHANCED AVAILABILITY OF NUTRIENTS TO PLANTS

Plants with mycorrhizal association can compensate for root damage and tolerate pathogen activity (Azcon-Aguilar and Barea, 1992; Declerck *et al.*, 2002) due to improved nutrition and plant growth. As Declerck *et al.* (2002) described that the isolates of glomus species and *Galium proliferum* are able to enhance the uptake of phosphorus (P) in the shoot of banana plant. He also described that AMF are able to reduce the root infection caused by root rot fungus (*Cylindrocladiumspathiphylli*). On the contrary, a few other reports indicated that AMF could be a biocontrol agent (Grey *et al.*, 1989; Rempel and Bernier 1990; Boyetchko and Bernier, 1990).

Mycorrhizal fungi have many hosts and cooperate with them, but the main host is the most preferable. Grey *et al.* (1989) stated that the mycorrhizal-associated cultivar (WI2291) of barley showed a greater effect against the root rot pathogen of barley (*Bipolarissorokiniana*) compared with other cultivars (Harmal) without mycorrhizal association and also gave higher yield. Boyetchko and Tewari (1992) described that AMF species cause reduction in infection of common root rot of barley caused by *B. sorokiniana*. This effect shows that improved nutrition in the host by AMF also has a significant effect against the pathogen (Thompson and Wildermuth, 1989). Modified root exudation is the result of mycorrhizal colonization, which improves P uptake and reduces the severity of disease, and its effect has also been described for the control of wheat diseases (Graham and Menge, 1982). The initial mechanism of AMF is to enhance the P nutrition in the host.

IMPROVE TOLERANCE IN HOST PLANT AGAINST PATHOGEN

AMF enhance host plant's tolerance to pathogens without any yield reduction. This reduction in yield is basically due to increased photosynthetic ability of plant (Heike *et al.*, 2001; Abdalla and Abdel-Fattah, 2000Karajeh and Al-Raddad, 1999) and delay in early maturity of a plant. These factors reduce disease severity and yield loss (Heike *et al.*, 2001). For example, soyabean seedlings planted in soil infested with *Rhizoctoniasolani, F. solani, or M. phaseolina* show reduction in shoot and root weight and plant height as compared to plants grown in non infested soil (Zambolin and Schenck, 1983).

The colonization of *G. mosseae* does not affect disease incidence. However, plants with the mycorrhizal association have greater ability to tolerate the pathogen infection than those with no mycorrhizal association, as described by Karajeh and Al-Raddad (1999) using seedlings of olive and by Hwang (1988) using alfalfa infested with *Pythiumparoecandrum*. It is not recognized that mycorrhizal alfalfa has additional mechanisms or tolerates the effect of *P. paroecandrum*. The benefit of AMF to susceptible plants is observed until pathogen inoculum is present at the threshold level (Stewart and Pfleger, 1977). Improved P level not only enhances plants' fitness and vigor but also modifies the root exudation to change pathogen dynamics (Kaye *et al.*, 1984). The tolerance ability of plants depends upon the species of mycorrhizal fungi and their ability to enhance growth and nutrition, although some species of AMF stimulate defense mechanisms in plants and reduce pathogen penetration (Davis and Menge, 1981). Matsubara *et al.* (2000) recognized that there is a wide dissimilarity in the ability of Glomus species R10, *G. mosseae, G. fasciculatum*, and *Gi. Margarita* to improve the growth of asparagus and to tolerate the severity of *Helicobasidiummompa* (violet root rot).

CHANGES IN PATHOGEN INOCULUM

All the alterations in the rhizosphere community of microflora are beneficial for the host plants because they create a friendly environment for the proliferation of microflora that show an anti-effect against the pathogen. For example, *Pythium* and *Phytophthora* spp. Malajczuk and McComb (1979) show this effect in eucalyptus seedlings. AMF colonization produced unfavorable conditions for the pathogen *Phytophthora cinnamomi* in tomato plants, which results in inducing qualitative changes to the mycorhizosphere and limiting the germination of its sporangia (Meyer and Linderman, 1986).

MECHANISM OF COMPETITION

The AMF spores in soil are not known to compete for nutrients as spore reserves are utilized for survival, until root contact is achieved. The spores of AMF are not capable of competing with nutrients in soil until physical contact with host. After getting into host's cells, competition may occur for the photosynthetic material of the host, root space, and sites for infection (Smith and Read, 1997). This competition between pathogen and AMF causes physical elimination of the pathogen (Davis and Menge, 1980; Smith, 1988; Hussey and Roncadori, 1982). If AMF and the pathogen are simultaneously colonized, this simultaneous colonization will not give competitive edge for AMF to build inoculum (Daniels and Menge, 1980), because of higher growth rate of the pathogen as compared to AMF (Sempavalan *et al.*, 1995).

Competition is not considered as significant method to suppress pathogen, because sometimes pathogens are suppressed by non-colonized roots which were regarded as induced resistance by AMF (Pozo *et al.*, 1999).

BIOCHEMICAL AND PHYSIOLOGICAL CHANGES OF HOST

Uptake of P level is enhanced by the AMF colonization, which changes the composition of a phospholipid; as a result, the root membrane becomes permeable and reduces the leakage of the net amount of sugars, amino acids, and carboxylic acid in the rhizosphere (Graham and Menge, 1982; Ratnayake *et al.*, 1978; Schwab *et al.*, 1983). These changes limit the chemotactic effect of the pathogen on the plant and prevent pathogen entry. If *G. mosseae* is inoculated early, it decreases the colony formation of *Alternaria alternate;* no effect was observed on the inoculum of pathogenic soil when both organisms inoculated simultaneously (McAllister *et al.*, 1996). It is also demonstrated that the quantity and quality of exudation substances of roots are reduced due to alteration in the permeability of the membrane of host roots infection caused by *G. mosseae* (Graham *et al.*, 1981), which restricts the germination of the pathogen propagule indicating that inoculation time enhances the activity of biocontrol.

SYSTEMIC-INDUCED RESISTANCE

Systemic-induced resistance (SIR) is a mechanism activated by inoculation of the pathogen or by chemical treatment stimulating tolerance to or resistance against plant disease by increasing the physical or chemical barrier of the host plant. (Handelsman and Stabb, 1996; Kuc, 1995). It is suggested that inoculation of AMF into plants may stimulate the SIR mechanism as a biocontrol (Benhamou *et al.*, 1994; Brendan *et al.*, 1996; Trotta *et al.*, 1996).

In mycorrhizal plants, the SIR mechanism against the pathogen is described as systemic and localized resistance (Cordier *et al.*, 1998). After AMF colonization, the concentration of lignin is increased in the cell wall, limiting the spread of pathogens (Dehne and Schonbeck, 1979). With the help of a split root system, it is described that the tomato plant is protected against *Phytophthoraparasitica* by the use of *G. mosseae,* because the development of the pathogen is reduced and the deposition of phenolic compound and appositions of the cell wall and defense response given by the plant cell are increased (Cordier *et al.* 1998). They further illustrated the SIR reaction in mycorrhozal plants, especially pea plant. In SIR, the host wall becomes thick and containspathogenesis-related (PR-1) protein and pectins in non-mycorrhizal root areas. Only the invaded pea tissues contain PR-1 protein. This effect was noted in those root tissues that were infected with pathogen and without mycorrhizal association. Bodker *et al.* in 1998 reported that pea plants inoculated with *G. intraradices* showed increased resistance to *A. euteiches*. SIR phenomenon mediated by AMFs is known to play an important part in managing disease of potato dry rot. The inoculation of AMF against the disease prolongs the reproduction and growth phases and increases the potato storage duration. It is observed that PR proteins mediated by AMF play an important role in disease control (Liu *et al.*, 1995). Cotton plants that have challenged *Verticillium dahliae*and were inoculated with *G.versiforme*or *G.mosseae* produced ten different PRproteins compared with plants with no pathogen challenges. The growth of *V. dahliae* is restricted by PR proteins as they kill fungal conidia.

PHYTOANTICIPINS AND PHYTOALEXINS

Phytoalexins are produced in plants on the stimulation of microbes (Paxton, 1981), and phytoanticipins are stored in plants before any infection (VanEtten *et al.*, 1995). Pathogen produced phytoalexins in higher amounts than symbiotic organisms (Wyss *et al.*, 1991). It was observed that after colonization of AMF, the level of soluble phenolics of the plant—such as flavonoids or isoflavonoids, syringic, coumaric acids or coumaric, and lignin—was enhanced (Harrison and Dixon, 1993; Morandi, 1989, 1996). Some flavonoids may respond to mycorrhizal colonized plant roots (Harrison and Dixon, 1993; Morandi and Le-Quere, 1991; Volpin *et al.*, 1995). When *G.mosseae*is inoculated into tomato plants, it causes more resistance against the pathogen *Fusarium oxysporum*; increased activity of b-glucosidase and phenylalanine is observed in the roots of such plants compared withthose that are not inoculated (Dehne and Schonbeck, 1979). In mycorrhizal plants, addition of fertilizer does not affect the production of phytoalexins (Caron *et al.*, 1986b).

HYDROLASES

Now a days, defense-related genes are manipulated to induce AMF-mediated biocontrol in plants with mycorrhizal association (Blee and Anderson, 1996; Dumas-Gaudot *et al.*, 1996; Lambais and Mehdy, 1995; Pozo *et al.*, 2002). It shows that entry of mycorrhizal fungi into the roots of the host stimulates a weak local but active defense mechanism against the pathogen (Pozo *et al.*, 2002). A high positive correlation has been established between pathogen resistance and the activity level of glucanase (Graham and Graham, 1991).

ANTIBIOSIS

Unidentified microbial substances were produced by the mychorihizal spp. *Glomus intraradices*, by which the conidial germination of *F.oxysporum*was reduced (Filion

et al., 1999). Budi *et al.* (1999) isolated a Paenibacillus species strain from the mycorrhizal sphere of sorghum that was inoculated early with *G. mosseae* and caused asig-
nificant antagonistic effect against *P.parasitica.*

MYCORRHIZA FUNGI: ROLE IN SUSTAINABLE AGRICULTURE

With the steady increase in world population, the need to supply food to a large community has become a challenge, specifically requiring the agriculture sector to adopt
farming practices that ensure better yields and crop production. For the sustainability
of agricultural land and better crop production management, efforts are required to
comprehend the complex processes related to soil biological systems and agro-ecosystems (Pimentel *et al.*, 2005; Moonen and Bàrberi, 2008). In this regard, AMF have
proved quite useful. AMF bring about considerable interaction among host plants
and the nutrients obtained from minerals present in the soil, serving as a source of
interest for ecosystem engineers and biofertilizers manufacturers (Gianinazzi *et al.*,
1990; Gianinazzi and Vosatka, 2004; Fitter *et al.*, 2011).

Arbuscular mycorrhizal fungi represent 10% of biomass related to soil microorganisms; they establish symbiotic relationships that cover almost 80% of land plants
species and agronomic crops (Smith and Read, 2008). In plants, the exchange of nutrients is done by AMF by providing phosphorous as the replacement for carbon that has
been fixed as a result of photosynthesis (Hodge *et al.*, 2010; Bago *et al.*, 2000). AMF,
in a mutualistic symbiotic relationship with plants, serve to enhance agricultural productivity by considerably enhancing the growth of plants (Koide, 1991), enhancing the
efficiency of seeds (Shumway and Koide, 1994), and providing resistance to plants,
specifically against pathogens of roots and drought. Moreover, AMF play a direct role
in the ecosystem by manipulating the structure of plant communities and enhancing
soil quality by improving aggregation and organic content such as carbon.

Long-term food production permits the plants to efficiently use mineral elements
such as nitrogen and phosphorous by using biofertilizers such as the inoculation
of microorganisms (Alori *et al.*, 2017). Phosphorous deficiency could reduce plant
development since it is regarded as an important mineral. Arbuscular mycorrhizal
associations are the means by which many plants receive enough phosphorous from
the soil surrounding the root. The phosphorous level can be enhanced by the arbuscular mycorrhizal associations, mainly in P-deficient soils, contributing significantly
to plants' enhanced growth. With gel electrophoresis and ultra-cytochemistry techniques, alkaline phosphatase has been accepted as a valuable enzyme for AMF.

Cultivating oilseed crops like soybean (*Glycine max* L.) on a vast area of reclaimed
soil tends to increase yield, thereby contributing to agricultural sustainability.
Improved phosphorous absorption is due to interaction of AMF which results in
enhanced productivity of soybean (Abdel-Fattah and Asrar, 2012). Sulphur can also
be made available to host plants through AMF, that can improve the sulfur nutritional
status of host by affecting the plant sulfate transporters.

Irrespective of potassium's importance in plants, the critical role of AMF for the
uptake of potassium has barely been demonstrated. The bioavailability of potassium
in the plant is low, but relatively higher in the soil. In plants, the overexpression of
potassium transporter protein has been demonstrated in the roots of *Lotus japonicas*

colonized with AMF (Berruti *et al.*, 2015). These fungi enhance the micronutrient levels in crops, thus playing a significant role in agricultural biofortification. For example, zinc level has been greatly impacted by the interaction of AMF in various agriculture crops in varying environmental conditions (Lehmann *et al.*, 2014). In contrast, increased copper, manganese, and iron have been reported in AMFungi.

The bioavailability of phosphorous can significantly affect arbuscular mycorrhizal symbiosis, causing an environmental issue. Enhanced phosphorous concentration significantly reduces the establishment of mutualistic symbiosis by AMF. Besides phosphorous, nitrate also has a negative impact on AMF colonization. However, it has been reported that magnesium, iron, and calcium ions positively affect AMF (Nouri *et al.*, 2014; Berruti *et al.*, 2015).

Because ofthe presence of high-affinity phosphate transporter (PT) in fungi, the processes involved in the uptake of nutrients in AM interactions have been studied extensively. Plant gene is stimulated by AM associations that express inorganic proteins such as a phosphorous (pi) transporter. Besides the expression of the Pi gene in plants, AM is also involved in forming the structure and shape of the arbuscule and maintaining a symbiotic relationship (Yang *et al.*, 2012; Xie *et al.*, 2013). Studies have demonstrated that as a result of AM associations, two Pi transporters, MtPT4 and LjPT4, are expressed by *Medicagotruncatula* and *Lotus japonicas*, respectively. These transporters are present on the root tip of plants (Volpe *et al.*, 2015).

AMF associations have also discovered ammonium transporters besides Pi transporters. The transmission of inorganic nutrients occurswhen there is aperi arbuscular membrane surrounding the arbuscule. Arbuscular mycorrhizal associations contribute to the maintenance of the arbuscule and supply nutrients to the root cells by transferring inorganic phosphorous and ammonium through such ammonium transporters (Berruti *et al.*, 2015; Breuillin-Sessoms *et al.*, 2015). Sulphateis acquired from arbuscules usually using sulphate base transporters.

AMF play a significant role in the ecosystem by considerably controlling soil erosion, focusing mainly on the structure and aggregation of soil and returning the products obtained from photosynthesis into the methyl furans biomass (Vosátka *et al.*, 2012). Using their mycelial networks, AMF resist the growth of other organisms. They act as potent environment protectors by significantly increasing phytoremediation (Bhargava *et al.*, 2012; Meier *et al.*, 2012).

AMF can lower accumulations of heavy metals on the upper surface of plant parts, ensuring the maintenance of edibility in food crops. A considerable quantity of sugars and valuable elements, such as Zn, Mg, and antioxidants, have also been well demonstrated in associations of AMF (Albrechtova *et al.*, 2012). It has recently been reported that lipo-chitooligosaccharides, a bioactive compound, can contribute to the establishment of AM interactions; thus, it involves the AMF potential for the efficient increase in the productivity of agricultural crops.

It is the need of the hour to establish a synergistic relationship or co-inoculation with other beneficial microorganisms for the effective enhancement of crop productivity. For instance, co-inoculation of AMF with rhizobacteria (Babalola, 2010) improves plant growth greatly compared with that of the single inoculation. The combination of AMF with saprophytic fungi has been reported to positively impact crop productivity and food quality in onion crops, with the ultimate aim of achieving agricultural sustainability (Albrechtova *et al.*, 2012). Agricultural sustainability through AMF can explicitly be achieved by controlling eroded soils, enhancing plant-based approaches, such as phytoremediation, and increasing utilization of mycelial networks, which eradicate organisms that can harm agricultural crops.

AMF AND ENVIRONMENTAL HEALTH

AMF play an important part in the maintenance of the ecosystem by forming mutualistic symbiotic relationships with a wide variety of plants. AMF have been regarded as potent agents to facilitate a wide range of sustainability-related programs, such as in agricultural, restoration, and natural resources depletion, and global climate change contexts (Babikova *et al.*, 2013). The world is facing severe challenges in terms of food shortages, overconsumption of natural resources, and ongoing climate change problems (Grassini *et al.*, 2013). Due to the increasing daily risk of such challenges, the United Nations has put climate action and land management at the top of its development goals to achieve sustainability in those areas that rank top (UN, 2015).

Mycorrhizal associations play a significant role in maintaining the environment, providing considerable benefits to the ecosystem concerning the host plants (Field andPressel, 2018; Treseder, 2004). The mutualistic symbiotic interaction of AMF with plants has benefited plant more in terms of nutrient availability compared with non-mycorrhizal plants of the same species in the natural environment (Chen *et al.*, 2018). The performance, receptivity, and compatibility of mycorrhiza among crop plants are crucial factors in achieving sustainability in the agricultural system. Resource exchange among the symbionts is greatly dependent on abiotic factors (including temperature, CO_2, and soil nutrients, in either inorganic or organic forms) as well as biotic factors.

Distribution of AMF over vast areas in almost all ecosystems (Varga *et al.*, 2015), from Arctic regions to the Himalayans, has made them highly adaptive in all conditions which are harsh for their survival (Rosendahl *et al.*, 2009). Due to such adaptabilities, AMF have the potential to withstand or combat abiotic stresses such as heat, drought, and nutrient starvation (Bunn *et al.*, 2009).

A highly branched network of hyphae in AMF has made them potent in ameliorating soil structures by protecting the soil with soil compaction. Due to the three-dimensional matrix of AMF that provides cross-linkage meshes to the soil particles efficiently. Glomalin, a soil glycoprotein, has been identified by AMF, and it provides stable soil aggregation (Rillig, 2004; Singh *et al.*, 2013). Glomalin-related soil protein provides 2–5% of total organic carbon fraction. In addition, it is responsible for protecting carbon from (Rillig *et al.*, 2001; Wilson *et al.*, 2009). By considering the extensive mycelial network of AMF and its beneficial effect on the growth and maturation of plants and roots, it can be well utilized to efficiently protect soil from wind erosion (Gutjahr and Paszkowski, 2013).

AMF reduce nutrient leaching from the soil in a way that AMF tends to integrate with nutrients flux by regulating nutrient cycles (Cavagnaro *et al.*, 2015). Nutrient leaching refers to the draining of nutrients from the soil, resulting in loss of soil fertility and pollution of surface and groundwater. Efficient retention of nutrients in an ecosystem exhibits good retention and absorption capacity of nutrients through soil and root (Cameron K.C. *et al.*, 2013). Asarbuscular mycorrhizal associations prevent nutrient leaching from the soil, they act as a potent source of environmentalprotection and sustainability.

Afforestation and renaturationare methods throughwhichthe stability of eroded and degraded ground surfaces can be achieved. Young trees are significantly more susceptible to abiotic stresses such as heat, drought, and nutrient deficiencies due to their underdeveloped root system and inefficiency in accessing sufficient groundwater reserves. Hence, inoculation with AMF is to be practiced in trees before planting to overcome this critical stage faced by young trees. For example, the establishment of the Moroccan argan tree requires AMF to be inoculated in nurseries that exhibit the potential to reduce dry climate exposure and enhance growth conditions, thereby enhancing the potential for survival and fitness in the native plants on planting (Sellal *et al.*, 2017).

The mineral scavenging capacities of AMF involve the phytoremediation technique, which involves the plants removing harmful contaminants, such as heavy metals, from the soil and the water. Heavy metal–tolerant AMF and highly tolerant host plantsare required in bioremediation. AMFeither accumulate and cloister the toxic metal ions to protect the host plant from the pollutant or they deliver these metal ions as an essential mineral nutrient to the plant, which results in accumulation of heavy metal in the host (Leyval *et al.*, 2002; Göhre and Paszkowski, 2006). Plant production can be enabled in the first case; however, in the second case, phytoextraction tends to be done, in which plants can be harvested, and their destruction, thereafter, tends to eliminate toxic heavy metal accumulation from the site (Burns *et al.*, 1996; Khan *et al.*, 2000). A relatively large shoot biomass is required to amass a substantial amount of heavy metals.

RESEARCH ON MYCHORRIZA AND FUTURE PROSPECTS

Agricultural sustainability heavily relies on the minimum use of chemical fertilizers and pesticides, primarily inorganic (Philippot *et al.*, 2013; Babalola, 2014; Alori *et al.*, 2017). The main area of research involves selecting and cultivating such plant cultivars that can grow well amid the available biotic and abiotic resources in the soil. Selection can be made on the genetic bases, where plants become resistant enough to withstand abiotic stresses and overcome diseases by adopting different mechanisms.

In Genetically modified plants showing resistance to *Pectobacterium carotovorum,* an elevated lebel of acyl-homoserinelactonase protein is observed, which tends to break quorum-sensing compounds (Zeller *et al.*, 2013).

Another factor that could be considered for improving agricultural sustainability and AMFinvolves identifying and recognizing such features of plants that are adversely affected either by the emission of greenhouse gases produced by soil microorganisms or by the nutrient leaching, mainly of nitrogen. We could develop a new cultivar with modified characteristics to detect emissions and release to alleviate such problems. Knowledge about root exudations is not much available, so better and deeper insights into the mechanisms of such exudations are required to analyze new features for plant breeding and genetic engineering (González-López, 2013).

As far as the problem of climate change is concerned, an immense amount of research should be geared up to enhance AMF to provide resistance against environmental stressors. The key is to accumulate ample information about AMF to determine the mechanistic pathway that leads to the resistance and tolerance against environmental factors. Genomics approaches such as transcriptomics, proteomics, and metabolomics are helpful for the detection of gene expression in AMF plants about the nutrient uptake and dimmish abiotic stresses such as drought and resistance against diseases. For the efficient utilization of AMF to ensure agricultural sustainability, management of AMF and AMF inoculants present in the soil rhizosphere should be prioritized (Schwartz *et al.*, 2006).

Energy inputs and fertilizers to improve plant growth prominently can be manipulated by using AMF, so the mycorrhizal application should not be neglected in forthcoming agricultural sustainability and development policies. In order to reduce and circumvent the issues related to food safety, yield, and ecological protection in the coming years, AMF inoculants will become more pertinent since the detrimental effects of chemicals during agricultural practices will shortly increase to affect the environment. Studies about the aftereffect of interactions between the soil microorganisms in the rhizosphere having AMF should be verified to assess their impact on the environment as a practicable replacement for land reclamation and agriculture (González-López, 2013).

Following Rodriguez and Sanders (2015) for the efficient exploitation of AMF in order to achieve sustainability in agriculture, ecologists should encounter these four aspects:

- Gaining insight into the persistence and infection potential of inoculated AMF in the already present AMF community
- Developing enough awareness about the adaptabilities of non-indigenous AMF to the foreign environment
- The importance of genetic variations exists between the arbuscular mycorrhizal species of fungi and their potential effect on the growth of plants
- Ascertaining the impacts on crop productivity, either indirect or direct, by non-indigenous AMF.

REFERENCES

Abdalla ME, and Abdel-Fattah GM (2000). Influence of the endomycorrhizal fungus Glomus mosseae on the development of peanut pod rot disease in Egypt. *Mycorrhiza* 10:29–35.

Abdel-Fattah GM, and Asrar AWA (2012) Arbuscular mycorrhizal fungal application to improve growth and tolerance of wheat (*Triticum aestivum* L.) plants grown in saline soil. *Acta Physiol Plant* 34:267–277.

Akiyama K, Matsuzakiand KI, and Hayashi H (2005). Plant sesquiterpenes induce hyphal branching in arbuscular mycorrhizal fungi. *Nature* 435:824–827.

Albrechtova J, Latr A, Nedorost L, PoKluda R, Posta K, and Vosatka M (2012). Dual inoculation with mycorrhizal and saprotrophic fungi applicable in sustainable cultivation improves the yield and nutritive value of onion. *Sci World J* 2012: 8, Article ID 374091 https://doi.org/10.1100/2012/374091.

Alexopoulos CJ, and Mims CW (1979). *Introducción a la micología* (No. QK603. A4318 3A ED.). Eudeba.

Alizadeh O (2011). Mycorrhizal symbiosis. *Adv Stud Biol* 6:273–281.

Alori ET, Dare MO, and Babalola OO (2017). Microbial inoculants for soil quality and plant health. *Sust Agric Rev*:281–307.

Assigbetse K, Gueye M, Thioulouseand J, and Duponnois R (2005). Soil bacterial diversity responses to root colonization by an ectomycorrhizal fungus are not root-growth-dependent. *Microb Ecol* 50:350–359.

Azcon-Aguilar C, and Barea JM (1992). Interactions between mycorrhizal fungi and other rhizosphere microorganisms. In: Allen MF (ed.) *Mycorrhizal Functioning: An Integrative Plant-Fungal Process*. New York: Chapman and Hall, pp. 163–198.

Babalola OO (2010). Beneficial bacteria of agricultural importance. *Biotechnol Lett* 32:1559–1570.

Babalola OO (2014). Does nature make provision for backups in the modification of bacterial community structures? *Biotechnol Genet Eng Rev* 30:31–48.

Babikova Z, Gilbert L, Bruce TJ, Birkett M, Caulfield JC, Woodcock C, . . . Johnson D (2013). Underground signals carried through common mycelial networks warn neighbouring plants of aphid attack. *Ecol Lett* 16(7):835–843.

Bago B, Pfeffer PE, and Shachar-Hill Y (2000). Carbon metabolism and transport in arbuscular mycorrhizas. *Plant Physiol* 124:949e957.

Bedini S, Cristani C, Avio L, Sbrana C, Turriniand A, and Giovannetti M (2008). *Influence of organic farming on arbuscular mycorrhizal fungal populations in a Mediterranean agro-ecosystem*. Modena: Organic World Congress.

Benhamou N, Fortin JA, Hamel C, St-Arnaud M, and Shatilla A (1994). Resistance responses to mycorrhizal Ri T-DNA transformed carrot roots to infection by *Fusarium oxysporum* f. sp. chrysanthemi. *Phytopathology* 84:958–968.

Berruti A, Lumini E, BalestRini R, and Bianciotto V (2015). Arbuscular mycorrhizal fungi as natural biofertilizers: let's benefit from past successes. *Front Microbiol* 6:1559.

Besserer A, Puech-Pagès V, Kiefer P, Gomez-Roldan V, Jauneau A, Roy S, Portais JC, Roux C, Bécardand G, and Séjalon-Delmas N (2006). Strigolactones stimulate arbuscular mycorrhizal fungi by activating mitochondria. *PLoS Biol* 4:e226.

Bhargava A, CarMona FF, Bhargava M, and Srivastava S (2012). Approaches for enhanced phytoextraction of heavy metals. *J Environ Manag* 105:103–120.

Blee KA, and Anderson AJ (1996). Defense-related transcript accumulation in Phaseolus vulgaris L. colonized by the arbuscular mycorrhizal fungus Glomus intraradices Schenck and Smith. *Plant Physiol* 110:675–688.

Bodker L, Kjoller R, and Rosendahl S (1998). Effect of phosphate and the arbuscular mycorrhizal fungus Glomus intraradices on disease severity of root rot of peas (*Pisum sativum*) caused by Aphanomyces euteiches. *Mycorrhiza* 8:169–174.

Bonfante P (2001). *At the interface between mycorrhizal fungi and plants: The structural organization of cell wall, plasma membrane and cytoskeleton, fungal associations*. Berlin, Heidelberg: Springer, pp. 45–61.

Bonfante P, and Genre A (2010). Mechanisms underlying beneficial plant—fungus interactions in mycorrhizal symbiosis. *Nat Commun* 1:1–11.

Boyetchko SM, and Tewari JP (1992). Interaction of VA mycorrhizal fungi with the common root rot of barley. *Proc Int Common Root Rot Workshop, Saskatoon, Saskatchewan*:166–169.

Brendan NA, Hammerschmidt R, and Safir GR (1996). Postharvest suppression of potato dry rot (*Fusarium sambucinum*) in prenuclearminitubers by arbuscular mycorrhizal fungal inoculum. *Am Potato J* 73:509–515.

Breuillin-Sessoms F, Floss DS, Gomez SK, Pumpli NN, Ding Y, LeVesque-Tremblay V, Noar RD, Daniels DA, Bravo A, and Eaglesham JB (2015). Suppression of arbuscule degeneration in Medicagotruncatula phosphate transporter4 mutants is dependent on the ammonium transporter 2 family protein AMT2; 3. *Plant Cell* 27:1352–1366.

Bucher M, Wegmueller S, and Drissner D (2009). Chasing the structures of small molecules in arbuscular mycorrhizal signaling. *Curr Opin Plant Biol* 12(4):500–507.

Budi SW, van Tuinen D, Martinotti G, and Gianinazzi S (1999). Isolation from the Sorghum bicolor mycorrhizosphere of a bacterium compatible with arbuscular mycorrhiza development and antagonistic towards soilborne fungal pathogens. *Appl Environ Microbiol* 65:5148–5150.

Bunn R, Lekberg Y, and Zabinski C (2009). Arbuscular mycorrhizal fungi ameliorate temperature stress in thermophilic plants. *Ecology* 90(5):1378–1388.

Burns RG, Rogers SL, and McGhee I (1996). Remediation of inorganics and organics in industrial and urban contaminated soils. In: Naidu R, Kookana RS, Oliver DP, and McLaughlin MJ (eds.) *Contaminants and the Soil Environment in the Australia Pacific Region.* London: Kluwer Academic Publishers, pp. 361–410.

Cameron KC, Di H, and Moir JL (2013). Nitrogen losses from the soil/plant system: A review. *Ann. Appl. Biol* 162:145–173.

Caron M, Fortin JA, and Richard C (1986b). Effect of phosphorus concentration and Glomus intraradices on Fusarium crown and root rot of tomatoes. *Phytopathology* 76:942–946.

Cavagnaro TR, Bender SF, Asghari HR, and van der Heijden MGA (2015). The role of arbuscular mycorrhizas in reducing soil nutrient loss. *Trends Plant Sci* 20:283–290.

Chabaud M, Venard C, Defaux-Petras A, Bécardand G, and Barker D (2002). Targeted inoculation of Medicagotruncatula in vitro root cultures reveals MtENOD11 expression during early stages of infection by arbuscular mycorrhizal fungi. *New Phytologist* 156:265–273.

Chen M, Arato M, Borghi L, Nouri E, and Reinhardt D (2018). Beneficial services of arbuscular mycorrhizal fungi—From ecology to application. *Front Plant Sci* 9:1270.

Cordier C, Pozo MJ, Barea JM, Gianinazzi S, and Gianinazzi Pearson V (1998). Cell defense responses associated with localized and systemic resistance to Phytophthora induced in tomato by an arbuscular mycorrhizal fungus. *Mol Plant-Microb Interact* 11:1017–1028.

Dalpé Y, and Monreal M (2004). Arbuscular mycorrhiza inoculum to support sustainable cropping systems. *Crop Manage* 3:1–11.

Daniels BA, and Menge JA (1980). Hyperparasitization by *Anquillosporapseudo longissima* and *Humicola fuscoatra* of vesicular-arbuscular mycorrhizal fungi *Glomus epigaeus* and *Glomus fasciculatus. Phytopathology* 70:584–588.

Davis RM, and Menge JA (1980). Influence of Glomus fasciculatus and soil phosphorus on Phytophthora root rot of citrus. *Phytopathology* 70:447–452.

Davis RM, and Menge JA (1981). Phytophthoraparasitica inoculation and intensity of vesicular-arbuscular mycorrhizae in citrus. *New Phytol* 87:705–715.

Declerck S, Risede JM, Ruflikiri G, and Delvaux B (2002). Effects of arbuscular mycorrhizal fungi on severity of root rot of bananas caused by *Cylindrocladium spathiphylli. Plant Pathol* 51:109–115.

Dehne HW, and Schonbeck F (1979). The influence of endotrophic mycorrhiza on plant diseases. II. Phenol metabolism and lignification Fusarium oxysporum. Untersuchungenzum Einfluss der endotrophen Mycorrhiza auf Pflanzenkrankheiten. II. Phenolstoffwechsel und Lignifizierung. *Phytopathol-Z* 95:210–216.

Dumas-Gaudot E, Slezack S, Dassi B, Pozo MJ, Gianinazzi-Pearson V, and Gianinazzi S (1996). Plant hydrolytic enzymes (chitinases and b-1, 3-glucanases) in root reactions to pathogenic and symbiotic microorganisms. *Plant Soil* 185:211–221.

Duponnois R, and Plenchette C (2003). A mycorrhiza helper bacterium enhances ectomycorrhizal and endomycorrhizal symbiosis of Australian Acacia species. *Mycorrhiza* 13:85–91.

Field KJ, and Pressel S (2018). Unity in diversity: Structural and functional insights into the ancient partnerships between plants and fungi. *New Phytologist* 220(4):996–1011.

Filion M, St. Arnaud M, and Fortin JA (1999). Direct interaction between the arbuscular mycorrhizal fungus Glomus intraradices and different rhizosphere microorganisms. *New Phytol* 141:525–533.

Finlay RD (2008). Ecological aspects of mycorrhizal symbiosis: With special emphasis on the functional diversity of interactions involving the extra-radical mycelium. *Journal of Experimental Botany* 59:1115–1126.

Fitter AH, Helgason T, and Hodge A (2011). Nutritional exchanges in the arbuscular mycorrhizal symbiosis: Implications for sustainable agriculture. *Fungal Biol Rev* 25:68e72.

Frank B (2005). On the nutritional dependence of certain trees on root symbiosis with below-ground fungi (an English translation of AB Frank's classic paper of 1885). *Mycorrhiza* 15:267–275.

Genre A, Chabaud M, Faccio A, Barkerand DG, and Bonfante P (2008). Prepenetration apparatus assembly precedes and predicts the colonization patterns of arbuscular mycorrhizal fungi within the root cortex of both Medicagotruncatula and Daucuscarota. *The Plant Cell* 20:1407–1420.

Genre A, Chabaud M, Timmers T, Bonfanteand P, and Barker DG (2005). Arbuscular mycorrhizal fungi elicit a novel intracellular apparatus in Medicagotruncatula root epidermal cells before infection. *The Plant Cell* 17:3489–3499.

Gianinazzi S, Trouvelot A, and Gianinazzi-Pearson V (1990). Role and use of mycorrhizas in horticultural crop production. *Adv Hortic Sci* 4:25e30.

Gianinazzi S, and Vosatka M (2004). Inoculum of arbuscular mycorrhizal fungi for production systems, science meets business. *Can J Bot* 82:1264e1271.

Göhre V, and Paszkowski U (2006). Contribution of the arbuscular mycorrhizal symbiosis to heavy metal phytoremediation. *Planta* 223:1115–1122.

González-López J (2013). *Beneficial plant-microbial interactions: Ecology and applications.* Boca Raton: CRC Press.

Graham JH, Leonard RT, and Menge JA (1981). Membrane mediated decrease in root exudation responsible for inhibition of vesicular-arbuscular mycorrhiza formation. *Plant Physiol* 68:548–552.

Graham JH, and Menge JA (1982). Influence of vesicular-arbuscular mycorrhizae and soil phosphorus on take-all disease of wheat. *Phytopathology* 72:95–98.

Graham TL, and Graham MY (1991). Cellular coordination of molecular responses in plant defense. *Mol Plant-Microb Interact* 4:415–422.

Grassini P, Eskridge KM, and Cassman KG (2013). Distinguishing between yield advances and yield plateaus in historical crop production trends. *Nat Commun* 4:2918.

Grey WE, van Leur JAG, Kashour G, and El-Naimi M (1989). The interaction of vesicular-arbuscular mycorrhizae and common root rot (*Cochliobolus sativus*) in barley. *Rachis* 8:18–20.

Gutjahr C, and Paszkowski U (2013). Multiple control levels of root system remodelling in arbuscular mycorrhizal symbiosis. *Front Plant Sci* 4:204.

Hacskaylo E (1972). Mycorrhiza: The ultimate in reciprocal parasitism? *BioScience* 22:577–583.

Handelsman J, and Stabb EV (1996). Biocontrol of soilborne plant pathogens. *Plant Cell* 8:1855–1869.

Harley JL (1984). *The mycorrhizal associations, cellular interactions.* Berlin, Heidelberg: Springer, pp. 148–186.

Harley JL (1969). *The biology of mycorrhiza,* 2nd ed. London: Leonard Hill.

Harrison MJ, and Dixon RA (1993). Isoflavonoid accumulation and expression of defense gene transcripts during the establishment of VA mycorrhizal associations in roots of Medicagotruncatula. *Mol Plant-Microb Interact* 6:643–654.

Hazledine S, Sun J, Wysham D, Downie JA, Oldroydand GE, and Morris RJ (2009). Nonlinear time series analysis of nodulation factor induced calcium oscillations: Evidence for deterministic chaos? *PLoS One* 4:e6637.

Heike G, von Alten H, and Poehling HM (2001). Arbuscular mycorrhiza increased the activity of a biotrophic leaf pathogen: Is a compensation possible? *Mycorrhiza* 11:237–243.

Hodge A, Helgason T, and Fitter A (2010). Nutritional ecology of arbuscular mycorrhizal fungi. *Fungal Ecol* 3:267e273.

Hussey RS, and Roncadori RW (1982). Vesicular-arbuscular mycorrhizae may limit nematode activity and improve plant growth. *Plant Dis* 66:9–14.

Hwang SF (1988). Effect of VA mycorrhizae and metalaxyl on growth of alfalfa seedlings in soils from fields with "alfalfa sickness" in Alberta. *Plant Dis* 72:448–452.

Janerette CA (1991). An introduction to mycorrhizae. *The American Biology Teacher* 13–19.

Johansson JF, Pauland LR, and Finlay RD (2004). Microbial interactions in the mycorrhizosphere and their significance for sustainable agriculture. *FEMS Microbiology Ecology* 48:1–13.

Karajeh M, and Al-Raddad A (1999). Effect of VA mycorrhizal fungus (Glomus mosseae Gerd & Trappe) on Verticillium dahlia Kleb of olive. *Dirasat Agric Sci* 26:338–341.

Kaye JW, Pfleger FL, and Stewart EL (1984). Interaction of *Glomus fasciculatum* and *Pythiumultimum* on greenhouse-grown poinsettia. *Can J Bot* 62:1575–1579.

Khan AG, Kuek C, Chaudhry TM, Khoo CS, and Hayes WJ (2000). Role of plants, mycorrhizae and phytochelators in heavy metal contaminated land remediation. *Chemosphere* 41:197–207.

Koide RT (1991). Nutrient supply, nutrient demand and plant response to mycorrhizal infection. *New Phytol* 117:365e386.

Kosuta S, Chabaud M, Lougnon G, Gough C, Dénarié J, Barkerand DG, and Bécard G (2003). A diffusible factor from arbuscular mycorrhizal fungi induces symbiosis-specific MtENOD11 expression in roots of Medicagotruncatula. *Plant Physiology* 131:952–962.

Kosuta S, Hazledine S, Sun J, Miwa H, Morris RJ, Downieand JA, and Oldroyd GE (2008). Differential and chaotic calcium signatures in the symbiosis signaling pathway of legumes. *Proceedings of the National Academy of Sciences* 105:9823–9828.

Kuc J (1995). Systemic induced resistance. *Asp Appl Biol* 42:235–242.

Kuhn H, Küsterand H, and Requena N (2010). Membrane steroid-binding protein 1 induced by a diffusible fungal signal is critical for mycorrhization in Medicagotruncatula. *New Phytologist* 185:716–733.

Lambais MR, and Mehdy MC (1995). Differential expression of defense-related genes in arbuscular mycorrhiza. *Can J Bot* 73:S533–S540.

Lehmann A, Veresoglou SD, Leifheit EF, Rillig MC (2014) Arbuscular mycorrhizal influence on zinc nutrition in crop plants—a meta-analysis. *Soil Biol Biochem* 69:123–131.

Leyval C, Joner EJ, del Val C, and Haselwandter K (2002). Potential of arbuscular mycorrhizal fungi for bioremediation. In: Gianinazzi S, Schiüepp H, Barea JM, and Haselwandter K (eds.) *Mycorrhizal Technology in Agriculture*. Basel: Birkhäuser.

Liu RJ, Liu HF, Shen CY, and Chiu WF (1995). Detection of pathogenesis-related proteins in cotton plants. *Physiol Mol Plant Pathol* 47:357–363.

Malajczuk N, and McComb AJ (1979). The microflora of unsuberized roots of Eucalyptus calophylla R. Br. and Eucalyptus marginata Donn ex Sm. seedlings grown in soil suppressive and conducive to Phytophthora cinnamomi Rands. I. Rhizosphere bacteria, actinomycetes and fungi. *Aust J Bot* 27:235–254.

Maronek DM, Hendrixand JW, and Stevens CD (1981). Fertility-mycorrhizal-isolate interactions in production of containerized pin oak seedlings. *Scientia Horticulturae* 15:283–289.

Martin F, and Nehls U (2009). Harnessing ectomycorrhizal genomics for ecological insights. *Curr Opin Plant Biol* 12:508–515.

Matsubara Y, Kayukawa Y, Yano M, and Fukui H (2000). Tolerance of asparagus seedlings infected with arbuscular mycorrhizal fungus to violet root rot caused by *Helicobasidiummompa. J Jpn Soc Hortic Sci* 69:552–556.

McAllister CB, GarcIA-Gairido JM, Garcia-Romera I, Godeas A, and Ocampo JA (1996). In vitro interactions between *Alternaria alternata, Fusarium equiseti* and *Glomus mosseae. Symbiosis* 20:163–174.

Meier S, Borie F, Bola N, and Cornejo P (2012). Phytoremediation of metal polluted soils by arbuscular mycorrhizal fungi. *Crit Rev Environ Sci Technol* 42:741–775.

Melin E (1948). Recent advances in the study of tree mycorrhiza. *Trans Br Mycol Soc* 30:92–99.

Meyer JR, and Linderman RG (1986). Response of subterranean clover to dual inoculation with vesicular-arbuscular mycorrhizal fungi and a plant growth promoting bacterium Pseudomonas putida. *Soil Biol Biochem* 18:185–190.

Mitra RM, Gleason CA, Edwards A, Hadfield J, Downie JA, Oldroydand GE, and Long SR (2004). A Ca^{2+}/calmodulin-dependent protein kinase required for symbiotic nodule development: Gene identification by transcript-based cloning. *Proc Natl Acad Sci* 101:4701–4705.

Moonen C, Bàrberi P (2008). Functional biodiversity: An agro-ecosystem approach. *Agric Ecosyst Environ* 127:7e21.

Morandi D (1989). Effect of xenobiotics on endomycorrhizal infection and isoflavonoid accumulation in soybean roots. *Plant Physiol Biochem* 27:697–701.

Morandi D (1996). Occurrence of phytoalexins and phenolic compounds in endomycorrhizal interactions, and their potential role in biological control. *Plant Soil* 185:241–251.

Morandi D, and Le-Quere JL (1991). Influence of nitrogen on accumulation of isosojagol (a newly detected coumestan in soybean) and associated isoflavonoids in roots and nodules of mycorrhizal and non-mycorrhizal soybean. *New Phytol* 117:75–79.

Nouri E, BreUillin-Sessoms F, Feller U, and Reinhardt D (2014). Phosphorus and nitrogen regulate arbuscular mycorrhizal symbiosis in *Petunia hybrida*. *PLoS One* 9:e90841.

Paszkowski U (2006). A journey through signaling in arbuscular mycorrhizal symbioses 2006. *New Phytologist* 172:35–46.

Paxton JD (1981). Phytoalexins—a working redefinition. *Phytopathol Z* 101:106–109.

Philippot L, Raaijmakers JM, Lemanceau P, and Van Der Putten WH (2013). Going back to the roots: The microbial ecology of the rhizosphere. *Nat Rev Microbiol* 11:789–799.

Pimentel D, Hepperly P, Hanson J, Douds D, Seidel R (2005). Environmental, energetic and economic comparisons of organic and conventional farming systems. *Bioscience* 55:573e582.

Pozo MJ, Azcon-Aguilar C, Dumas-Gaudot E, and Barea JM (1999). b-1,3-glucanase activities in tomato roots inoculated with arbuscular mycorrhizal fungi and/or Phytophthora-parasitica and their possible involvement in bioprotection. *Plant Sci* 141:149–157.

Pozo MJ, CordiEr C, Dumas-Gaudot E, Gianinazzi S, Barea JM, and Azcon-Aguilar C (2002). Localized versus systemic effect of arbuscular mycorrhizal fungi on defence responses to Phytophthora infection in tomato plants. *J Exp Bot* 53:525–534.

Pumplin N, and Harrison MJ (2009). Live-cell imaging reveals periarbuscular membrane domains and organelle location in Medicagotruncatula roots during arbuscular mycorrhizal symbiosis. *Plant physiology* 151:809–819.

Ramanankierana N, Ducousso M, Rakotoarimanga N, Prin Y, Thioulouse J, Randrianjohany E, Ramaroson L, Kisa M, Galianaand A, and Duponnois R (2007). Arbuscular mycorrhizas and ectomycorrhizas of *Uapacabojeri* L. (*Euphorbiaceae*): Sporophorediversity, patterns of root colonization, and effects on seedling growth and soil microbial catabolic diversity. *Mycorrhiza* 17:195–208.

Ratnayake M, Leonard RT, and Menge JA (1978). Root exudation in relation to supply of phosphorus and its possible relevance to mycorrhiza formation. *New Phytol* 81:543–552.

Rempel CB, and Bernier CC (1990). Glomus intraradices and Cochliobolussativus interactions in wheat grown under two moisture regimes. *Can J Plant Pathol* 12:338, [Abstr].

Rillig MC (2004). Arbuscular mycorrhizae, glomalin, and soil aggregation. *Can J Soil Sci* 84:355–363.

Rillig MC, Wright SF, Nichols KA, Schmidt WF, and Torn MS (2001). Large contribution of arbuscular mycorrhizal fungi to soil carbon pools in tropical forest soils. *Plant Soil* 233:167–177.

Rodriguez A, Sanders IR (2015). The role of community and population ecology in applying mycorrhizal fungi for improved food security. *ISME J* 9:1053–1061.

Rosendahl S, McGee P, and Morton JB (2009). Lack of global population genetic differentiation in the arbuscular mycorrhizal fungus *Glomus mosseae* suggests a recent range expansion which may have coincided with the spread of agriculture. *Mol Ecol* 18:4316–4329.

Schwab SM, Leonard RT, and Menge JA (1983). Quantitative and qualitative comparison of root exudates of mycorrhizal and non-mycorrhizal plant species. *Can J Bot* 62:1227–1231.

Schwartz MW, Hoeksema JD, Gehring CA, Johnson NC, Klironomos JN, Abbott LK, and Pringle A (2006). The promise and the potential consequences of the global transport of mycorrhizal fungal inoculum. *Ecol Lett* 9(5):501–515.

Sellal Z, Ouazzani Touhami A, Mouden N, El Ouarraqi M, Selmaoui K, Dahmani J, et al (2017). Effect of an endomycorrhizal inoculum on the growth of Argan tree. *Int J Environ Agric Biotechnol* 2:928–939.

Selosse MA, and Roy M (2009). Green plants that feed on fungi: Facts and questions about mixotrophy. *Trends Plant Sci* 14:64–70.

Sempavalan J, Wheeler CT, and Hooker JE (1995). Lack of competition between Frankia and Glomus for infection and colonization of roots of *Casuarina equisetifolia*. *New Phytol* 130:429–436.

Shumway DL, and Koide RT (1994). Reproductive responses to mycorrhizal colonization of Abutilon theophrasti Medic. Plants grown for two generations in the field. *New Phytol* 128:219e224.

Singh PK, Singh M, and Tripathi BN (2013). Glomalin: An arbuscular mycorrhizal fungal soil protein. *Protoplasma* 250:663–669.

Smith GS (1988). The role of phosphorus nutrition in interactions of vesicular-arbuscular mycorrhizal fungi with soilborne nematodes and fungi. *Phytopathology* 78:371–374.

Smith SE, and Read DJ (1997). *Mycorrhizal Symbiosis*, 2nd ed. Toronto: Academic Press.

Smith SE, and Read DJ (2008). *Mycorrhizal Symbiosis*, 3rd ed. Amsterdam, The Netherlands: Academic Press.

Smith SE, and Read DJ (2010). *Mycorrhizal Symbiosis*. Amsterdam: Academic Press. Elsevier Ltd.

Stewart EL, and Pfleger FL (1977). Development of poinsettia as influence by endomycorrhizae, fertilizer, and root rot pathogens, *Pythiumultimum* and *Rhizoctoniasolani*. *Flor Rev* 159:79–80.

Strack D, and Fester T (2006). Isoprenoid metabolism and plastid reorganization in arbuscular mycorrhizal roots. *New Phytologist* 172:22–34.

Thompson JP, and Wildermuth GB (1989). Colonization of crop and pasture species with vesicular-arbuscular mycorrhizal fungi and negative correlation with root infection by *Bipolarissorokiniana*. *Can J Bot* 69:687–693.

Tollot M, Wong Sak Hoi J, Van Tuinen D, Arnould C, Chatagnier O, Dumas B, Gianinazzi-Pearsonand V, and Seddas PM (2009). An STE12 gene identified in the mycorrhizal fungus *Glomus intraradices* restores infectivity of a hemibiotrophic plant pathogen. *New Phytologist* 181:693–707.

Treseder KK (2004). A meta-analysis of mycorrhizal responses to nitrogen, phosphorus, and atmospheric CO_2 in field studies. *New Phytologist* 164(2):347–355.

Trotta A, Varese GC, Gnavi E, Fusconi A, Sampo S, and Berta G (1996). Interactions between the soilborne pathogen *Phytophthora nicotianae* var. parasitica and the arbuscular mycorrhizal fungus *Glomus mosseae* in tomato plants. *Plant Soil* 185:199–209.

Umehara M, Hanada A, Yoshida S, Akiyama K, Arite T, Takeda-Kamiya N, Magome H, Kamiya Y, Shirasuand K, and Yoneyama K (2008). Inhibition of shoot branching by new terpenoid plant hormones. *Nature* 455:195–200.

UN (2015). Resolution adopted by the general assembly, transforming our world: The 2030 Agenda for Sustainable Development, 70th session, agenda items 15 and 16. Retrieved from www.un.org/ga/search/view_doc.asp?symbol=A/RES/70/1&Lang=E.

Van Etten HD, Mansfield JW, Bailey JA, and Farmer EE (1995). Two classes of plant antibiotics: Phytoalexins versus "phytoanticipins." *Plant Cell* 6:1191–1192.

Varga S, Finozzi C, Vestberg M, and Kytoviita MM (2015). Arctic arbuscular mycorrhizal spore community and viability after storage in cold conditions. *Mycorrhiza* 25:335–343.

Volpe V, Giovannetti M, Sun XG, Fiorilli V, and Bonfante P (2015). The phosphate transporters LjPT4 and MtPT4 mediate early root responses to phosphate status in non mycorrhizal roots. *Plant Cell Environ* 39:660–670.

Volpin H, Phillips DA, Okon Y, and Kapulnik Y (1995). Suppression of an isoflavonoidphytoalexin defense response in mycorrhizal alfalfa roots. *Plant Physiol* 108:1449–1454.

Vosátka M, Látr A, Gianinazzi S, and Albrechtová J (2012). Development of arbuscular mycorrhizal biotechnology and industry: Current achievements and bottlenecks. *Symbiosis* 58:29–33.

Wang B, and Qiu YL (2006). Phylogenetic distribution and evolution of mycorrhizas in land plants. *Mycorrhiza* 16:299–363.

Wilson GWT, Rice CW, Rillig MC, Springer A, and Hartnett DC (2009). Soil aggregation and carbon sequestration are tightly correlated with the abundance of arbuscular mycorrhizal fungi: Results from long-term field experiments. *Ecol Lett* 12:452–461.

Wyss P, Boller T, and Wiemken A (1991). Phytoalexin response is elicited by a pathogen (Rhizoctoniasolani) but not by a mycorrhizal fungus (Glomus mosseae) in soybean roots. *Experientia* 47:395–399.

Xie X, Huang W, Liu F, TaNg N, Liu Y, Lin H, and Zhao B (2013). Functional analysis of the novel mycorrhiza-specific phosphate transporter AsPT1 and PHT1 family from *Astragalussinicus* during the arbuscular mycorrhizal symbiosis. *New Phytol* 198:836–852.

Yang SY, Grønlund M, Jakobsen I, Grotemeyer MS, Rentsch D, MiyAo A, Hirochika H, Kumar CS, Sundaresan V, and Salamin N (2012). Nonredundant regulation of rice arbuscular mycorrhizal symbiosis by two members of the phosphate transporter1 gene family. *Plant Cell* 24:4236–4251.

Zambolin L, and Schenck NC (1983). Reduction of the effects of pathogenic root-infecting fungi on soybean by the mycorrhizal fungus, Glomus mosseae. *Phytopathology* 73:1402–1405.

Zeller SL, Kalinina O, and Schmid B (2013). Costs of resistance to fungal pathogens in genetically modified wheat. *J Plant Ecol* 6:92–100.

12 Multiomics Approaches and Plant–Fungal Interactions

Qaiser Shakeel, Rabia Tahir Bajwa, Guoqing Li,
Yang Long, Mingde Wu, Jing Zhang
and Ifrah Rashid

CONTENTS

INTRODUCTION

The ecology of plant diseases is multidisciplinary and depends on many domains, such as pathogen ecology, pathogen epidemiology, host plant physiology, and plant genetics, to elucidate a critical observation of infected plants. Researchers have observed for a century that pathogenic microorganisms, including fungi, viruses, bacteria, and oomycetes, induce disease dynamics in susceptible plants under suitable environmental conditions (Agrios 2005; Crandall *et al.* 2020). Most phytopathogens form antagonistic interactions with their host plants in natural ecosystems,

DOI: 10.1201/9781003162742-12

hence, plant disease is a significant force regulating plant populations (Burdon 1991; Jeger *et al.* 2014; Bever *et al.* 2015). Plant health is vital to maintain plant production in agricultural systems, including agro-ecosystems and forest plantations, with farmers and land managers having a priority of actively controlling plant diseases. However, the management and understanding of such ecosystems are conventionally based on a reductionist technique, i.e., observing the individual plant–microbe interactions associated with a particular phytopathology compared with the complicated ecological interactions among populations of microorganisms, host plants, and the environment. Over the past few decades, technological improvements have led to a unique and broader technique for understanding ecology of phyto-pathogens, enabling a more holistic method for analyzing pathogenesis and its fundamental processes.

Fungi are one of the most prevalent phyto-pathogens, and the incidences of plant–fungal infections are increasing day by day. It is important to understand in depth the mechanism of interaction of a fungus with its host plant. An interaction of plant–pathogen is required for pathogenesis. Several technologies help elucidate the plant–fungal interaction in terms of pathogens causing phyto-diseases. Such technologies include multiomics techniques, which allow for a thorough investigation of plants and features of the pathogen across the genotype–phenotype range. In-depth understanding of such traits has aided in identifying mechanisms of plant–fungal interaction considering the ecology of the host plant defense system and creating more efficient strategies (from plant breeding to cultural management practices) to limit the spread of phyto-diseases. These advancements in omics-based approaches for the ecology of fungal phyto-diseases are especially significant in the context of the present rapidly changing environment. Adaptation and minimizing the impact of emerging biotic and abiotic stresses are crucial for maintaining viable agricultural ecosystems and healthy terrestrial systems. Globalization, for instance, and the movement of humans, plants, and microorganisms throughout the world may significantly alter environmental diversity and ecology.

Invasive fungal infections and soil-borne pathogens, particularly with a broad range of plant hosts, can severely damage millions of terrestrial trees (Dell and Malajczuk 1989; Rizzo *et al.* 2002). Fungal pathogens and their habitats are altering due to climate change and ecological incursions (Gilbert and Parker 2010; Santini and Ghelardini 2015). For example, changing temperatures and the length and frequency of weather conditions within a particular duration cause the fast development of fungal diseases or abiotic stresses that can damage host plants (Santini and Ghelardini 2015).

However, in its initial phases, this combination has a broad potential for making basic and functional discoveries regarding the ecology of fungal phyto-diseases. To manage fungal phyto-diseases, it is necessary to understand the vulnerability of the phytopathogenic fungus, expression of the phytopathogenic fungal community, pathways of volatilomic and metabolomics cascades occurring due to fungal infection, and spectranomic foliage properties providing several opportunities for the intrusion to be altered and adjusted for management of phyto-diseases. In this chapter, such opportunities are discussed in detail with an overview of multiomics techniques and the particular obstacles that every sector faces. We present the latest studies that illustrate the assimilation of multiomics findings. Next, we discuss other opportunities and concerns for incorporating such techniques and concisely discuss the projected advantages. These opportunities and concerns will be examined by concentrating on the primary objective, i.e., promoting the growth of healthy plants and their associated fungal communities in vibrant environmental conditions.

THE OVERVIEW OF MULTIOMICS APPROACH

GENOMICS

Since 2001, after publishing the first human genome draft, the genomics study, including the study of the organism's complete genetic composition, has evolved tremendously (Santini and Ghelardini 2015; Weissenbach 2016). While the term "genomics" is comparatively new, its origins can be traced to the early 1900s, when Wihelm Johannsen proposed the notion of the gene, and subsequently to 1920, when the term "genome" was developed by Hans Winkler (McKusick and Ruddle 1987). The microbial genomes were the first to be sequenced by developing signal tools and exponential creation of a complete genome sequence. The first free-living prokaryote with a completely sequenced genome was the *Haemophilus influenza* bacterium, which was sequenced through shotgun sequencing (Fleischmann *et al.* 1995). The *Saccharomyces cerevisiae* fungus was the first eukaryotic organism with a completely sequenced genome (Goffeau *et al.* 1996), while the *Caenorhabditis elegans* nematode was the first multicellular eukaryotic living organism to have a completely sequenced genome (elegans Sequencing Consortium TC 1998). The cost of sequencing has been minimized significantly. The accessibility of whole genome sequences has increased, including integrating de novo sequence and re-sequencing various strains of an individual species by evolving next-generation sequencing techniques.

Microbial genomics, an interdisciplinary sector, concentrates on the genomic characteristics, structure, properties, functions, development, representation, and editing for fungi and other microorganisms. Ecological problems related to environmental change can be better understood by assimilating genomics. For example, understanding the history of genomic evolution is crucial for deep insights in case of rapid change in several components under fluctuating temperatures (i.e., quick evolution). Occasionally, for in-depth understanding of the microbial functions, the comparison of genomes with transcriptomics, proteomics, and other multi-omics techniques can be required. The genomic information is used for the study of gene and protein function and communication linked with transcriptomics and proteomics on a genome-wide scale. The assimilation of genomics and transcriptomics has led to enhanced insights into several features associated with the biology of plant fungus, the interaction of fungus with the host plant, and the plant's health (Lindeberg 2012; Sundin *et al.* 2016). It is helpful to deliberate the two major genomic groups (i.e., structural genomica and functional genomics) of microbes while investigating the ecology of the fungal phyto-pathogen and the health of the host plant.

Structural genomics has concentrated on allocating and plotting the markers and genes to specific chromosomes, leading to the whole genomic physical chart. In functional genomics, the genome sequence is assimilated with the transcriptome of a microbe (transcripts that are developed by a specific microbe, including fungus) and a proteome (which are encoded proteins) for the description of interactions and functions of the genes (Hieter and Boguski 1997; Barone *et al.* 2008). The annotation and prediction of genes have been considerably improved by integrating genome information and next-generation sequence-based RNA sequencing (Yandell and Ence 2012). Regarding the modern ecology of phytopathogen, based on the research attention and perspective, several kinds of genomics are considered. Relative genomics has the purpose of detecting functional and structural genomic components preserved within respective species or across various species (Nobrega and Pennacchio 2004). This technique makes it clear that the architecture of several fungal phytopathogenic genomes is very dynamic and diverse, including the fields of quick evolution, ranging from the chromosome of transposon-rich sections to the whole accessory chromosomes (Möller and Stukenbrock 2017). For instance, a comparison of four *Fusarium* species clearly has been recognized for the sectioning of genomes that has been recognized at both the functional and structural levels: a fundamental genomic constituent encoding the microbial functions is essential for the development and persistence of *Fusarium* species and is shared commonly among tested *Fusarium* species and is an accessory element (Ma *et al.* 2010, 2013). The accessory genome consists of a small variable (<2Mb) excessively or tentatively expendable chromosomes augmented for identical compounds, and horizontally transfer the genes. Some of such genes are involved in the pathogenicity of the host plant. Correspondingly, the genomes of Verticillium vary in four zones, each of which is about 300–350 kb. Such genome sizes can be found in *V. alfalfae* and *V. dahliae* (Ls.17 and Ms.102, respectively). Such four zones comprise genes that differ significantly between fungal strains with the function of involvement in plant pathogenicity and pathogen virulence (Klosterman *et al.* 2011; de Jonge *et al.* 2013). Comparing with more genomes, it is evident that the dynamics of these chromosomes are predominant in several other phytopathogenic fungi as well (Poppe *et al.* 2015; Croll and McDonald 2017).

Similar queries regarding plant–fungal interactions have been addressed. Such sorts of addressed queries involve genome-based evolutionary and demographic systems, and the former affects the specification and deviation of closely associated taxa and the specificity of the locus on particular chromosomes or genes having an impact on definite phenotypes, while the latter seems to affect the structure of the population (Stukenbrock *et al.* 2011; de Vries *et al.* 2020; Grunwald 2016). Such information provides a basis for understanding the phytopathogens' evolutionary potential. Due to the complex nature of genome dynamics and the quantity of measured horizontal gene transfer, the major issue that is faced in the genomics of microbes is the difficulty in determining and understanding the limitations of microbial species and their communities (Aylward *et al.* 2017; Ceresini *et al.* 2019; Valent *et al.* 2019; VanInsberghe *et al.* 2020). It cannot be predicted how these dynamics will be influenced by fluctuating environmental conditions in which more leaps by the host and extreme environments are foreseen (Friesen *et al.* 2006; Ma LJ *et al.* 2010, 2013).

METAGENOMICS

Metagenomics extends devices and speculations of molecular genomics to explore DNA directly derived from the atmosphere. The microbial DNA is extracted and amplified after isolation from animals, plants, air, soil, and water (Handelsman *et al.* 2007). Although microorganisms are thought to be universal and found in every environment worldwide, just about 1% of all microbes can be cultured and identified through the traditional 1st Sanger sequencing techniques (Solden *et al.* 2016). The metagenomics domain emerged from a desire to comprehend the diversification and biological functions of unculturable microorganisms. In the late 1990s, phytopathologists and microbiologists primarily initiated the various tools and approaches of molecular assessment (Chen and Pachter 2005). This domain has now progressed to address questions about environmental dynamics, including the population of microorganisms, regulation of several species, microorganism–microorganism and host–microorganism interaction mechanisms, and coevolution of communities (Segata *et al.* 2013).

The term "metagenomics" has been derived from the Greek parafix "meta," which means to "exemplify": metagenomics explains how it goes beyond conventional genomic technologies for the identification of genomic diversification and functions of microbial genes and the unculturable microbes that have been referred to as the dark matter by some researchers (Solden *et al.* 2016). With the latest developments in high-throughput DNA sequencing technologies, we can sequence either a fraction or a whole genome of microbes to address questions about structure, composition, phylogenetic similarity, and activity at the community level of microbial species (Caporaso *et al.* 2012). A traditional metagenomic process currently includes the collection and sequencing of ecological DNA, preprocessing of bioinformatic to read the DNA sequence, determining the phylogenetic resume and each functional or genomic aspect of concern, statistical analyses, validation of data, and ultimately simulation and conversation of the findings (Quince *et al.* 2017).

The definition of the microbial species core idea is itself a massive issue in metagenomics. The concept of living species perform well with more derived taxonomies which clearly define reproductive limitations and obstacles as in case of *Homo sapiens*. While microorganisms are typically attached to a common species when reassociating, their mutual, pairwise DNA attributes throughout hybridization of DNA experimentations are greater than 70% (Staley 2006). The typical problem of microorganisms' vertical DNA transfer can further complicate the issue. Typically, strains of the same species must have some phenotypic reliability, and species characterizations would have to include the strain of more than one type, like genetic mutation or subtype. A name must be allocated to the microbial species depending on the difference of its members from other species having at least one diagnostic trait of its phenotypes. Furthermore, it is challenging to assign phenotypes to unculturable taxa of microorganisms (Staley 2006).

Another barrier to allocating taxonomic neologisms is the lack of a systematic social database sequence to which genetic readings can be mapped. Recent sequence databases—such as GenBank and Greengenes for prokaryotes and UNITE—are good starters, but genomic data about the diverse microbial array is still absent and remains undiscovered in the world. Until now, the studies concerning metagenomics in plant health have focused on two main sections. The first relates to plant growth regulation by acknowledging the endophytic microbial population in host plant shoots or roots (Fadiji and Babalola 2020). Large proportions of this research are currently being undertaken using model plants such as maize (*Zea mays*), and the research must be broadened to other crop plants and more biological ecosystems. Others have gone further than roots leading into the plant rhizoplane and rhizosphere to investigate plant and soil health's functional genes' activities (Yurgel *et al.* 2019). Determining how soil-borne microorganisms reduce plant diseases is the second type of plant health study (Larkin 2015). The studies that have been undertaken in this chapter are under crop fungal pathoecosystems, and it is necessary to investigate the mechanism of suppressive soils in other environments, such as controlled and uncontrolled forests and meadows or grasslands.

METABOLOMICS

At the phenotypic edge of the multiomics scale, metabolomics collects the outcomes of data sequence starting from the genome and moving via the transcriptome and proteome (Liu and Locasale 2017). However, the metabolomics discipline is strongly embedded in the chemical assessment of individual components of each substance; it has quickly advanced in the last 20 years, having the capacity to examine metabolites in several geographical and temporal domains (Irie *et al.* 2014; Bartels and Svatoš 2015; Sumner *et al.* 2018). Nevertheless, some constraints still exist to fully comprehending the immense diversity of metabolites found inside a particular plant. Targeted and untargeted techniques offer the possibility of profound understanding that, when combined with alternative multiomics techniques, can produce a bulk of information (Oksman-Caldentey and Saito 2005; Dettmer and Aronov 2007; Johnson *et al.* 2016). In the 1970s, metabolomics initiated medical investigations of interested substances by employing gas chromatography–mass spectrometry (GS-MS) (Sumner *et al.* 2003). With the advancement of technologies, the profile of the medical field concerning human metabolites continued to progress for the diagnostics and drugs development. In the 1990s, studies on herbicide mechanisms or modes of action initiated such profiling approaches in plants (Sauter *et al.* 1991). In the mid-1990s, the association of metabolomics with functional genomics emerged and extended parallel to enhancing capacities of these technologies in mass spectrometry, chromatography, and imaging techniques (Oliver *et al.* 2002; Ward *et al.* 2007).

The recent technology of metabolomics is traditionally the combination of liquid chromatography at an ultra high pressure with mass spectrometry at high resolution, or resonance spectroscopy based on nuclear magnetism delivers an unprecedented resolution into the phenotypic traits of chemical in particular microbes, enabling the acquisition and profiling of thousands of elements (De Vos *et al.* 2007; Ward *et al.* 2007; Allwood *et al.* 2008). However, such advancements also have some drawbacks. The high diversity of potential metabolites available in any specimen is a significant issue in metabolomics. An individual species of plants may create more than 5,000 distinct molecules, whereas animals and microbes synthesize approximately 2,500 and 1,500 unique metabolic compounds, respectively.

Overall, the kingdom Plantae has approximately 0.2 million unique metabolites (Hall *et al.* 2002; Oksman-Caldentey and Saito 2005). Focused metabolomics techniques solve these issues by limiting the profile to a typical pattern of identified molecules (Griffiths *et al.* 2010). Comparing these chemicals to existing databases of analytical standards allows for accurate quantitative measurement (Lu *et al.* 2008; Roberts *et al.* 2012).

Targeted techniques can yield detailed data on a wide range of chemical compounds (usually in hundreds) on the basis of the applied approach (Lu *et al.* 2008; Roberts *et al.* 2012). Targeted or selective techniques of metabolomics are specifically useful for analyzing defined activities and responses, since they enable optimization at higher resolution into established biological systems. Contrary, untargeted or nonselective techniques of metabolomics are used for unspecific activities (Dunn *et al.* 2013; Schrimpe *et al.* 2016).

Untargeted metabolomics, which comes close to globalized metabolomics, aims to characterize many chemical compounds found in biological samples by comparing the respective intensities. These chemicals may be labeled and supposedly detected in certain circumstances. By avoiding limitations, it is possible to make broad comparisons inside and across experimental groups to find routes and substances that drive biological activities (Johnson *et al.* 2016). Untargeted metabolomics improves the detection of interaction across genomic, metagenomic, spectromic, and volatilomic techniques, which is specifically relevant for multiomics techniques (Oksman-Caldentey and Saito 2005; Johnson *et al.* 2016).

Volatilomics

In strict terms, volatilomics might be regarded as a subgroup of metabolomics (Bicchi and Maffei 2012). However, volatilomics is not only a subgroup of metabolomics but also an implement to evaluate food authenticity as it involves the systematic profile of high-vapor pressure substances generated by microorganisms (Rosenkranz and Schnitzler 2001). Volatilomics evaluates the chemicals emitted by a microbe—the essential constituents of chemically induced intermicrobial communication (Majchrzak *et al.* 2020) whereas metabolomics attempts detailed profiling of mostly intramicrobial substances. As volatilomics is primarily essential for communication, it offers unique prospects for understanding population dynamics when combined with a multiomics technique (Cumeras and Correig 2018). Plants communicate through a diverse range of substances, including phytohormones, terpenoids, volatile compounds of green leaf, and volatile compounds synthesized in response to pest attacks (predator-mediated phytovolatiles). Such volatile chemicals have a variety of functions in plants, including regulation of pollination, dispersion of pests, and protection against biotic (pathogens) and abiotic (environmental) stress (Rosenkranz and Schnitzler 2001).

Plants are estimated to emit approximately 30,000 distinct volatile compounds into the environment and rhizosphere (Majchrzak *et al.* 2020). Similarly, in an ecosystem lacking visual data integration, the volatile plant compounds induce everything from plant–microbial pathogen interaction to the interaction between phytopathogen and natural antagonist or biocontrol agents (Insam and Seewald 2010; van Dam and

Bouwmeester 2016). Developments in gas chromatography, combined with in analytical chemistry laboratories mass spectrometry, have enabled deeper profiling of volatile gaseous chemicals, resulting in the area of volatilomics (Cumeras and Correig 2018). Developments in volatile chemical–collecting technologies have been essential to expanding volatilomics as a discipline. Mindfulness gathering or high sorption items are frequently utilized to collect volatile chemicals involved in the plants communication with anticipation of the introduction of GC-MS analysis (Bicchi and Maffei 2012). In a nondestructive multiomics technique, these more recent techniques have emerged to be applied in high-through phenotyping amount, where the quick evaluation of volatilomic profiles of the plant is associated with the gene (Jud et al. 2018). Adapting this technique to the plant-associated microbiota might lead to the identification of new ways of diagnosing and managing plant pests and diseases (Bailly and Weisskopf 2017; Parthasarathy et al. 2017).

SPECTRANOMICS

In the agricultural ecosystem, the characterization of foliar functional trait has developed as a fundamental paradigm for a deeper understanding of the dynamic nature of plant function and its variability in response to environment alteration and stress conditions. The spatial data sector has endorsed the idea of foliar functional traits as several attributes have been demonstrated to closely associate with normal and stress-mediated deviation in the functioning of the plant (Wright et al. 2004) that may be identified and measured using remote sensing information (Townsend et al. 2003; Ustin et al. 2004; Ustin and Gamon 2010; Singh et al. 2015). Spectroscopy based on remote sensing of noninvasive, adjacent, and foliar functional traits might block gaps in space and time across a farmer's field observations, limiting the risk of downstream assessment and decision-making, allowing for a better evaluation of assumptions about plant functioning responding to biotic and abiotic stress. Researchers in the subject has developed the term "spectranomics" to describe the utilization of spectroscopy in conjunction with biochemistry, taxonomy, and population ecology. The spectranomics technique is based on the following ideas:

- Plants have biochemical footprints that grow increasingly distinctive as more components are added (Ustin et al. 2004).
- Spectroscopic fingerprints identify a portfolio of available phytochemicals (Jacquemoud et al. 1995).

Plant pathologists have just started to reap the benefits of spectranomics pathway.

Current research has established in-situ (foliar) and image-processing spectroscopy (formerly known as "hyperspectral imaging") as valuable techniques that are noninvasive and adaptable for detecting pathogenic attacks in natural and agricultural ecosystems (Mahlein 2016; Mahlein et al. 2018; Zarco-Tejada et al. 2018; Fallon et al. 2020). Beneficial (Fisher et al. 2016) and pathogenic plant–fungal interactions (Mahlein et al. 2019) have noninvasive effects on many plant characteristics. Fungi can harm, degrade, or modify foliar activity either directly or indirectly, affecting the chemical properties of host plant by producing systemic effector proteins or secondary metabolic compounds or the presence of physical fungal structures such as hyphae, mycelium, and spores (Agrios 2009). With local and remote spectroscopy, the total number of modifications that the pathogen causes to plant health may be evaluated using a statistical method to the spectroscopic analysis of information (Mahlein 2016; Mahlein *et al.* 2018). In agricultural plant–breeding systems, this strategy has proved beneficial in enhancing efficacy and precision (Ge *et al.* 2019; Meacham-Hensold *et al.* 2019). Therefore, the spectral identification of foliar functional characteristics enables in identifying, tracing, and simulating physiological and biochemical pathogenic processes that enable the employment of spectroscopy to determine plant–fungal interactions (Arens *et al.* 2016; Couture *et al.* 2018; Fallon *et al.* 2019; Meacham-Hensold *et al.* 2020; Gold et al. 2019, 2020). Since the 1980s, hyperspectral and wideband approaches based on visible and near-infrared reflectance indicators, such as the normalized difference index, have been employed to detect late-stage photo disease (Jackson 1986; Hatfield and Pinter 1993; Nilsson 1995). Due to the sensitivity of short-wavelength infrared to many foliage traits, such as the content of nutrients (Gillon *et al.* 1999; Serbin *et al.* 2012; Zhai *et al.* 2013; Singh *et al.* 2015), water (Gao 1996), the activity of photosynthesis (Oren *et al.* 1986), physiology (Serbin *et al.* 2019), polyphenolic compounds, and secondary metabolites (Kokaly and Skidmore *et al.* 2015; Couture *et al.* 2016), it has proven beneficial in sensing the interaction between plant and microorganism. With its recently proven potential for vital pre visual diagnosis of phyto-disease, sensing by visible to short-wavelength infraRed (400–2,500 nm) has revived the study of interaction sensing between the plant and the microorganism (Rumpf *et al.* 2010; Xie *et al.* 2017; Couture *et al.* 2018; Bienkowski *et al.* 2019).

CASE STUDIES ON THE ECOLOGY OF PHYTOPATHOGENS UNDER VARYING ENVIRONMENTS

MULTIOMICS APPROACH IN THE DEFENSE SYSTEM OF PLANT: SALICYLIC ACID PATHWAY

Plant defense mechanisms regulate responses to environmental and pathogenic stresses below and above ground by mediating the interactions between plant and microorganism under fluctuating environments (Bezemer and van Dam 2005; Stout *et al.* 2006; Kumar *et al.* 2015a). A proposed multiomics strategy is required to understand the processes behind this capacity and the nature of impacts mediated by host plants on microbial populations. The involvement of the signaling pathway, i.e., salicylic acid, is an excellent example of this mechanism, and there are several effective plant defense mechanisms including other key plant hormones, such as jasmonic acid and abscisic acid. For instance, the salicylic acid pathway regulates systemic acquired resistance SAR of the plant and mediates plant–microbial interactions (Filgueiras *et al.* 2019). The discovery that this secondary metabolite phenol was practically ubiquitous in the kingdom Plantae and implicated everywhere, from plant reproduction to photosynthetic processes to modulating responses and to environmental and pathogen stressors, prompted primary molecular research (Mahdi *et al.* 2006). The study of plant metabolite process associated with salicylic acid perform function as chemical-signaling molecules originated with the metabolomics field; focus on these chemical signals prompted research into routes of phytometabolites associated with salicylic acid and their activity as signaling molecules (Chen *et al.* 2009; Dempsey *et al.* 2011). At first, biochemical pathways associated with the development of salicylic acid were formed, collecting the left off-pathway of shikimic acid and progressing via concurrent phenylalanine ammonia-lyase and isochorismate synthase pathways to produce the real molecule of salicylic acid that can be adapted further downward by adding molecular tags, including glycosylation, methylation, and conjugation of an amino acid (Chen *et al.* 2009; Dempsey *et al.* 2011; Kumar *et al.* 2015b). Recent selective and nonselective metabolomics techniques have described signaling cascades of salicylic acid from plant stresses after explication of the pathway (Noctor *et al.* 2015; Mhlongo *et al.* 2017; Mhlongo *et al.* 2018). Such metabolomics techniques have provided insight into how the plant reactions are produced and impacted and on induction of the plant's salicylic acid cascades that influence other metabolic activities such as cross-communication and hormonal modulation (Kunkel and Brooks 2002; Caarls *et at.* 2015). Metabolomics techniques are emerging to clarify how plant–pathogen interactions might affect microbial diversity in diverse environments, specifically significant for understanding microbial modifications in the rhizosphere (van Dam and Bouwmeester 2016). Salicylic acid has a role in arbitrating sequence and triggering immune responses via effectors, generating particular defense substances, including antimicrobial peptides, antimicrobial terpenoids, and hypersensitive response, and likely inducing systemic-acquired resistance in terms of defense against phytopathogens (Tsuda and Katagiri 2010; Sudisha *et al.* 2012; Lu *et al.* 2016; Klessig *et al.* 2018). However, metabolomics also started the genetics-based investigation for observed phenomena, which swiftly led to questions about systemic-acquired resistance, i.e., plants' potential to resist fungal and other microbial pathogens (Durrant and Dong 2004). Significant nonexpressor of pathogenesis-related genes implicated in systemic- acquired resistance has been discovered through genomic research and how such genes can be regulated transcriptionally and systemically in contexts of induced resistance (Fu *et al.* 2012; Ali *et al.* 2018; Innes 2018). A broad gene expression range of modulating transcription systems influences the responses of plant defense implicated in systemic-acquired resistance, which is impacted by pathogenic microorganisms and stimulation of the salicylic pathway, in addition to such nonexpressor of pathogenesis-related genes (Wang *et al.* 2006; Blanco *et al.* 2009). Such genomics investigations are significant for plant breeding since targeted genes that can encode systemic-acquired resistance against pathogenic fungi might be introduced into plants to ensure those fungal communities can be manipulated in a beneficial way (Li *et al.* 2013; Xu *et al.* 2017a, 2017b).

The signaling pathway of salicylic acid and SAR do not just impact the phytopathogenic population but a wide range of other associated organisms (Mahdi *et al.* 2006; Filgueiras *et al.* 2019). Such pathways utilize mobile, volatile-based signals, such as methyl salicylate (a methylated derivative of salicylic acid) (Park *et al.* 2007). In volatilomics techniques, the plant volatile profiles vary when the defense mechanism of salicylic acid is induced (Filgueiras *et al.* 2019). The activation of salicylic acid responses in surrounding plants and the induction of natural antagonists above and below ground in the rhizosphere can occur as a consequence of these alterations in profiles of volatile signaling (Baldwin *et al.* 2006; Holopainen and Blande 2012; Filgueiras *et al.* 2016a; Filgueiras *et al.* 2016b). Exogenous stimulation of salicylic acid responses in annual and perennial plants can attract natural antagonists, e.g., entomopathogenic nematodes, over vast distances, changing community composition and interactions with other nematodes (Filgueiras *et al.* 2017).

MULTIOMICS APPROACH IN DETECTION OF PLANT STRESS

The primary history of multiomics techniques is closely linked to how various salicylic acid–mediated ecologies of phytopathogens were identified. Our understanding of the mechanisms behind plant defense signals of salicylic acid has increased as multiomics has evolved, along with our capacity to employ this understanding for beneficial manipulation in controlled natural ecosystems. Protein, nitrogen, pigments, structural components (i.e., lignin and cellulose), and mineral compounds were the primary targets of initial vegetative spectroscopic investigations regarding capturing foliar biochemicals (Kokaly *et al.* 2009; Thulin *et al.* 2014). This capability was quickly extended to involve phenolic compounds and other secondary metabolic compounds related to salicylic acid and plant defense pathways (Kokaly and Skidmore 2015; Couture *et al.* 2018). In the last decade, the variety of preferred traits has developed significantly to involve particular photosynthetic indices, including maximal carboxylation and starch, sucrose, fructose, free amino acids, and rate of photosynthesis in leaves (Yendrek *et al.* 2017; Couture *et al.* 2016; Meacham-Hensold *et al.* 2019; Serbin *et al.* 2019). To accurately analyze and comprehend plant–fungal interactions, spectroscopic approaches, including Fourier-transform infrared spectroscopy, fluorescence spectroscopy, and chemometrics, are applied, although these approaches need sample collection and processing (Ni *et al.* 2011). The capacity to measure and forecast phytometabolites via passive screening using vegetative spectroscopy, particularly under *in vivo* situations, would allow field metabolomics, providing a way for new plant generation and plant–microbial interaction phenotyping techniques. Numerous subtypes in plant pathology have been theorized recently (Martins *et al.* 2019). *In vivo* metabolomics have significantly less precision than conventional metabolomics regarding multispectral sensors and significance of precision farming. The capacity of multispectral has been demonstrated initially *in vivo*-sensors to determine profiles of foliar metabolite non-invasively. This study discovered that nearly a quarter of metabolic compounds observed via the profile of GC-MS in leaves and ear rachis of the wheat plant might be effectively estimated with multispectral information, with about a 50% accuracy rate. Most of the metabolites might be predicted by them, such as amino acids, sugars, and organic acids; these are important for primary and secondary metabolism. Despite the metabolite assessment, it was discovered that the blue area was the most important waveband to predict metabolite in both examined wheat organs (leaf and ear). Similarly, a consistency was observed in the proposed canopy model, revealing that the best predictors to predict metabolites ranged from 1300 to 1400 and 2200 to 2400 nm zones, coinciding with the absorption spectrum related to nitrogen and sugars. Integrating spectranomics with metabolomics would not lack difficulties: There would be issues associated with the structure or framework of the canopy, and the consequent effects on the reflection of light restrict measuring the traits of plant biochemicals at the canopy level (Vergara 2019).

Furthermore, historical understanding of the visible-near infrared-short wave infraRed (VIS–NIR–SWIR) spectral characteristics (spatial patterns in plant reflectivity) involved in phytometabolites are restricted since MIR, UV, X-ray, FTIR, or Raman spectroscopy have traditionally been employed (Ni *et al.* 2011). Such sensitive methods can estimate the concentrations of isolated metabolites more accurately, but they do not take into consideration how micro- and macro-physical features of plants affect the interaction of light with these linkages. Previous understanding of VIS–NIR–SWIR spectroscopic properties considerably facilitates generalization and scaling and helps users to understand spectral information and data about basic plant processes. Spectroscopic procedures detect actual but minute variations in plant morphology, physiology, and biochemistry; all these are affected by plant–fungal interactions and give rise to our capacity to employ spectroscopic approaches to detect changes within the plant (Carter and Knapp 2001; Crandall *et al.* 2020).

MULTIOMICS APPROACH IN SUPPRESSIVE SOILS TO REDUCE PHYTOPATHOGENS

In agricultural contexts, the ability of soils to reduce phytodiseases has been known since long, likely close to a century or more (Chandrashekara *et al.* 2012). Soil scientists are yet to determine which soil microbes are responsible, under which combination, and in which environmental situations they evoke an inhibitory response (Pascale *et al.* 2020). However, it is now easy to comprehend the ecosystem underlying the inhibitory impact due to the advancement of multiomics techniques (Schlatter *et al.*, 2017). Disease suppressive soils include a group of microorganisms, usually fungi or bacteria, that inhibit a phytopathogen from invading roots or developing the disease. Suppressive soils might have a broad or a selective effect. Instead of employing a single taxon or a few taxa of microorganisms, extensive soil suppression is achieved by exploiting the suppressive capacities of diverse microbial populations (Hennessy *et al.* 2017). A metagenomics technique can be used to determine taxa associated with soil suppression, which might disclose their diverse activities and provide information on the biological process of disease suppression. A multipronged strategy is required to adapt fundamental understanding of the microbial population in the soil, functions of soil identities and challenges of soil identities in botanical studies, including restoration, conservation, agriculture, and forestry. Plant nutrient deficits, for example, have historically been recognized to influence crop production and overall disease resistance. Scientists separated the interaction between the content of plant nutrients and host defense on model plants, e.g., Arabidopsis (Castrillo *et al.* 2017). The scientists discovered that the core microbiota is shaped by the phosphate starvation response (PSR). A combination of amplicon sequencing to detect the host plant–related microbiota, genomic modeling, and multiple metagenomic assays to uncover the roles of individual genes that influenced the defensive response was employed to arrive at this conclusion (Castrillo *et al.* 2017). The researchers revealed that PSR could shape specific natural microbial antagonistic populations that aid defense against phytopathogens. The identification of more specific micro-organisms that might be used as natural antagonists against soil-borne phytopatho-gens and the development of synthesized community are two other applications of a multiomics technique. The study of biocontrol has traditionally followed a focused approach, emphasizing the use of an individual or a few microorganisms to reduce pathogenicity. Whole microbiota has been found and is still being found due to the advent of high throughput sequencing methods like amplicon sequencing. It is now necessary to move away from characterization of microbial communities to a func-tional knowledge of these communities' ability to not only reduce the phytopathogens but also have a single microorganism or a group of microorganisms inducing antago-nistic relationships to prevent phytopathogens (Massart *et al.* 2015; Köhl *et al.* 2019).

Research exploring predicted patterns and outcomes is still required to effectively understand the ecology of microorganisms concerning soil suppression and the abil-ity to implement biological activities to reduce plant disease. Notably, it is vital to determine the spatial-temporal pattern of soil suppression, as either particular micro-bial taxonomy or various functional groupings of soil microbiota suppress several phyto-diseases, specifically fungal diseases. For example, compared with monocul-tures, many sustainable agricultural strategies, such as cover cropping (Vukicevich *et al.* 2016), organic farming (van Bruggen *et al.* 2016; Bonanomi *et al.* 2018), and diverse cropping systems, have been known to prevent phytopathogens with vary-ing degrees of effectivity based on the cropping season and environment (Hiddink *et al.* 2010). Further difficulties for plant pathologists include diverging microbial biocontrol agent identification and activity, pathogen individuality and activity, and environmental conditions, and it is important to do this under *in vivo* conditions because environmental influences are critically required.

MULTIOMICS FRONTIER CONCERNING THE ECOLOGY OF PHYTO-DISEASES

Knowing the impact of endophytes in photodisease is a potential field of multiomics study. Endophytes are microbes that reside within plant tissue and do not produce symptoms; they are necessary for the development and health of the plant (Khare *et al.* 2018). Endophytes have been shown to make a significant contribution in limiting plant predation both above and below ground (Ahmad *et al.* 2020), influencing the pathways of plant immune response, and in certain instances, systemic acquired resis-tance as well (Kusajima *et al.* 2018), preventing limited gene expressions and oxidative bursts and attempting to fix biotic and abiotic issues in plants via specified abscisic acid deregulation. Most studies to date have concentrated on the activity of a single or a few endophytes compared with that of whole microbial communities. Outstanding research concentrated on the individual endophyte of potato (*Burkholderia phy-tofirmans* PsJN) and how several extra-cytoplasmatic functional grouping com-ponents (i.e., sigma factorsin group 4) were critical in allowing other microbes to detect variations around their environment, including humidity or temperature, and switching their metabolic activities to persist in unfavourable environment. This study adopted a dual-omic technique, linking high-throughput sequencing to iden-tify the primary participants—Burkholderia—other microbes, and fundamental metabolomics (Sheibani-Tezerji *et al.* 2015). Recently, research has been conducted to comprehend better the biodiversity and structuring of foliar fungal endophytes in natural forests (Jia *et al.* 2020). However, the ecology of endophytic microbes in other ecosystems and at the microbiome level is still poorly known (Harrison and Griffin 2020). Integrating spectranomics with genomics to generate multidimensional phylogenies reflects the dynamic resource of leaf biochemistry and morphology which is another intriguing frontier of multiomics. It has demonstrated that spectroscopy may reliably capture evolutionary signals, and such reflectance spectra may be used to identify large plant groupings, orders, and respective families (Meireles *et al.* 2020). It has been discovered that distinct spectral bands developed at varying paces and under various constraints, reflecting the evolution of the primary qualities they relate to and employing evolutionary simulations. This groundbreaking discovery demonstrates that spectranomics and spectroscopy can provide new perspectives regarding the evolution of leaf and phylogenetic diversification of plants on a large scale. This method might be used to study the ecology of fungal diseases to better comprehend plant evolutionary dynamics and interaction with beneficial and pathogenic fungi (Crandall *et al.* 2020).

POTENTIAL, DRAWBACKS, AND FUTURE PERSPECTIVES ON MULTIOMICS APPROACHES

Multiomic techniques help us figure out which microorganisms are participants in the ecosystem, how they disseminate over time, at what distance, and what are their solitary and collective roles. Multiomics can give a deeper perspective of the systems underpinning important themes in the ecology of phytodisease, including plant defense mechanism, response to stress, and efficiency to inhibit phytodisease, as outlined in the case studies. This is particularly true if compared with depending just on a single-omic approach. When adequately combined at the correct level and for the relevant research issue, these techniques have the potential to offer a multidimensional perspective of phytodisease. However, researchers, when drafting a study plan or project, need to keep in mind that there are several difficulties and restrictions in integrating and applying these techniques. The benefits and drawbacks of multiomics-based research and the prospects for applying such methods in phytopathogen studies are discussed next.

Research publications have increased significantly over the last decade due to discoveries in high-throughput sequencing, only a few approaches have been followed to experimental problems in the disciplines of sustainability and environmental genomics (Galla *et al.* 2016). The science of ecology for fungal diseases might take advantage by moving away from research hypotheses that assess interaction to those that decode actual causes for the ecological phenomenon. For instance, designing a research-based hypothesis to test the ecological model is critical in understanding how microbial community and dynamics alter in response to changing environmental conditions. Studies that characterize the microbiological participants in an ecosystem are significant; such studies facilitate a scientist to fine-tune their thoughts and queries after they have explored the available taxa and specific environmental factors. Amplicon sequencing methods are frequently used in such early research. While employing amplicon sequencing to detect taxa is a potential initial step toward identifying a system's taxonomic and correlative constraints, descriptive research has limited ability to disclose basic processes or causes for the observed phenomena.

For example, undertaking basic statistical studies to uncover patterns regarding similarities in microbes in drought-affected versus non-stressed plant roots might help us obtain a thorough understanding of the absence or presence of microbial taxonomy under stressful conditions. The only way to figure out why we observe these patterns is to look at them closely. Designing and building sector, greenhouse, and *in vitro* experimental models illuminating functional activities of microbial diversity and dynamics at diverse temporal and spatial scales are crucial for comprehending the mechanisms that produce these patterns, which are detected through amplicon information (Fierer *et al.* 2012; Talbot *et al.* 2014).

Another significant drawback of multiomics research is the lack of chemical compounds or taxonomic datasets, which are still in their infancy, as more microorganisms are sequenced and identified worldwide. Even so, a few well-organized databases constitute the basis for undertaking multiomics research. Silva, UNITE, Greengenes, and RDP (Mcdonald *et al.* 2011; Quast *et al.* 2012; R Cole's *et al.* 2013; Nilsson *et al.* 2019) are open-source datasets for both metagenomic and amplicon-based investigations. FUNGuild is a program based on (Nguyen *et al.* 2015) Python for parsing taxonomic units of fungal operations by ecological association, regardless of *in situ* analysis or platform. The latest database for functional features of fungi is FunFun, which may be used to examine and anticipate how a functional fungal community changes by taxa, groups, and other groupings on the basis of evolution or ecology (Zanne *et al.* 2020). Among databases with more specificity, metabolomics Workbench and MassBank are databases applied in volatilomics and metabolomics research (Horai *et al.* 2010; Johnson and Lange 2015; Sud *et al.* 2016). Furthermore, most researchers keep personal device-specific databases due to the constraints in scope and capacity to access them as actual benchmark datasets for component comparison. The databases ecosis.org and ecosml.org (i.e., Ecological Spectral Information System and Ecological Spectral Model Library, respectively) are spectroscopy (spectranomics) databases that may be used to locate spectral data and simulations on leaf nutritional content, cellulose, and other physical responses.

The sampling expense (particularly while considering different assessment methodologies) and difficulty in combining diverse datasets are further constraints in integrating multiomics techniques in the ecology of phyto-diseases and plant–fungal interaction. The multiomics data arises in various forms, each preprocessing method analyzing and summarizing the outcomes. It can be challenging to integrate such findings throughout the genotype–phenotype continuum.

Interaction is required to address these obstacles and achieve the promise of a multiomics study. This writing aims to encourage such interactions. No single individual can be a specialist in all multiomics disciplines; efficiently executing multiomics studies requires integrating information and datasets. This feature of interaction is possibly the most significant characteristic of the multiomics technique. Integration of several disciplines leads to a convergence of viewpoints. Novel views from other disciplines seem to help solve problems in one discipline.

Integration of these techniques will enable investigators to address previously unanswerable issues and complex challenges. Understanding the mechanism of plant defense response, sensing plant stress, and management of phytodisease through suppressive soils have all benefited from a multiomics technique to tackle challenges in the ecology of phytopathogens. This method also appears to have the potential to revolutionize our awareness of endophytic fungi. However, the potential of multiomics in phytopathogenic ecology and plant–fungal interaction goes beyond these domains. It has the potential to make significant advances in our knowledge of how microbial populations will adapt to change in the future while interacting with plants (Sarrocco *et al.* 2020).

CONCLUSION

The rapid emergence of multiomics techniques has made it possible to study microbial diversity and host plant–fungal interactions over a wide range of ecosystems and spatial and temporal levels. Meanwhile, the environment is altering swiftly. Fungal species and the changing climatic significantly influence emerging phytodiseases and the management of present outbreaks. Considering the dynamics of shifting environmental circumstances, it is vital in managing phytodiseases to adopt a comprehensive strategy for comprehending why and how pathogenesis arises. A multiomics technique provides a detailed image of the interaction between host plants and associated fungi that might be used to form prediction models for how plants and associated fungi adapt to stress due to environmental shifts. This chapter plays the role of a primer for people interested in adopting multiomics techniques while studying phytodiseases due to fungal pathogens. However, additional drawbacks and restrictions should be discussed before initiating a project based on a multiomics integration study.

REFERENCES

Agrios G (2005) *Plant Pathology*, 5th ed. Burlington, MA: Elsevier Academic Press, 79–103.

Agrios G (2009) *Plant Pathogens and Disease: General Introduction,* 613–646. https://doi.org/10.1016/B978-012373944-5.00344-8

Ahmad I, del Mar Jime'nez-Gasco M, Luthe DS, Shakeel SN and Barbercheck ME (2020) Endophytic Metarhizium robertsii promotes maize growth, suppresses insect growth, and alters plant defense gene expression. *Biological Control*:104167.

Ali A, Shah L, Rahman S, Riaz MW, Yahya M, Xu YJ, Liu F, Si W, Jiang H and Cheng B (2018) Plant defense mechanism and current understanding of salicylic acid and NPRs in activating SAR. *Physiological and Molecular Plant Pathology* 104:15–22.

Allwood JW, Ellis DI and Goodacre R (2008) Metabolomic technologies and their application to the study of plants and plant—host interactions. *Physiologia plantarum* 132(2):117–135.

Arens N, BackhauS A, Döll S, Fischer S, Seiffert U and Mock HP (2016) Non-invasive presymptomatic detection of *Cercospora beticola* infection and identification of early metabolic responses in sugar beet. *Frontiers in Plant Science* 7:1377.

Aylward FO, Boeuf D, Mende DR, Wood-Charlson EM, VislovA A, Eppley JM, Romano AE and DeLong EF (2017) Diel cycling and long-term persistence of viruses in the ocean's euphotic zone. *Proceedings of the National Academy of Sciences* 114(43):11446–11451.

Bailly A and Weisskopf L (2017) Mining the volatilomes of plant-associated microbiota for new biocontrol solutions. *Frontiers in Microbiology* 8:1638.

Baldwin IT, HAlitschke R, Paschold A, Von Dahl CC and Preston CA (2006) Volatile signaling in plant-plant interactions: "talking trees" in the genomics era. *Science* 311(5762):812–815.

Barone A, Chiusano ML, Ercolano MR, Giuliano G, Grandillo S and Frusciante L (2008) Structural and functional genomics of tomato. *International Journal of Plant Genomics* 2008:820274. https://doi.org/10.1155/2008/820274.

Bartels B and Svatoš A (2015) Spatially resolved in vivo plant metabolomics by laser ablation-based mass spectrometry imaging (MSI) techniques: LDI-MSI and LAESI. *Frontiers in Plant Science* 6:471.

Bever JD, Mangan SA and Alexander HM (2015) Maintenance of plant species diversity by pathogens. *Annual Review of Ecology, Evolution, and Systematics* 46:305–325.

Bezemer TM and van Dam NM (2005) Linking aboveground and belowground interactions via induced plant defenses. *Trends in Ecology & Evolution* 20(11):617–624.

Bicchi C and Maffei M (2012) The plant volatilome: Methods of analysis. In: *High-Throughput Phenotyping in Plants*. Totowa, NJ: Springer, 289–310.

Bienkowski D, Aitkenhead MJ, Lees AK, Gallagher C and Neilson R (2019) Detection and differentiation between potato (*Solanum tuberosum*) diseases using calibration models trained with non-imaging spectrometry data. *Computers and Electronics in Agriculture* 167:105056.

Blanco F, Salinas P, Cecchini NM, Jordana X, Van Hummelen P, and Holuigue L (2009) Early genomic responses to salicylic acid in Arabidopsis. *Plant Molecular Biology* 70(1–2):79–102.

Bonanomi G, Lorito M, Vinale F and Woo SL (2018) Organic amendments, beneficial microbes, and soil microbiota: Toward a unified framework for disease suppression. *Annual Review of Phytopathology* 56:1–20.

Burdon J (1991) Fungal pathogens as selective forces in plant populations and communities. *Australian Journal of Ecology* 16(4):423–432.

Caarls L, Pieterse CM and Van Wees S (2015) How salicylic acid takes transcriptional control over jasmonic acid signaling. *Frontiers in Plant Science* 6:170.

Caporaso JG, Lauber CL, WAlters WA, Berg-Lyons D, Huntley J, Fierer N, Owens SM, Betley J, Fraser L, Bauer M and Gormley N (2012) Ultra-high-throughput microbial community analysis on the Illumina HiSeq and MiSeq platforms. *The ISME Journal* 6(8):1621–1624.

Carter GA and Knapp AK (2001) Leaf optical properties in higher plants: linking spectral characteristics to stress and chlorophyll concentration. *American Journal of Botany* 88(4):677–684.

Castrillo G, Teixeira PJPL, Paredes SH, Law TF, de Lorenzo L, Feltcher ME and Finkel OM (2017) Root microbiota drive direct integration of phosphate stress and immunity. *Nature* 543(7646):513–518.

Ceresini PC, Castroagudín VL, Rodrigues FA', Rios JA, Aucique-Pe'rez CE, Moreira SI, Croll D, AlvEs E, Maciel JL and McDonald BA (2019) Wheat blast: from its origins in South America to its emergence as a global threat. *Molecular Plant Pathology* 20(2):155–172.

Chandrashekara C, Bhatt J, Kuma R and Chandrashekara K (2012) Suppressive soils in plant disease management. In: *Eco-Friendly Innovative Approaches in Plant Disease Management* (ed.) A Singh. New Delhi: International Book Distributors, 241–256.

Chen K, and Pachter L (2005) Bioinformatics for whole-genome shotgun sequencing of microbial communities. *PLoS Computational Biology* 1(2):e24.

Chen Z, Zheng Z, Huang J, Lai Z and Fan B (2009) Biosynthesis of salicylic acid in plants. *Plant Signaling & Behavior* 4(6):493–496.

Couture JJ, Singh A, ChaRkowski A, Groves R, Gray SM, Bethke PC and Townsend PA (2018) Integrating spectroscopy with potato disease management. *Plant Disease* 102(11):2233–2240.

Couture JJ, Singh A, Rubert-Nason KF, Serbin SP, Lindroth RL and Townsend PA (2016) Spectroscopic determination of ecologically relevant plant secondary metabolites. *Methods in Ecology and Evolution* 7(11):1402–1412.

Crandall SG, Gold KM, Jiménez-Gasco MD, Filgueiras CC and Willett DS (2020) A multi-omics approach to solving problems in plant disease ecology *PLoS One* 15(9):e0237975.

Croll D and McDonald BA (2017) The genetic basis of local adaptation for pathogenic fungi in agricultural ecosystems. *Molecular Ecology* 26(7):2027–2040.

Cumeras R and Correig X (2018) The volatilome in metabolomics. In *Volatile Organic Compound Analysis in Biomedical Diagnosis Applications*. Toronto, ON: Apple Academic Press, 23–50.

de Jonge R, Bolton MD, Kombrink A, van den Berg GC, Yadeta KA and Thomma BP (2013) Extensive chromosomal reshuffling drives evolution of virulence in an asexual pathogen. *Genome Research* 23(8):1271–1 282.

De Vos RC, Moco S, Lommen A, Keurentjes JJ, Bino RJ and Hall RD (2007) Untargeted large-scale plant metabolomics using liquid chromatography coupled to mass spectrometry. *Nature Protocols* 2(4):778.

de Vries S, Stukenbrock EH and Rose LE (2020) Rapid evolution in plant—microbe interactions—an evolutionary genomics perspective. *New Phytologist* 226(5):1256–1262.

Dell B and Malajczuk N (1989) Jarrah dieback—a disease caused by *Phytophthora cinnamomi*. In: *The Jarrah Forest*. Dordrecht: Springer, 67–87.

Dempsey DA, Vlot AC, Wildermuth MC and Klessig DF (2011) Salicylic acid biosynthesis and metabolism. *The Arabidopsis book/American Society of Plant Biologists* 9:124.

Dettmer K and Aronov PA (2007) Hammock BD. Mass spectrometry-based metabolomics. *Mass Spectrometry Reviews* 26(1):51–78.

Dunn WB, Erban A, Weber RJ, Creek DJ, Brown M, BreiTling R Hankemeier T, Goodacre R, Neumann S, Kopka J and Viant MR (2013) Mass appeal: Metabolite identification in mass spectrometry-focused untargeted metabolomics. *Metabolomics* 9(1):44–66.

Durrant WE and Dong X (2004) Systemic acquired resistance. *Annual Review Phytopathology* 42:185–209.

elegans Sequencing Consortium TC (1998) Genome sequence of the nematode *C. elegans*: A platform for investigating biology. *Science*:2012–2018.

Fadiji AE and Babalola OO (2020) Elucidating mechanisms of endophytes used in plant protection and other bioactivities with multifunctional prospects. *Frontiers in Bioengineering and Biotechnology* 8:467.

Fallon B, Yang A, Lapadat C, Armour I, Juzwik J, Montgomery RA and Cavender-Bares J (2020) Spectral differentiation of oak wilt from foliar fungal disease and drought is correlated with physiological changes. *Tree Physiology* 40(3):377–390.

Fallon B, Yang A, Nguyen C, Armour I, Juzwik J, Montgomery RA and Cavender-Bares J (2019) Leaf and canopy spectra, symptom progression, and physiological data from experimental detection of oak wilt in oak seedlings. Data Repository for the University of Minnesota. https://doi.org/10.13020/cgy7-2564

Fierer N, Leff JW, Adams BJ, Nielsen UN, Bates ST, Lauber CL, OwenS S, Gilbert JA, Wall DH and Caporaso JG (2012) Cross-biome metagenomic analyses of soil microbial communities and their functional attributes. *Proceedings of the National Academy of Sciences* 109(52):21390–21395.

Filgueiras CC, Martins AD, Pereira RV and Willett DS (2019) The ecology of salicylic acid signaling: Primary, secondary and tertiary effects with applications in agriculture. *International Journal of Molecular Sciences* 20(23):5851.

Filgueiras CC, Willett DS, Junior AM, Pareja M, El Borai F, Dickson DW, Stelinski LL and Duncan LW (2016a) stimulation of the salicylic acid pathway aboveground recruits entomopathogenic nematodes belowground. *PLoS One* 11(5).

Filgueiras CC, Willett DS, Pereira RV, Junior AM, Pareja M and Duncan LW (2016b) Eliciting maize defense pathways aboveground attracts belowground biocontrol agents. *Scientific Reports* 6:36484.

Filgueiras CC, Willett DS, Pereira RV, Sabino PHdS, Junior AM, Pareja M and Dickson DW (2017) Parameters affecting plant defense pathway mediated recruitment of entomopathogenic nematodes. *Biocontrol Science and Technology* 27(7):833–843.

Fisher JB, Sweeney S, Brzostek ER, Evans TP, Johnson DJ, Myers JA, Bourg NA, Wolf AT, Howe RW and Phillips RP (2016) Tree-mycorrhizal associations detected remotely from canopy spectral properties. *Global Change Biology* 22(7):2596–2607.

Fleischmann RD, Adams MD, White O, Clayton RA, Kirkness EF, Kerlavage AR, Bult CJ, Tomb JF, Dougherty BA, Merrick JM and McKenney K (1995) Whole-genome random sequencing and assembly of *Haemophilus influenzae* Rd. *Science* 269(5223):496–512.

Friesen N, Fritsch RM and Blattner FR (2006) Phylogeny and new intrageneric classification of Allium (Alliaceae) based on nuclear ribosomal DNA ITS sequences. Aliso: *A Journal of Systematic and Evolutionary Botany* 22(1):372–395.

Fu ZQ, YAn S, Saleh A, Wang W, Ruble J, Oka N, Mohan R, Spoel SH, Tada Y, ZheNg N and Dong X (2012) NPR3 and NPR4 are receptors for the immune signal salicylic acid in plants. *Nature* 486(7402):228–232.

Galla SJ, Buckley TR, Elshire R, Hale ML, Knapp M, McCallum J, MoRaga R, Santure AW, Wilcox P and Steeves TE (2016) Building strong relationships between conservation genetics and primary industry leads to mutually beneficial genomic advances. *Molecular Ecology* 25(21):5267–5281.

Gao BC (1996) NDWI—A normalized difference water index for remote sensing of vegetation liquid water from space. *Remote Sensing of Environment* 58(3):257–266.

Ge Y, Atefi A, ZHang H, Miao C, Ramamurthy RK, Sigmon B, Yang J and Schnable JC (2019) High-throughput analysis of leaf physiological and chemical traits with VIS—NIR—SWIR spectroscopy: A case study with a maize diversity panel. *Plant Methods* 15(1):66.

Gilbert GS and Parker IM (2010) Rapid evolution in a plant-pathogen interaction and the consequences for introduced host species. *Evolutionary Applications* 3(2):144–156.

Gillon D, Houssard C and Joffre R (1999) Using near-infrared reflectance spectroscopy to predict carbon, nitrogen and phosphorus content in heterogeneous plant material. *Oecologia* 118(2):173–182.

Goffeau A, Barrell BG, Bussey H, Davis R, Dujon B, Feldmann H, Galibert F, Hoheisel JD, JaCq C, Johnston M and Louis EJ (1996) Life with 6000 genes. *Science* 274(5287):546–567.

Gold KM, Townsend PA, Chlus A, Herrmann I, Couture JJ, Larson ER and Gevens AJ (2020) Hyperspectral measurements enable pre-symptomatic detection and differentiation of contrasting physiological effects of late blight and early blight in potato. *Remote Sensing* 12(2):286.

Gold KM, Townsend PA, Herrmann I and Gevens AJ (2019) Investigating potato late blight physiological differences across potato cultivars with spectroscopy and machine learning. *Plant Science* 110316.

Griffiths WJ, Koal T, Wang Y, Kohl M, Enot DP and Deigner HP (2010) Targeted metabolomics for biomarker discovery. *Angewandte Chemie International Edition* 49(32):5426–5445.

Grunwald S (Ed) (2005) *Environmental Soil-Landscape Modeling: Geographic Information Technologies and Pedometrics*, 1st ed. Boca Raton, FL: CRC Press. https://doi.org/10.1201/9781420028188

Hall R, Beale M, Fiehn O, Hardy N, Sumner L and Bino R (2002) Plant metabolomics: The missing link in functional genomics strategies. *Plant Cell* 14(7):1437–1440.

Handelsman J, Tiedje J, Alvarez-Cohen L, Ashburner M, Cann IK and DeLong E (2007) *The new science of metagenomics*. Committee on metagenomics: challanges and functional group applications, National Research Council (US) of National Academy of Sciences.

Harrison JG and Griffin EA (2020) The diversity and distribution of endophytes across biomes, plant phylogeny and host tissues: how far have we come and where do we go from here? *Environmental Microbiology* 22(6):2107–2123.

Hatfield P and Pinter P Jr (1993) Remote sensing for crop protection. *Crop Protection* 12(6):403–413.

Hennessy RC, Glaring MA, OlSson S, and Stougaard P (2017) Transcriptomic profiling of microbe—microbe interactions reveals the specific response of the biocontrol strain *P. fluorescens* In5 to the phytopathogen *Rhizoctonia solani*. *BMC Research Notes* 10(1):376.

Hiddink GA, Termorshuizen AJ and van Bruggen AH (2010) Mixed cropping and suppression of soil-borne diseases. In: *Genetic Engineering, Biofertilisation, Soil Quality and Organic Farming*. Dordrecht, Heidelberg, London, New York: Springer, 119–146.171.

Hieter P and Boguski M (1997) Functional genomics: it's all how you read it. *Science* 278(5338):601–602.

Holopainen JK, Blande JD (2012) Molecular Plant Volatile Communication. In: *Sensing in Nature. Advances in Experimental Medicine and Biology*, vol 739. New York, NY: Springer. https://doi.org/10.1007/978-1-4614-1704-0_2

Horai H, Arita M, Kanaya S, Nihei Y, Ikeda T and Suwa K (2010) MassBank: A public repository for sharing mass spectral data for life sciences. *Journal of Mass Spectrometry* 45(7):703–714.

Innes R (2018) The positives and negatives of NPR: A unifying model for salicylic acid signaling in plants. *Cell* 173(6):1314–1315.

Insam H and Seewald MS (2010) Volatile organic compounds (VOCs) in soils. *Biology and Fertility of Soils* 46(3):199–213.

Irie M, FujiMura Y, Yamato M, Miura D and Wariishi H (2014) Integrated MALDI-MS imaging and LC—MS techniques for visualizing spatiotemporal metabolomic dynamics in a rat stroke model. *Metabolomics* 10(3):473–483.

Jackson R (1986) Remote sensing of biotic and abiotic plant stress. *Annual Review of Phytopathology* 24(1):265–287.

Jacquemoud S, Verdebout J, Schmuck G, Andreoli G and Hosgood B (1995) Investigation of leaf biochemistry by statistics. *Remote Sensing of Environment* 54(3):180–188.

Jeger M, Salama N, Shaw MW, Van Den Berg F and Van Den Bosch F (2014) Effects of plant pathogens on population dynamics and community composition in grassland ecosystems: two case studies. *European Journal of Plant Pathology* 138(3):513–527.

Jia Q, Qu J, Mu H, Sun H and Wu C (2020) Foliar endophytic fungi: diversity in species and functions in forest ecosystems. *Symbiosis* 1–30.

Johnson CH, Ivanisevic J and Siuzdak G (2016) Metabolomics: beyond biomarkers and towards mechanisms. *Nature Reviews Molecular Cell Biology* 17(7):451–459.

Johnson SR and Lange BM (2015) Open-access metabolomics databases for natural product research: present capabilities and future potential. *Frontiers in Bioengineering and Biotechnology* 3:22.

Jud W, Winkler JB, Niederbacher B, Niederbacher S and Schnitzler JP (2018) Volatilomics: A non-invasive technique for screening plant phenotypic traits. *Plant Methods* 14(1):1–18.

Khare E, Mishra J and Arora N (2018) Multifaceted interactions between endophytes and plant: Developments and prospects. *Frontiers in Microbiology* 9:2732.

Klessig DF, Choi HW and Dempsey DA (2018) Systemic acquired resistance and salicylic acid: Past, present, and future. *Molecular Plant-Microbe Interactions* 31(9):871–888.

Klosterman SJ, Subbarao KV, Kang S, Veronese P, Gold SE, Thomma BP, Chen Z, Henrissat B, Lee YH, Park J and Garcia-Pedrajas MD (2011) Comparative genomics yields insights into niche adaptation of plant vascular wilt pathogens. *PLoS Pathogens* 7(7):e10021 37.

Köhl J, Kolnaar R and Ravensberg WJ (2019) Mode of action of microbial biological control agents against plant diseases: relevance beyond efficacy. *Frontiers in Plant Science* 10:845.

Kokaly RF, Asner GP, Ollinger SV, Martin ME and Wessman CA (2009) Characterizing canopy biochemistry from imaging spectroscopy and its application to ecosystem studies. *Remote Sensing of Environment* 113:S78–S91.

Kokaly RF and Skidmore AK (2015) Plant phenolics and absorption features in vegetation reflectance spectra near 1.66 μm. *International Journal of Applied Earth Observation and Geoinformation* 43:55–83.

Kumar D, ChaPagai D, Dean P and Davenport M (2015a) Biotic and abiotic stress signaling mediated by salicylic acid. In: *Elucidation of Abiotic Stress Signaling in Plants*. New York: Springer, 329–346.

Kumar D, Haq I, Chapagai D, Tripathi D, Donald D, Hossain M and Devaiah S (2015b) Hormone signaling: current perspectives on the roles of salicylic acid and its derivatives in plants. In: *The Formation, Structure and Activity of Phytochemicals*. Berlin: Springer, 115–136.

Kunkel BN and Brooks DM (2002) Cross talk between signaling pathways in pathogen defense. *Current Opinion in Plant Biology* 5(4):325–331.

Kusajima M, ShiMa S, Fujita M, Minamisawa K, Che FS, Yamakawa H and Nakashita H (2018) Involvement of ethylene signaling in *Azospirillum* sp. B510-induced disease resistance in rice. *Bioscience, Biotechnology, and Biochemistry* 82(9):1522–1526.

Larkin RP (2015) Soil health paradigms and implications for disease management. *Annual Review of Phytopathology* 53:199–221.

Li Y, Huang F, Lu Y, Shi Y, Zhang M, Fan J and Wang W (2013) Mechanism of plant—microbe interaction and its utilization in disease-resistance breeding for modern agriculture. *Physiological and Molecular Plant Pathology* 83:51–58.

Lindeberg M (2012) Genome-enabled perspectives on the composition, evolution, and expression of virulence determinants in bacterial plant pathogens. *Annual Review of Phytopathology* 50:111–132.

Liu X and Locasale JW (2017) Metabolomics: A primer. *Trends in Biochemical Sciences* 42(4):274–284.

Lu H, Greenberg JT and Holuigue L (2016) Salicylic acid signaling networks. *Frontiers in Plant Science* 7:238.

Lu W, Bennett BD, Rabinowitz JD (2008) Analytical strategies for LC—MS-based targeted metabolomics. *Journal of Chromatography B* 871(2):236–242.

Ma LJ, Geiser DM, Proctor RH, Rooney AP, O'Donnell K, Trail F, Gardiner DM, Manners JM and Kazan K (2013) Fusarium pathogenomics. *Annual Review of Microbiology* 67:399–416.

Ma LJ, Van Der Does HC, Borkovich KA, Coleman JJ, Daboussi MJ, Di Pietro A, Dufresne M, Freitag M, Grabherr M, Henrissat B and Houterman PM (2010) Comparative genomics reveals mobile pathogenicity chromosomes in Fusarium. *Nature* 464(7287):367–373.

Mahdi J, MAhdi A, Mahdi A and Bowen I (2006) The historical analysis of aspirin discovery, its relation to the willow tree and antiproliferative and anticancer potential. *Cell Proliferation* 39(2):147–155.

Mahlein AK (2016) Plant disease detection by imaging sensors—parallels and specific demands for precision agriculture and plant phenotyping. *Plant Disease* 100(2):241–251.

Mahlein AK, Kuska MT, Behmann J, Polder G and Walter A (2018) Hyperspectral sensors and imaging technologies in phytopathology: State of the art. *Annual Review of Phytopathology* 56:535–558.

Mahlein AK, KuSka MT, Thomas S, Wahabzada M, Behmann J, Rascher U and Kersting K (2019) Quantitative and qualitative phenotyping of disease resistance of crops by hyperspectral sensors: seamless interlocking of phytopathology, sensors, and machine learning is needed. *Current Opinion in Plant Biology* 50:156–162.

Majchrzak T, Wojnowski W, Rutkowska M, Wasik A (2020) Real-time volatilomics: A novel approach for analyzing biological samples. *Trends in Plant Science* 25(3):302–312.

Martins RC, Magalhães S, Jorge P, Barroso T and Santos F (2019) Metbots: Metabolomics robots for precision viticulture. In: *EPIA Conference on Artificial Intelligence*. Cham: Springer, 156–166.158.

Massart S, Martinez-Medina M and Jijakli MH (2015) Biological control in the microbiome era: challenges and opportunities. *Biological Control* 89:98–108.

Mcdonald D, Price M, Goodrich J, Nawrocki E, DeSanTis T and Probst A (2011) An improved GreenGenes taxonomy with explicit ranks for ecological and evolutionary analyses of Bacteria and Archaea. *The ISME Journal* 6:610–618.

McKusick VA and Ruddle FH (1987) A new discipline, a new name, a new journal. *Genomics* 1(1):1–2.

Meacham-Hensold K, Fu P, Wu J, Serbin S, Montes CM, Ainsworth E, Guan K, Dracup E, Pederson T, Driever S and Bernacchi C (2020) Plot-level rapid screening for photosynthetic parameters using proximal hyperspectral imaging. *Journal of Experimental Botany* 71(7):2312–2328.

Meacham-Hensold K, Montes CM, Wu J, Guan K, Ainsworth EA, Pederson T, Moore CE, Brown KL, Raines C and Bernacchi C (2019) High-throughput field phenotyping using hyperspectral reflectance and partial least squares regression (PLSR) reveals genetic modifications to photosynthetic capacity. *Remote Sensing of Environment* 231:111176.

Meireles JE, Cavender-Bares J, Townsend PA, UStin S, Gamon JA, Schweiger AK, Schaepman ME, Asner GP, Martin RE, Singh A and Schrodt F (2020) Leaf reflectance spectra capture the evolutionary history of seed plants. *New Phytologist* 485–493.

Mhlongo MI, Piater LA, Madala NE, LabuschagNe N, Dubery IA (2018) The chemistry of plant—microbe interactions in the rhizosphere and the potential for metabolomics to reveal signaling related to defense priming and induced systemic resistance. *Frontiers in Plant Science* 9:112.

Mhlongo MI, Tugizimana F, Piater L, Steenkamp PA, Madala N and Dubery I (2017) Untargeted metabolomics analysis reveals dynamic changes in azelaic acid- and salicylic acid derivatives in LPS-treated *Nicotiana tabacum* cells. *Biochemical and Biophysical Research Communications* 482(4):1498–1503.

Möller M and Stukenbrock EH (2017) Evolution and genome architecture in fungal plant pathogens. *Nature Reviews Microbiology* 15(12):756.

Nguyen N, Song Z, BateS S, Branco S, Tedersoo L and Menke J (2015) FUNGuild: An open annotation tool for parsing fungal community datasets by ecological guild. *Fungal Ecology* 20.

Ni J, Tian Y, Yao X, Zhu Y and Cao W (2011) Application of monitoring system about plant growth information based on spectroscopy technique. In: *PIAGENG 2010: Photonics and Imaging for Agricultural Engineering* (Vol. 7752). International Society for Optics and Photonics, 77521E. https://doi.org/10.1117/12.890951

Nilsson HE (1995) Remote sensing and image analysis in plant pathology. *Canadian Journal of Plant Pathology* 17(2):154–166.

Nilsson RH, LArsson KH, Taylor A, Bengtsson-Palme J, Jeppesen T and Schigel D (2019). The UNITE database for molecular identification of fungi: Handling dark taxa and parallel taxonomic classifications. *Nucleic Acids Research* 47:D259–D264.

Nobrega MA and Pennacchio LA (2004) Comparative genomic analysis as a tool for biological discovery. *The Journal of Physiology* 554(1):31–39.

Noctor G, Lelarge-Trouverie C and Mhamdi A (2015) The metabolomics of oxidative stress. *Phytochemistry* 112:33–53.

Oksman-Caldentey KM and Saito K (2005) Integrating genomics and metabolomics for engineering plant metabolic pathways. *Current Opinion in Biotechnology* 16(2):174–179.

Oliver DJ, Nikolau B and Wurtele ES (2002) Functional genomics: High-throughput mRNA, protein, and metabolite analyses. *Metabolic Engineering* 4(1):98–106.

Oren R, Schulze ED, Matyssek R and Zimmermann R (1986) Estimating photosynthetic rate and annual carbon gain in conifers from specific leaf weight and leaf biomass. *Oecologia* 70(2):187–193.

Park SW, Kaimoyo E, Kumar D, MoSher S and Klessig DF (2007) Methyl salicylate is a critical mobile signal for plant systemic acquired resistance. *Science* 318(5847):113–116.

Parthasarathy S, Thiribhuvanamala G, Subramanian K, Paliyath G, JayaSankar S and Prabakar K (2017) Volatile metabolites fingerprinting to discriminate the major post harvest diseases of mango caused by *Colletotrichum gloeosporioides* Penz. and *Lasiodiplodia theobromae* Pat. *Annals of Phytomędicine* 6(2):55–62.75.

Pascale A, Proietti S, Pantelides IS and Stringlis IA (2020) Modulation of the root microbiome by plant molecules: The basis for targeted disease suppression and plant growth promotion. *Frontiers in Plant Science* 10:1741.

Poppe S, Dorsheimer L, HaPpel P and Stukenbrock EH (2015) Rapidly evolving genes are key players in host specialization and virulence of the fungal wheat pathogen *Zymoseptoria tritici* (*Mycosphaerella graminicola*). *PLoS Pathogens* 11(7):e1005055.

Quast C, Pruesse E, Yilmaz P, Gerken J, Schweer T and Yarza P (2012) The SILVA ribosomal RNA gene database project: Improved data processing and web-based tools. *Nucleic Acids Research* 41(D1):D590–D596.

Quince C, Walker AW, Simpson JT, Loman NJ, and Segata N (2017) Shotgun metagenomics, from sampling to analysis. *Nature Biotechnology* 35(9):833–844.

R Cole's J, Wang Q, Fish J, Chai B, Mcgarrell D and Sun Y (2013) Ribosomal DATABASE PROject: data and tools for high throughput rRNA analysis. *Nucleic Acids Research* 42.

Rizzo D, Garbelotto M, Davidson J, SlauGhter G and Koike S (2002) *Phytophthora ramorum* as the cause of extensive mortality of *Quercus* spp. and *Lithocarpus densiflorus* in California. *Plant Disease* 86(3):205–214.

Roberts LD, Souza AL, Gerszten RE and Clish CB (2012) Targeted metabolomics. Current Protocols in *Molecular Biology* 98(1):30–2.

Rosenkranz M and Schnitzler JP (2001) Plant volatiles. *eLS* 1–9.

Rumpf T, Mahlein AK, Steiner U, Oerke EC, Dehne HW and Plümer L (2010) Early detection and classification of plant diseases with support vector machines based on hyperspectral reflectance. *Computers and Electronics in Agriculture* 74(1):91–99.

Santini A and Ghelardini L (2015) Plant pathogen evolution and climate change. *CABI Rev* 10.

Sarrocco S, Herrera-Estrella A and Collinge DB (2020) Plant disease management in the postgenomic era: From functional genomics to genome editing. *Frontiers in Microbiology* 11:107.

Sauter H, Lauer M and Fritsch H (1991) Metabolic profiling of plants: A new diagnostic technique. *ACS Publications* 443:288–299.

Schlatter D, Kinkel L, Thomashow L, Weller D and Paulitz T (2017) Disease suppressive soils: New insights from the soil microbiome. *Phytopathology* 107(11):1284–1297.

Schrimpe-Rutledge AC, Codreanu SG, Sherrod SD and McLean JA (2016) Untargeted metabolomics strategies—challenges and emerging directions. *Journal of the American Society for Mass Spectrometry* 27(12):1897–1905.

Segata N, Börnigen D, Morgan XC and Huttenhower C (2013) PhyloPhlAn is a new method for improved phylogenetic and taxonomic placement of microbes. *Nature Communications* 4(1):1–11.

Serbin SP, Dillaway DN, Kruger EL and Townsend PA (2012) Leaf optical properties reflect variation in photosynthetic metabolism and its sensitivity to temperature. *Journal of Experimental Botany* 63(1):489–502.

Serbin SP, Wu J, Ely KS, Kruger EL, Townsend PA, Meng R, Wolfe BT, Chlus A, Wang Z and Rogers A (2019) From the Arctic to the tropics: multibiome prediction of leaf mass per area using leaf reflectance. *New Phytologist* 224(4):1557–1568.

Sheibani-Tezerji R, RAttei T, Sessitsch A, Trognitz F and Mitter B (2015) Transcriptome profiling of the endophyte Burkholderia phytofirmans PsJN indicates sensing of the plant environment and drought stress. *MBio* 6(5).

Singh A, Serbin SP, McNeil BE, Kingdon CC and Townsend PA (2015) Imaging spectroscopy algorithms for mapping canopy foliar chemical and morphological traits and their uncertainties. *Ecological Applications* 25(8):2180–2197.

Solden L, Lloyd K and Wrighton K (2016) The bright side of microbial dark matter: lessons learned from the uncultivated majority. *Current Opinion in Microbiology* 31:217–226.

Staley JT. (2006). The bacterial species dilemma and the genomic—phylogenetic species concept. *Philosophical Transactions of the Royal Society B: Biological Sciences* 361(1475):1899–1909. https://doi.org/10.1098/rstb.2006.1914

Stout MJ, Thaler JS and Thomma BP (2006) Plant-mediated interactions between pathogenic microorganisms and herbivorous arthropods. *Annual Review of Entomology* 51:663–689.

Stukenbrock EH, Bataillon T, Dutheil JY, Hansen TT, Li R, Zala M, McDonald BA, Wang J and Schierup MH (2011) The making of a new pathogen: insights from comparative population genomics of the domesticated wheat pathogen *Mycosphaerella graminicola* and its wild sister species. *Genome Research* 21(12):2157–2166.

Sud M, Fahy E, Cotter D, Azam K, VadIvelu I and Burant C (2016) Metabolomics workbench: An international repository for metabolomics data and metadata, metabolite standards, protocols, tutorials and training, and analysis tools. *Nucleic Acids Research* 44(D1):D463–D470.

Sudisha J, Sharathchandra R, Amruthesh K, Kumar A and Shetty HS (2012) Pathogenesis related proteins in plant defense response. In: *Plant Defence: Biological Control.* Dordrecht, New York, NY: Springer, 379–403.

Sumner LW, Mendes P and Dixon RA (2003) Plant metabolomics: large-scale phytochemistry in the functional genomics era. *Phytochemistry* 62(6):817–836.

Sumner LW, Yang DS, Bench BJ, Watson BS, Li C and Jones AD (2018) Spatially resolved plant metabolomics. *Annual Plant Reviews Online*:343–366.

Sundin GW, Wang N, Charkowski AO, Castiblanco LF, Jia H and Zhao Y (2016) Perspectives on the transition from bacterial phytopathogen genomics studies to applications enhancing disease management: From promise to practice. *Phytopathology* 106(10):1071–1082.

Talbot JM, Bruns TD, Taylor JW, Smith DP, Glassman SI, ErlandSon S, Vilgalys R, Liao HL (2014) Endemism and functional convergence across the North American soil mycobiome. *Proceedings of the National Academy of Sciences* 111(17):6341–6346.

Thulin S, Hill MJ, Held A, JoneS S and Woodgate P (2014) Predicting levels of crude protein, digestibility, lignin and cellulose in temperate pastures using hyperspectral image data. *American Journal of Plant Sciences* 5:997–1019. https://doi.org/10.4236/ajps.2014.57113.

Townsend PA, Foster JR, Chastain RA and Currie WS (2003) Application of imaging spectroscopy to mapping canopy nitrogen in the forests of the central Appalachian Mountains using Hyperion and AVIRIS. *IEEE Transactions on Geoscience and Remote Sensing* 41(6):1347–1354.

Tsuda K and Katagiri F (2010) Comparing signaling mechanisms engaged in pattern-triggered and effector-triggered immunity. *Current Opinion in Plant Biology* 13(4):459–465.

Ustin SL and Gamon JA (2010) Remote sensing of plant functional types. *New Phytologist* 186(4):795–816.

Ustin SL, Roberts DA, Gamon JA, Asner GP and Green RO (2004) Using imaging spectroscopy to study ecosystem processes and properties. *Bioscience* 54(6):523–534.

Valent B, Farman M, Tosa Y, Begerow D, FourniEr E, Gladieux P, Islam MT, Kamoun S, KeMler M, Kohn LM and Lebrun MH (2019) *Pyricularia graminis-triticiis* not the correct species name for the wheat blast fungus: response to Ceresini *et al.*(MPP 20: 2). *Molecular Plant Pathology* 20(2):173.

van Bruggen AH, Gamliel A and Finckh MR (2016) Plant disease management in organic farming systems. *Pest Management Science* 72(1):30–44.

van Dam NM and Bouwmeester HJ (2016) Metabolomics in the rhizosphere: tapping into belowground chemical communication. *Trends in Plant Science* 21(3):256–265.

VanInsberghe D, Arevalo P, Chien D and Polz MF (2020) How can microbial population genomics inform community ecology? *Philosophical Transactions of the Royal Society B* 375(1798):20190253.

Vergara Díaz O (2019) High-throughput phenotyping in cereals and implications in plant eco-physiology. (Doctoral dissertation, Universitat de Barcelona).

Vukicevich E, Lowery T, Bowen P, U' rbez-Torres JR and Hart M (2016) Cover crops to increase soil microbial diversity and mitigate decline in perennial agriculture. *A Review. Agronomy for Sustainable Development* 36(3):48.

Ward JL, Baker JM and Beale MH (2007) Recent applications of NMR spectroscopy in plant metabolomics. *The FEBS Journal* 274(5):1126–1131.

Weissenbach J (2016) The rise of genomics. *Comptes Rendus Biologies* 339(7–8):231–239.

Wright IJ, Reich PB, Westoby M, Ackerly DD, Baruch Z, Bongers F, Cavender-Bares J, Chapin T, Cornelissen JH, DieMer M and Flexas J (2004) The worldwide leaf economics spectrum. *Nature* 428(6985):821–827.

Xie C, Yang C and He Y (2017) Hyperspectral imaging for classification of healthy and gray mold diseased tomato leaves with different infection severities. *Computers and Electronics in Agriculture* 135:154–162.

Xu G, Greene GH, Yoo H, Liu L, Marque's J, Motley J and Dong X (2017a) Global translational reprogramming is a fundamental layer of immune regulation in plants. *Nature* 545(7655):487–490.

Xu G, Yuan M, Ai C, Liu L, Zhuang E, Karapetyan S, Wang S and Dong X (2017b) uORF-mediated translation allows engineered plant disease resistance without fitness costs. *Nature* 545(7655):491–494.

Yandell M and Ence D (2012) A beginner's guide to eukaryotic genome annotation. *Nature Reviews Genetics* 13(5):329–342.

Yendrek CR, Tomaz T, Montes CM, Cao Y, Morse AM, Brown PJ, McIntyre LM, Leakey AD and Ainsworth EA (2017) High-throughput phenotyping of maize leaf physiological and biochemical traits using hyperspectral reflectance. *Plant Physiology* 173(1):614–626.

Yurgel ME, Kakad P, Zandawala M, Nässel DR, Godenschwege TA and Keene AC (2019) A single pair of leucokinin neurons are modulated by feeding state and regulate sleep—metabolism interactions. *PLoS Biology* 17(2):e2006409.

Zanne AE, Abarenkov K, Afkhami ME, Aguilar-Trigueros CA, BateS S and Bhatnagar JM (2020) Fungal functional ecology: Bringing a trait-based approach to plant-associated fungi. *Biological Reviews* 95(2):409–433.

Zarco-Tejada P, Camino C, Beck P, CaldeRon R, Hornero A, HeRna'ndez-ClemenTe R, Kattenborn T, Montes-Borrego M, Susca L, Morelli M and Gonzalez-Dugo V (2018) Previsual symptoms of *Xylella fastidiosa* infection revealed in spectral plant-trait alterations. *Nature Plants* 4(7):432–439.

Zhai S, Chen H, Ding C and Zhao X (2013) Double-negative acoustic metamaterial based on meta-molecule. *Journal of Physics D: Applied Physics* 46(47):475105.

13 Metabolic Engineering of Plant Fungi for Industrial Applications

*Maria Babar, Barbaros Çetinel,
Nabeeha Aslam Khan
and Ebtihal Alsadig Ahmed Mohamed*

CONTENTS

INTRODUCTION TO METABOLIC ENGINEERING

Jay Bailey discussed the emergence of a new field of biological science in 1991, which he called "metabolic engineering." He defined this science as "the improvement of cellular activities by altering the cell's enzymatic, transport, and regulatory functions through recombinant DNA technology." Primarily, metabolic engineering was considered part of applied molecular biology, but later advances showed that it combines genetic engineering and cellular function analysis. Metabolic and cellular engineering are interconnected fields of great importance and broad- term applications. Metabolic engineering is a multidisciplinary field that draws techniques from genetics, chemistry, system analysis, molecular biology, and biochemistry (Cameron and Tong, 1993). Metabolic engineering involves the genetic alteration of pathways involved in various cellular functions to produce valuable products. In order to improve the final product of any pathway, it is imperative to study and analyze the steps involved in the synthesis of the desired product. For studying the intermediate steps, one must

DOI: 10.1201/9781003162742-13

gain significant knowledge about cellular functions and their interaction with the respective steps of particular pathways. As mentioned, metabolic engineering is not an isolated discipline; it is also associated with functional genomics. Additionally, metabolic engineering requires appropriate "tools" for cellular function analysis and genetic modifications. Furthermore, recent advances in genomics, genetic engineering, and related fields have paved new avenues for metabolic engineering.

The field of metabolic engineering has progressed significantly in the last few years. Its success stories include the introduction of genes for protein synthesis or alteration in pathways through gene disruption. The primary purpose of metabolic engineering is to increase the production of commercially vital natural products and produce new products. The engineering of metabolic pathway is done on various points as modification of a single gene does not give the desirable product, because the innate cellular system absorbs the effect of alteration of a single enzymatic reaction to maintain cell homeostasis (Gutteridge et al., 2007). It is possible to simultaneously regulate a pathway at many sites using transcriptional regulators to control numerous endogenous genes. In a nutshell, utilizing molecular biology methodologies, metabolic engineering involves implementing genetic alterations to improve certain products in a cell (Barve et al., 2012). In recent years, metabolic engineering has been applied to optimize various fermentation procedures, which has extensively increased the market value of fermented products. Nowadays, metabolic engineering is not limited to fermentation; instead, the metabolic engineering examples are classified into various categories on the basis of the required final product (Nielsen, 2001).

AN INSIGHT INTO FUNGAL METABOLITES

Plants and microbes are major kingdoms of life with well developed secondary metabolism (also called "natural products"). Approximately 500,000 secondary metabolites have been discovered to date. About 350,000 metabolites are derived from plants, approximately 100,000 from animals, and 70,000 from microbes (Nett et al., 2009; Bérdy, 2012). About 33,500 microbial metabolites have been described in total, out of which 41% are Actinomycete fermentation products, and approximately 47% of metabolites are of fungal origin (Bérdy, 2012). Furthermore, in the last 20 years, the rate of research and discoveries of novel fungal metabolites has been much more significant than for other microbes (Nett et al., 2009). In filamentous Ascomycota and Basidiomycota, the system of secondary metabolism is highly complex, while in unicellular forms of Ascomycota, Basidiomycota, Chytridiomycota, and Zygomycota, this system is underdeveloped. A high potential for metabolic variation exists in diverse species of Ascomycota and Basidiomycota due to the high diversification level of biosynthetic genes and gene clusters. These diversifications of secondary metabolites have made it possible for these fungi to survive and colonize themselves in various ecological conditions (Bills and Gloer, 2016).

Different fungi species produce multiple secondary metabolites that help their survival in diverse environmental conditions. A wide range of biological activities are associated with secondary metabolites; some positively impact ecology and are used for beneficial purposes, while some negatively impact surrounding living beings. Approximately 1,000 fungal compounds are notorious for their adverse effects, including cytochalasins, fumonisins, aflatoxins, and trichothecenes (Bräse et al., 2009; Bräse et al., 2013). Toxins are secondary metabolites of many organisms, and some of Botulinum toxin (Nigam and Nigam 2010) are even more toxic than fungi; however, mycotoxins (toxins produced by fungi) are considered more problematic due to their wide range of occurrence in the form of contaminants in food and forages as well as mold in damp environments (Miller and McMullin, 2014). The metabolites produced by phytofungi are also of great concern due to their deteriorating effects on economically important crops. Understanding mycotoxins is essential to reduce their adverse impact and decrease their exposure to the environment (Pusztahelyi et al., 2015).

On the contrary, several novel pharmaceuticals are discovered through studies of various fungal metabolites. Several therapeutic agents and major compounds of commercially important medicines comprise natural products used in therapies for autoimmune disorders, cancer, microbial infections, neurological disorders, and cardiovascular diseases (Newman and Cragg, 2016). Some agriculture-related chemicals are also natural by-products (Rimando and Duke, 2006; Asolkar et al., 2013). Mainly, fungi are a significant source of bioactive secondary metabolites (Peláez, 2005) that have immensely improved animal and human health. A spectacular example is β-lactam antibiotics, which include cephalosporins and penicillin. These compounds have had a tremendous impact on human health due to their potential against bacterial diseases. Additionally, penicillin has paved new avenues for modern pharmaceutical development by providing the basis for technological advances in biochemistry, chemistry, and microbiology. Several fungal metabolites are known to date, possess vital activities, and improve multiple disciplines of life.

The typical definition of secondary metabolites reveals that these compounds are not indispensable for microorganisms' growth and development. In laboratory experimentations, it has been observed that silencing of genes that encode enzymes (for catalysis) for secondary metabolites results only in a minute effect on vegetative growth (Wilkinson et al., 2004; Chiang et al., 2016). However, some secondary metabolites are essential for survival, e.g., sunscreens, pheromones, pigments, and siderophores. The importance of secondary metabolites is underestimated due to their partial inventory and ambiguous functions. It has been argued by various scientists that secondary metabolites are important for the organism to the extent that they should be included in the central dogma of biology as a fourth pillar (Schreiber, 2005). The line of demarcation between primary and secondary metabolites is not quite clear; in fact, sometimes primary metabolites show behavior similar to secondary metabolites, e.g., oxalic acid (Dutton and Evans, 1996; de Oliveira Ceita et al., 2007; Schmalenberger et al., 2015). Secondary metabolites are simple molecules, such as sugar, organic acid, and alcohol, directly derived from primary metabolism (Albuquerque and Casadevall, 2012). The secondary metabolites of fungi are excreted as volatiles into the environment or incorporated as structural units into cells. Secondary metabolites are categorized as biosynthesized compounds through the pathways in which primary metabolites are utilized as building blocks to synthesize highly complex molecules such as non-ribosomal peptides, polyketides (PKs), and nuclear-encoded ribosomal peptides and terpenoids. Hundreds of secondary metabolites have been identified and compiled in different databases (CRC Press,

2016). Many secondary metabolites are identified due to their association with a biological process or activity sensed outside laboratory conditions, e.g., pigments, mycotoxins (fungal toxins), phytotoxins. However, secondary metabolites of some fungi are characterized and correlated with phylogenetic classifications (Röhrich et al., 2014; Stadler et al., 2014).

The biosynthesis of secondary metabolites is highly controlled, and their expression is activated by external stimuli and coordinated with the organism's developmental stages (Brakhage, 2013; Lu et al., 2014; Calvo and Cary, 2015). For instance, early studies showed that bioactive secondary metabolites were significantly associated with sclerotia of Aspergillus, which was further supported by a recent study of gene expression in sclerotia versus mycelium (Calvo and Cary, 2015). Additionally, expression may also be stimulated in association with other organisms (Lamacchia et al., 2016). Several secondary metabolites are products of specific pathways that express during a specific stage of life. Some life stage–specific pathways respond to culture media, but some remain silent in an artificial environment. The history of fermentation technology has shown that any strain grown in changed nutrient and environmental conditions has brought noticeable changes in the composition of the resultant metabolites (Yarbrough et al., 1993). Efforts to address this phenomenon have remained in experimental stages. This knowledge gap has been a critical element that decreased the identification rate of new secondary metabolites, their natural functions, and their novel applications in medicine and agriculture (Bills and Gloer, 2016).

ORIGIN OF SECONDARY METABOLITES AND THEIR EVOLUTION

The origins of secondary metabolites are most likely related to the origins of self-replicating biological molecules (Davies and Ryan, 2012; Strachan and Davies, 2016). Simple amino acids interacted with each other and withered other organic molecules to form new molecules; these primitive secondary metabolites were retained as a new cellular process when they positively affected macromolecular processes and cell reproduction. Many people believe that the initial chemical and structural interaction sites of primitive amino acids, which developed into polypeptides over time, still have some specificity today (Bellezza et al., 2014). Antibiotics' antagonistic activity, along with their numerous diverse small-molecule interactions, might constitute just one feature of their primordial beginnings. While the ubiquity of secondary metabolites does not prove their function in nascent biochemical development, the assumption that small bioactive molecules have played a crucial part in the diversification of all organisms throughout molecular history is constant (Davies and Ryan, 2012). Amino acid derivatives or their by-products' biosynthesis accounts for many primitive metabolites. Nonribosomal peptides may be the earliest secondary metabolites, as they are important amino acids in cellular composition and function, and they play an indispensable role as building blocks for numerous secondary metabolites (Strachan and Davies, 2016).

One of the simple fungal secondary metabolites is Hadacidin (N-formyl hydroxy-aminoacetic acid), discovered in *Penicillium frequentans* and other *Penicillium* species (Dulaney and Gray, 1962). Its name comes from the fact that it inhibits human adenocarcinoma (Shigeura and Gordon, 1962). Furthermore, it has a strong inhibitory effect on plant growth and development (Gray et al., 1964). Hadacidin inhibited purine production by inhibiting adenylosuccinate synthetase. According to isotope labeling experiments, it is generated via N-oxygenation of glycine to yield N-hydroxyglycine, followed by N-formylation to yield hydroxamic acid. Another fascinating example of primordial secondary metabolites generated by the linking of two amino acids is cyclic dipeptide (CDP, diketopiperazines or dioxopiperazines). CDPs are thought to be one of the oldest signaling molecules (Bellezza et al., 2014). These chemicals have been discovered in meteorites and are conserved in biological systems (Shimoyama and Ogasawara, 2002). CDPs are produced in abiotic systems from unprotected linear dipeptides and are relatively stable when synthesized. CDPs are promising templates for protein-based therapeutics because they lack free C- and N-terminal groups, making them resistant to the human digestion process.

Apart from spontaneous production, CDPs are synthesized by using two primary biosynthetic routes: as products of non-ribosomal peptides synthases (NRPSs) or through a recently discovered enzymatic pathway involving CDPSs (Belin et al., 2012; Jacques et al., 2015). CDPSs seize aminoacyl-tRNAs and use them to establish a cyclodipeptide connection between two aminoacyl transfer RNAs via a ping-pong process. An aminoacyl moiety of the first aminoacyl tRNA is transferred to a conserved serine residue, resulting in an aminoacyl enzyme production. Meanwhile, CDPSs divert the passage of loaded tRNAs away from the ribosomal machinery. The covalently attached intermediate binds the homologous aminoacyl tRNA's aminoacyl-binding site. It interacts with the second aminoacyl tRNA's aminoacyl moiety to produce a dipeptidyl enzyme that functions as a bridge between amino acid tRNA synthetase-based protein synthesis and other NRP bond-forming activities. Genome-mining studies have revealed CDPs genes present in bacteria and Fusarium species (Jacques et al., 2015); these indicate that fungal peptide synthesis through this pathway is present in these fungal species. Some CDPs display biological actions such as preventing microbial growth and human tumor cell proliferation. A fermentation round of *A. fumigatus*, for example, may yield at least seven CDPs, including cyclo (L-Phe-L-Pro), cyclo (L-Pro-L-Val), cyclo (L-Leu-L-Pro), cyclo(L-Leu-L-trans 4-OH-Pro), cyclo (Gly-L-Pro), and cyclo (L-Pro-L-Pro). Many of these CDPs possess antibacterial potential. Cyclo (L-Phe-L-Pro) is a phytotoxin produced by *Rosellinia necatrix* (Chen, 1960) that inhibits morphine addiction development (Walter et al., 1979). CDPs also serve as structural cores for various fungal secondary metabolites that contain diketopiperazine units such as chaetocin, gliotoxin, and verticillin B (Bills and Gloer, 2016).

One of the first fungal metabolites to be identified was kojic acid, primarily extracted from koji cultures of *Aspergillus oryzae* (Bentley, 2006). Kojic acid is a mild antibacterial (Morton et al., 1945) antioxidant and tyrosinase inhibitor; however, its function in its life cycle is unclear. Initial isotope labeling experiments revealed that two or three enzymes converted glucose into kojic acid without PK or amino acid pathways (Arnstein and Bentley, 1953, 1956). Statistical analysis identified its biosynthetic genes (Terabayashi et al., 2010), and the biosynthesis was associated with coregulated gene cluster, which comprised oxidoreductase (kojA), a transcription factor (kojR), and a transporter (kojT) (Bills and Gloer, 2016)

Dipicolinic acid (pyridine-2,6-dicarboxylic acid), a by-product of the lysine biosynthetic pathway, is another example of a simple amino acid–derived metabolite. Dipicolinic acid is produced in fungi by condensation of aspartyl-β-semialdehyde with pyruvate to make 2-keto-6-aminopimelic acid, which is then further cyclized to form 3,5-dihydro-4-hydroxyl-dipicolinc acid, considered to be a hypothetical precursor to dipicolinic acid (Kalle and Khandekar, 1983). It is a critical component of bacterial endospores, *Penicillium* spp., and several entomopathogenic fungi belonging to order Hypocreales (Asaff et al., 2005). It functions as an extracellular metabolite in *Penicillium* spp. It is assumed to play a scavenging role for external metal ligands, e.g., Mg2+, while it may possess insecticidal potential in fungi belonging to order Hypocreales (Asaff et al., 2005).

For preserving essential macromolecules from degradation (caused by high ultraviolet radiation), cyanobacteria and fungi have developed pathways that synthesize hybrid amino acid–containing compounds termed "mycosporines" (Bhatia et al., 2011). Mycosporine-like amino acids were primarily identified in fungi, involving UV-induced sporulation activation (Leach, 1965). Mycosporine-like amino acids are colorless and water-soluble small compounds derived from the intermediate pentose phosphate pathway. They are found in both the Ascomycota and Basidiomycota families (Oren and Gunde-Cimerman, 2007). Mycosporines absorb UV radiation and provide photoprotection to the organism. Moreover, they play a role in osmotic regulation (Miyamoto et al., 2014) and oxidative stress defense (CRC Press, 2016). Mycosporine and mycosporine-like amino acid biosynthesis pathways have recently been identified in cyanobacteria (Balskus and Walsh, 2010) and Actinomycetes (Miyamoto et al., 2014). Various genomic analyses reveal that fungi possess orthologous biosynthetic potential, suggesting that mycosporines are encoded by a similar gene cluster (Balskus and Walsh, 2010).

STRUCTURAL CLASSES OF FUNGAL SECONDARY METABOLITES

Different structural classes of secondary metabolites are produced by fungi, which include NRPs), terpenoids, PKs, and shikimate-derived metabolites; almost all the genes involved in the production of many types of secondary metabolites are grouped in a gene cluster (Calvo and Cary, 2015).

NRPs

Among all organisms, fungi have one of the most diverse and well-developed pools of NRPs. In the filamentous Ascomycota, this diversity reaches its pinnacle (Wang et al., 2015; Bushley and Turgeon, 2010). The NRPS mega enzymes involved in this process produce a broad spectrum of physiologically active natural peptide molecules. Iron transport, storage, and homeostasis, as well as pathogenicity toward plants and animals, are all regulated by NRPS-derived siderophores (Bushley et al., 2008; Haas, 2012). The biosynthesis of the well-known natural product–based therapeutics, i.e., penicillin, cephalosporins, cyclosporine A, and echinocandins, is carried out by NRPSs. The modular structure of NRPSs is identical to that of polyketide synthase (PKS), and they both use an assembly line thiotemplate method to accomplish peptide synthesis (Fischbach and Walsh, 2006). NRPS modules consist of three primary functional domains: the adenylation (A) domain, which identifies as well as activates specific amino acids through adenylation with ATP; a peptidyl carrier protein domain, which binds activated substrate with 4'-phosphopantetheine through thioester bond and transports the substrate; and the condensation (C) domain, which catalyzes peptide bond synthesis between monomers of NRPS complex.

During the 1940s, one of the most astonishing breakthroughs in the history of drug discovery research was the rapid development of therapeutic approaches and industrial-scale processes for the synthesis of lactam antibiotic penicillin (Lax, 2005). The development of -lactam antibiotics penicillin and cephalosporin and their contemporary variants resulted in significant improvements in human life duration. The biosynthetic gene clusters for penicillin and cephalosporin are among the most well studied of all NRPSs (Ozcengiz and Demain, 2013). The penicillin gene cluster from *Penicillium chrysogenum* was recently reconstructed as a single synthetic three open reading frame polycistronic gene under a single promoter (Unkles et al., 2014). The open reading frames were split by viral 2A peptide sequence, which caused cleavage and synthesis of distinct enzymes, and its random integration of this construct in genome resulted in penicillin synthesis in the heterologous host *Aspergillus nidulans* (Unkles et al., 2014).

TERPENOIDS

Terpenoids are abundantly produced by Ascomycota and Basidiomycota fungi. Terpenoids, particularly diterpenoids and sesquiterpenoids, are the most abundant secondary metabolites in the Basidiomycota. Recent reviews provide much information about fungal terpenoids (Wawrzyn et al., 2012; Schmidt-Dannert, 2014). Isopentenyl diphosphate and/or dimethylallyl diphosphate are five-carbon precursor molecules used to make terpenoids. These two isomers are produced in fungi via the mevalonate pathway from acetyl-CoA. (Miziorko, 2011). Geranyl pyrophosphate (C10), farnesyl pyrophosphate (C15), geranylgeranyl pyrophosphate (C20), squalene (C30), and phytoene/lycopene (C30) are branching hydrocarbons made up of isoprene or prenyl units formed via condensation of isopentenyl diphosphate and dimethylallyl diphosphate monomers (C40). Loss of a few methyl groups during biosynthesis results in the formation of Norterpenoids. "Meroterpenoid" is a term used to describe a hybrid terpenoid–PK (Bills and Gloer, 2016).

The genetic basis of the biosynthesis of trichothecenetype mycotoxins, which share a common tricyclic 12,13-epoxytrichothec-9-ene core structure, has been identified in fusaria and other species of the Hypocreales (McCormick et al., 2011; Semeiks et al., 2014). Indole diterpenoid metabolites have been extensively studied due to their biological effects, and they are occasionally found in fungi belonging to Sordariomycetes and Eurotiomycetes. The metabolites comprise the diterpene backbone, which is derived from geranylgeranyl diphosphate, and an indole moiety derived from indole-3-glycerol phosphate (Singkaravanit et al., 2010). The heterologous expression of many synthetic sesquiterpene and monoterpene synthases has been achieved, resulting in the *in vitro* synthesis of commercially novel terpenoids, including caryophyllene and eucalyptol, gurjunene, and guaiene, widely used as fragrances, fuel additives, and other products (Bills and Gloer, 2016).

PKs

PKs are a structurally diverse collection of secondary metabolites found in fungi. Their carbon scaffolds are produced from the polymerization of carboxylic acid units and have a shared biochemical origin. PKs are biosynthesized by PKSs, which are complex multifunctional enzymes that may be divided into four categories on the basis of catalytic domains and enzymatic pathways involved in their production (Shen, 2003). Fatty acid synthases and fungal PKS enzymes are closely linked. The most common forms of PKSs in Ascomycota and Basidiomycota are iterative type I PKSs. These fungal iterative PKSs are categorized into non-reducing, partial-reducing, and highly reducing PKSs. Among them, non-reducing PKSs are well studied compared with partial-reducing PKSs, which are far less characterized.

In comparison to the non-reducing PKSs, the highly reducing PKSs are highly complex. A broad spectrum of hybrid PK–amino acid metabolites, known as "iterative PKS–NRPSs," are closely related to the products of HR–PKSs. They are generated by fusing an HR–PK into an amino acid–binding domain (Hertweck, 2009; Fisch, 2013). The production of tetramic acid–containing metabolites, e.g., nematocidal thermolides, equisetin, the immunosuppressive agent pseurotin A, and the mycotoxins fusarin C, has been associated with functionally characterized hybrid gene clusters of fungal iterative PKS–NRPS (Bills and Gloer, 2016).

SHIKIMATE-DERIVED METABOLITES

Shikimate is the deprotonated version of shikimic acid (trihydroxy cyclohexene carboxylic acid) that provides the carbon backbone for the aromatic amino acids (Knaggs, 2003). As a result, all metabolites that contain aromatic amino acids in their structure are shikimate derived. The shikimate pathway has been extensively researched as mammals do not produce these imperative aromatic amino acids that meet necessary dietary needs (Tohge et al., 2013). This primitive eukaryotic pathway is highly conserved, because it is found in fungi, bacteria, and plants, and is to a substantial degree an important process involved in primary amino acid metabolism. The pathway has faced several evolutionary events, resulting in branches that create a varied range of secondary metabolites, including several aromatics and quinones. Chorismate and its derivatives are precursors for producing key isoprenoid quinones such as ubiquinones and menaquinones, which play critical roles in electron transport and antioxidant activity. Complex shikimate-derived metabolites are involved in multiple functions, e.g., the metabolite named Involutin, produced by *Paxillus involutus*, plays a key role as an Fe^{3+} reductant in Fenton-based decomposition of organic matter by mobilizing embedded nutrients, thereby increasing their accessibility to the host plant (Braesel et al., 2015).

ENGINEERING FUNGAL METABOLITES FOR THE SYNTHESIS OF BENEFICIAL INDUSTRIAL PRODUCTS

Plant pathogenic fungi/Phytofungi synthesize a wide range of secondary metabolites used to produce various medicines and agrochemical products. Analysis of whole-genome sequences has revealed that fungi contain many genes that encode for secondary metabolism, but several such genes remain silent under *in vitro* conditions. The secondary metabolism of fungi has been enhanced through various methodologies such as overexpressing the transcription factors of pathways, epigenetic approach, alteration in global regulators, ribosomal engineering, and chemical induction (Macheleidt et al., 2016). The biosynthesis of secondary metabolites is highly regulated, and various external stimuli activate their expression. A recent study revealed the effect of rice extract on gene expression of secondary metabolites in *Aspergillus nidulans* (Lacriola et al., 2020). *Pyricularia oryzae* synthesizes dihydroxynaphthalene melanin, biosynthesized through PK compound 1,8 DHN (1,8-dihydroxynaphthalene) polymerization (Butler and Day, 1998). The biosynthetic genes of melanin were identified, and their synthesis pathways have been described in detail (Eliahu et al., 2007). Melanin was primarily considered a penta-ketide compound, but ALB1 homolog analysis of Colletotrichum lagenarium revealed that melanin is a hexa-ketide compound with its backbone compound 1,3,6,8-THN (1,3,6,8-tetrahydroxynaphthalene) biosynthesized through acetyl-CoA (Vagstad et al., 2012). The 1,3,6,8-THN compound is converted into 1,8 DHN, which is then converted into DHN melanin through oxidative polymerization. The melanin biosynthesis can be triggered by modifying the factors involved in epigenetic control (Maeda et al., 2017). The biosynthetic enzymes of DHN melanin have been targeted for the production of agrochemicals, which has resulted in the production of three commercially important melanin biosynthesis inhibitors (MBIs). These MBIs are ecofriendly agrochemicals as they can inhibit fungal infection without inhibiting fungal growth (Motoyama et al., 2021).

Citric acid is an important product in the organic acid industry, mainly produced by *Aspergillus niger*. It is commonly used as a chelating and flavoring agent, as an acidifier, and also utilized in the pharma industry. About 80% of citric acid production is through *Aspergillus niger* fermentation. This production capacity of *Aspergillus niger* is possible due to its multiomics data. Two glucose transporters, i.e., MstH and MstG, and a rhamnose transporter, i.e., RhtA, have been identified through proteomics, which can be used for substrate uptake optimization in *Aspergillus niger*. Apart from citric acid, *A. niger* is also famous for the biosynthesis of various organic acids, including galactaric acid and itaconate. The former can replace petroleum-based polyethylene terephthalate used in the production of plastic, while the later has the potential to replace polyacrylic acid. The metabolic engineering of *A. niger* using the CRISPR approach is quite fascinating. In this process, the *A. niger* strain was developed, which could hydrolyze pectin into d-galacturonate, which was then converted into galactarate through oxidation.

Interestingly, d-galacturonate can be converted into I-ascorbate (vitamin C) through genetic engineering. It is a breakthrough for metabolic engineered fungi as it involves only a one-step fermentation of vitamin C production. *A. niger* is considered a cell factory for producing various enzymes and proteins due to its high secretion potential. Glucoamylase is abundantly secreted by *A. niger*, and Glucoamylase has starch-based industrial applications. Additionally, *A. niger* produces enzymes like phytases, cellulases, proteases, and pectinases (Meyer et al., 2015; Cairns et al., 2018).

Over many centuries, *Aspergillus oryzae* has been used for beverage production and Asian food and cuisine production; furthermore, in 2006, it was given the title of Japan's national microorganism by the Brewing Society of Japan (Ichishima, 2012). The transcription machinery is responsible for its efficient protein production (Gomi, 2019). It has become famous as a cell factory for the production of biofuels due to the presence of several xylanolytic and cellulolytic enzymes in its genomes. Furthermore, *A. oryzae* is biotechnologically significant since it is the sole producer of the secondary metabolite kojic acid, a by-product of soybean and rice fermentation. It is used as a skin lightener in cosmetics due to its strong biocompatibility and antioxidant action. It inhibits tyrosinase in melanin production (Leyden et al., 2011). Enzymes required for kojic acid synthesis are thought to be encoded in a gene cluster of 14 genes (Marui et al., 2011). *A. oryzae* was able to manufacture kojic acid directly from cellulose rather than starch by combining overexpression of KojR with overexpression of three cellulolytic genes (Yamada et al., 2014). In the future, *A. oryzae* will be an appealing cell factory for kojic acid–based biodegradable plastic. Furthermore, malic acid is a promising chemical that can be generated using *A. oryzae* and might help to strengthen the bioeconomy. This organic acid is derived from the citric acid cycle and has several uses in food as an acidulant and flavor enhancer, in chemicals as polyester resins, and in the medical field as an acidulant (Dai et al., 2018). Several bacterial, yeast, and filamentous fungal cell factories have been genetically modified to generate this chemical in the past few years. *A. oryzae* is one of the most promising strains, and it has sparked various metabolic engineering studies. A recently produced strain had a high malate titer (127 g 11) and malate yield, while producing significantly less succinate, which is an undesirable by-product in this procedure (Liu et al., 2018). This was accomplished by synergistically targeting carbon and redox metabolism, with 12 genetic changes introduced into *A. oryzae*. Overexpression of amylolytic genes, malate-producing enzyme fumarase, and glyoxylate helped bypass succinate (Liu et al., 2018). A considerable increase in the desired product (malate) with a decrease in the by-product (succinate) provides a sustainable way for the production of the desired product in *A. oryzae* (Meyer, 2021)

In 1939, Griseofulvin was first isolated from *Penicillium griseofulvum* and later identified for its fungicidal properties, known as "curling factor," as it induces the curling of fungal hyphae (Petersen et al., 2014). Griseofulvin was developed at the Glaxo Laboratories to treat ringworm; fungal skin ailments; andother dermatophyte diseases. Griseofulvin fungal metabolite is commercially available as a Fulcin, Fulsovin, and Grisovin.

Kojic acid, produced by *Aspergillus oryzae*, is commercially formulated in kojic dipalmitate to enhance shelf life. It suppresses the effect of melanogenesis, usually used to treat hyperpigmentation and as an antioxidant in cosmetics. Another fungal metabolite, α-Zearalanol, is produced by *Fusarium* spp., which is biosynthesized by zearalenone fermentation and hydrogenized to produce α-Zearalanol, commercially known as "Ralgro implants." It acts as mammalian estrogen to enhance weight gain in cattle (Bills and Gloer, 2016).

Retapamulin is a topical antibiotic produced by Clitopilus passeckerianus (mushroom) and other related species (Hartley et al., 2009; Jacobs, 2010). First, the pleuromutilin pathway was described by Sandoz AG as consisting of six transcriptionally coregulated genes: a putative geranylgeranyl diphosphate synthase gene, putative acyltransferase encoding gene, labdane-type diterpene synthase gene, and three cytochrome P450 monooxygenase genes.

In 1893, Bartolomeo Gosio discovered a mycophenolic acid (MPA) metabolite from *Penicillium brevicompactum* and other *Penicillium* species (Bentley, 2001). It inhibits the growth of *Bacillus anthracis* and the inosine monophosphate dehydrogenase enzyme essential for guanine production. In the 1960s and 1970s, the biological activity of MPA against viruses and some tumor types was recognized. Commercially, it is used to synthesize mycophenolate mofetil (Cellcept) and mycophenolate sodium (Myfortic) (Epinette et al., 1987).

Strobilurins A–D metabolite is produced by *Strobilurus tenacellus*. Commercially, it is used as a template for a synthetic family of β-methoxyacrylate strobilurin, which includes fluoxastrobin, azoxystrobin, kresoxim-methyl, fenamidone, and trifloxystrobin (Bills and Gloer, 2016).

FUTURE PERSPECTIVE

Genome-sequencing initiatives involving many fungi provide fruitful ground for identifying directed secondary metabolites. Several novel methodologies have revolutionized the field of fungal product discovery, particularly in biosynthetic mechanism, identification of gene clusters involved in biosynthesis, and their links to novel fungal secondary metabolites. A collaborative community initiative would make recognizing orthologous pathways in novel genomes easier and map broadly distributed fungal metabolite families. At the same time, the identification of new secondary metabolite pathways would be faster. It is reasonable to suppose that these novel platforms for discovering fungal natural products will be used in pharmaceutical, medical, and agrochemical industries at some time. Trials to develop industrial-scale Actinomycete genome-mining systems are also in progress (Doroghazi et al., 2014).

REFERENCES

Albuquerque P, Casadevall A. 2012. Quorum sensing in fungi: A review. *Med Mycol* 50: 337–345. http://dx.doi.org/10.3109/13693786.2011.652201.

Arnstein HRV, Bentley R. 1953. The biosynthesis of kojic acid. II. The occurrence of aldolase and triosephosphate isomerase in *Aspergillus* species and their relationship to kojic acid biosynthesis. *Biochem J* 54:508–516. http://dx.doi.org/10.1042/bj0540508.

Arnstein HRV, Bentley R. 1956. The biosynthesis of kojic acid. Production from pentoses and methyl pentoses. *Biochem J* 62:403–411. http://dx.doi.org/10.1042/bj0620403.

Asaff A, Cerda-García-Rojas C, de la Torre M. 2005. Isolation of dipicolinic acid as an insecticidal toxin from *Paecilomyces fumosoroseus*. *Appl Microbiol Biotechnol* 68:542–547. http://dx.doi.org/10.1007/s00253-005-1909-2.

Asolkar RN, Cordova-Kreylos AL, Himmel P, Marrone PG. 2013. Discovery and development of natural products for pest management. *ACS Symp Ser* 1141:17–30.

Balskus EP, Walsh CT. 2010. The genetic and molecular basis for sunscreen biosynthesis in cyanobacteria. *Science* 329:1653–1656http://dx.doi.org/10.1126/science.1193637.

Barve A, Rodrigues JF, Wagner A. 2012. Superessential reactions in metabolic networks. *Proc Natl Acad Sci* 109:1121–1130. https://doi.org/10.1073/pnas.1113065109.

Belin P, Moutiez M, Lautru S, Seguin J, Pernodet JL, Gondry M. 2012. The non-ribosomal synthesis of diketopiperazines in tRNAdependent cyclodipeptide synthase pathways. *Nat Prod Rep* 29:961–979. http://dx.doi.org/10.1039/c2np20010d.

Bellezza I, Peirce MJ, Minelli A. 2014. Cyclic dipeptides: From bugs to brain. *Trends Mol Med* 20:551–558. http://dx.doi.org/10.1016/j.molmed.2014.08.003.

Bentley R. 2001 Bartolomeo Gosio, 1863–1944: An appreciation pp. *Advances in Applied Microbiology* 48:229–250.

Bentley R. 2006. From miso, saké and shoyu to cosmetics: a century of science for kojic acid. *Nat Prod Rep* 23:1046–1062. http://dx.doi.org/10.1039/b603758p.

Bérdy J. 2012. Thoughts and facts about antibiotics: Where we are now and where we are heading. *J Antibiot (Tokyo)* 65:385–395 http://dx.doi.org/10.1038/ja.2012.27.

Bhatia S, Garg A, Sharma K, Kumar S, Sharma A, Purohit AP. 2011. Mycosporine and mycosporine-like amino acids: A paramount tool against ultra violet irradiation. *Pharmacogn Rev* 5:138–146. http://dx.doi.org/10.4103/0973-7847.91107.

Bills GF, Gloer JB. 2016. Biologically active secondary metabolites from the fungi. *Microbiol Spect* 4:4–6.

Braesel J, Götze S, Shah F, Heine D, Tauber J, Hertweck C, Tunlid A, Stallforth P, Hoffmeister D. 2015. Three redundant synthetases secure redox-active pigment production in the basidiomycete Paxillus involutus. *Chem Biol* 22:1325–1334. http://dx.doi.org/10.1016/j.chembiol.2015.08.016.

Brakhage AA. 2013. Regulation of fungal secondary metabolism. *Nat Rev Microbiol* 11: 21–32. http://dx.doi.org/10.1038/nrmicro2916.

Bräse S, Encinas A, Keck J, Nising CF. 2009. Chemistry and biology of mycotoxins and related fungal metabolites. *Chem Rev* 109:3903–3990. http://dx.doi.org/10.1021/cr050001f.

Bräse S, Gläser F, Kramer C, Lindner S, Linsenmeier AM, Masters K-S, Meister AC, Ruff BM, Zhong S. 2013. *The Chemistry of Mycotoxins*. Springer Verlag, Vienna, Austria. http://dx.doi.org/10.1007/978-3-7091-1312-7.

Bushley KE, Ripoll DR, Turgeon BG. 2008. Module evolution and substrate specificity of fungal non-ribosomal peptide synthetases involved in siderophore biosynthesis. *BMC Evol Biol* 8:328. http://dx.doi.org/10.1186/1471-2148-8-328.

Bushley KE, Turgeon BG. 2010. Phylogenomics reveals subfamilies of fungal non-ribosomal peptide synthetases and their evolutionary relationships. *BMC Evol Biol* 10:26. http://dx.doi.org/10.1186/1471-2148-10-26.

Butler MJ, Day AW. 1998 Fungal melanins: A review. *Can J Microbiol* 44:1115–1136. https://doi.org/10.1038/nrmicro2916

Cairns TC, Nai C, and Meyer V. 2018. How a fungus shapes biotechnology: 100 years of *Aspergillus niger* research. *Fungal Biol Biotechnol* 5:1. https://doi.org/10.1186/s40694-018-0054-5.

Calvo AM, Cary JW. 2015. Association of fungal secondary metabolism and sclerotial biology. *Front Microbiol* 6:62. http://dx.doi.org /10.3389/fmicb.2015.00062.

Cameron DC, Tong IT. 1993. Cellular and metabolic engineering. *App Biochem Biotech* 38:105–140.

Chen YU. 1960. Studies on the metabolic products of Rosellinia necatrix Berlese: Part I. isolation and characterization of several physiologically active neutral substances. *Bull Agric Chem Soc Jpn* 24:372–381.

Chiang Y-M, Ahuja M, Oakley CE, Entwistle R, Asokan A, Zutz C, Wang CCC, Oakley BR. 2016. Development of genetic dereplication strains in *Aspergillus* nidulans results in the discovery of aspercryptin. *Angew Chem Int Ed Engl* 55:1662–1665. http://dx.doi.org/10.1002/anie.201507097.

CRC Press. 2016. *The Chapman & Hall/CRC Dictionary of Natural Products Version 25.1*. London: CRC Press, Taylor & Francis Group, an Informa Group company.

Dai Z, Zhou H, Zhang S, et al. 2018. Current advance in biological production of malic acid using wild type and metabolic engineered strains. *Bioresour. Technol* 258:345–353.

Davies J, Ryan KS. 2012. Introducing the parvome: Bioactive compounds in the microbial world. *ACS Chem Biol* 7:252–259. http://dx.doi.org/10.1021/cb200337h.

de Oliveira Ceita G, Macêdo JNA, Santos TB, Alemanno L, da Silva Gesteira A, Micheli F, Mariano AC, Gramacho KP, da Costa Silva D, Meinhardt L, Mazzafera P, Pereira GAG, de Mattos Cascardo JC. 2007. Involvement of calcium oxalate degradation during programmed cell death in Theobroma cacao tissues triggered by the hemibiotrophic fungus *Moniliophthora perniciosa*. *Plant Sci* 173:106–117. http://dx.doi.org /10.1016/j. plantsci.2007.04.006.

Doroghazi JR, Albright JC, Goering AW, Ju KS, Haines RR, Tchalukov KA, Labeda DP, Kelleher NL, Metcalf WW. 2014. A roadmap for natural product discovery based on large-scale genomics and metabolomics. *NatChem Biol* 10:963–968. http://dx.doi. org/10.1038/nchembio.1659.

Dulaney EL, Gray RA. 1962. Penicillia that make N-formyl hydroxyaminoacetic acid, a new fungal product. *Mycologia* 54:476–480. http://dx.doi.org/10.2307/3756317.

Dutton MV, Evans CS. 1996. Oxalate production by fungi: Its role in pathogenicity and ecology in the soil environment. *Can J Microbiol* 42:881–895. http://dx.doi.org/10.1139/m96-114.

Eliahu N, Igbaria A, RoSe MS, Horwitz BA, Lev S. 2007. Melanin biosynthesis in the maize pathogen *Cochliobolus heterostrophus* depends on two mitogen-activated protein kinases, Chk1 and Mps1, and the transcription factor Cmr1. *Eukaryotic Cell* 6:421–429. https://doi.org/10.1128/ec.00264-06.

Epinette WW, Parker CM, Jones EL, Greist MC. 1987 Mycophenolic acid for psoriasis: A review of pharmacology, long-term efficacy, and safety. *Journal of the American Academy of Dermatology* 17:962–971.

Fisch KM. 2013. Biosynthesis of natural products by microbial iterative hybrid PKS-NRPS. *RSC Adv* 3:18228–18247 http://dx.doi.org/10.1039/c3ra42661k.

Fischbach MA, Walsh CT. 2006. Assembly-line enzymology for polyketide and non-ribosomal peptide antibiotics: Logic, machinery, and mechanisms. *Chem Rev* 106:3468–3496. http://dx.doi.org/10.1021/cr0503097.

Gomi K. 2019. Regulatory mechanisms for amylolytic gene expression in the koji mold Aspergillus oryzae. *Biosci. Biotechnol. Biochem* 83:1385–1401.

Gray RA, Gauger GW, Dulaney EL, Kaczka EA, Woodruff HB. 1964. Hadacidin, a new plant-growth inhibitor produced by fermentation. *Plant Physiol* 39:204–207. http://dx.doi. org/10.1104/pp.39.2.204.

Gutteridge A, Kanehisa M, Goto S. 2007. Regulation of metabolic networks by small molecule metabolites. *BMC Bioinform* 8:1–7. https://doi.org/10.1186/1471-2105-8-88.

Haas H. 2012. Iron: A key nexus in the virulence of Aspergillus fumigatus. *Front Microbiol* 3:28. http://dx.doi.org/10.3389/fmicb.2012.00028.

Hartley AJ, de Mattos-Shipley K, Collins CM, Kilaru S, Foster GD, Bailey AM. 2009. Investigating pleuromutilin-producing Clitopilus species and related basidiomycetes. *FEMS Microbiol Lett* 297:24–30.

Hertweck C. 2009. The biosynthetic logic of polyketide diversity. *Angew Chem Int Ed Engl* 48:4688–4716. http://dx.doi.org/10.1002/anie.200806121.

Ichishima E. 2012. Unique Enzymes of Aspergillus Fungi Used in Japanese Bioindustries. Novinka: *Nova Science*. ISBN: 978-1-61209-719-0

Jacobs MR. 2010 Retapamulin: Focus on its use in the treatment of uncomplicated superficial skin infections and impetigo. *Expert Review of Dermatology* 5:505–517.

Jacques IB, Moutiez M, Witwinowski J, Darbon E, Martel C, Seguin J, Favry E, Thai R, Lecoq A, Dubois S, Pernodet J-L, Gondry M, Belin P. 2015. Analysis of 51 cyclodipeptide synthases reveals the basis for substrate specificity. *Nat Chem Biol* 11:721–727. http:// dx.doi.org/10.1038/nchembio.1868.

Kalle GP, Khandekar PS. 1983. Dipicolinic acid as a secondary metabolite in *Penicillium citreoviride*. *J Biosci* 5:43–52. http://dx.doi.org/10.1007/BF02702592.

Knaggs AR. 2003. The biosynthesis of shikimate metabolites. *Nat Prod Rep* 20:119–136. http://dx.doi.org/10.1039/b100399m.

Lacriola CJ, Falk SP, WeisBlum B. 2020. Rice-induced secondary metabolite gene expression in Aspergillus nidulans. *J Indust Microbio Biotech* 47:1109–1116. https://doi.org/10.1007/s10295-020-02328-x

Lamacchia M, Dyrka W, Breton A, Saupe SJ, Paoletti M. 2016. Overlapping *Podospora anserina* transcriptional responses to bacterial and fungal non self indicate a multi-layered innate immune response. *Front Microbiol* 7:471. http://dx.doi.org/10.3389/fmicb.2016.00471.

Lax E. 2005. *The Mold in Dr. Florey's Coat: The Story of the Penicillin Miracle.* New York, NY: Henry Holt and Co. http://dx.doi.org/10.1172/JCI24342

Leach CM. 1965. Ultraviolet-absorbing substances associated with light-induced sporulation in fung. *Can J Bot* 43:185–200. http://dx.doi.org/10.1139/b65-024.

Leyden JJ, Shergill B, Micali G, et al. 2011. Natural options for the management of hyperpigmentation. *J. Eur. Acad. Dermatol. Venereol* 25(10):1140–1145.

Liu J, Li J, Liu Y, et al. 2018. Synergistic rewiring of carbon metabolism and redox metabolism in cytoplasm and mitochondria of *Aspergillus oryzae* for increased l-malate production. *ACS Synth. Biol* 7:2139–2147.

Lu MYJ, Fan WL, Wang WF, Chen T, Tang YC, Chu FH, Chang TT, Wang SY, Li MY, Chen YH, Lin ZS, Yang KJ, Chen SM, Teng YC, Lin YL, Shaw JF, Wang TF, Li WH. 2014. Genomic and transcriptomic analyses of the medicinal fungus *Antrodia cinnamomea* for its metabolite biosynthesis and sexual development. *Proc Natl Acad Sci USA* 111:E4743–E4752 http://dx.doi.org/10.1073/pnas.1417570111.

Macheleidt J, Mattern DJ, Fischer J, NeTzker T, Weber J, Schroeckh V, VAliante V, Brakhage AA. 2016. Regulation and role of fungal secondary metabolites. *Annual Review of Genetics* 50:371–392. https://doi.org/10.1146/annurev-genet-120215-035203.

Maeda K, Izawa M, Nakajima Y, Jin Q, Hirose T, Nakamura T, KosHino H, Kanamaru K, OhSaTo S, Kamakura T, Kobayashi T, Yoshida M, KiMura M. 2017. Increased metabolite production by deletion of an HDA1-type histone deacetylase in the phytopathogenic fungi, *Magnaporthe oryzae* (*Pyricularia oryzae*) and Fusarium asiaticum. *Lett Appl Microbiol* 65:446–452. https://doi.org/10.1111/lam.12797

Marui J, Yamane N, Ohashi-Kunihiro S, et al. 2011. Kojic acid biosyn thesis in *Aspergillus oryzae* is regulated by a Zn(II)(2)Cys(6) transcriptional activator and induced by kojic acid at the transcriptional level. *J Biosci Bioeng* 112:40–43.

McCormick SP, Stanley AM, Stover NA, Alexander NJ. 2011. Trichothecenes: from simple to complex mycotoxins. *Toxins (Basel)* 3:802–814. http://dx.doi.org/10.3390/toxins3070802.

Meyer V. 2021. Metabolic engineering of filamentous fungi. *Metab Eng Concepts Appl* 13:765–801.

Meyer V, Fiedler M, Nitsche B, and King R. 2015. The cell factory Aspergillus enters the big data era: Opportunities and challenges for optimising product formation. *Adv Biochem Eng Biotechnol* 149. http://dx.doi.org/10.1007/10_2014_297

Miller JD, McMullin DR. 2014. Fungal secondary metabolites as harmful indoor air contaminants: 10 years on. *Appl Microbiol Biotechnol* 98:9953–9966. http://dx.doi.org/10.1007/s00253-014-6178-5.

Miyamoto KT, Komatsu M, Ikeda H. 2014. Discovery of gene cluster for mycosporine-like amino acid biosynthesis from Actinomycetales microorganisms and production of a novel mycosporine-like amino acid by heterologous expression. *Appl Environ Microbiol* 80:5028–5036. http://dx.doi.org/10.1128/AEM.00727-14.

Miziorko HM. 2011. Enzymes of the mevalonate pathway of isoprenoid biosynthesis. *Arch Biochem Biophys* 505:131–143. http://dx.doi.org/10.1016/j.abb.2010.09.028.

Morton HE, Kocholaty W, Junowicz-Kocholaty R, Kelner A. 1945. Toxicity and antibiotic activity of kojic acid produced by Aspergillus luteo-virescens. *J Bacteriol* 50:579–584.

Motoyama, Takayuki, Choong-Soo Yun, Hiroyuki Osada. 2021. Biosynthesis and biological function of secondary metabolites of the rice blast fungus Pyricularia oryzae. *J Indus Microbiol Biotech.* https://doi.org/10.1093/jimb/kuab058.

Nett M, Ikeda H, Moore BS. 2009. Genomic basis for natural product biosynthetic diversity in the actinomycetes. *Nat Prod Rep* 26:1362–1384. http://dx.doi.org/10.1039/b817069j.

Newman DJ, Cragg GM. 2016. Natural products as sources of new drugs from 1981 to 2014. *J Nat Prod* 79:629–661. http://dx.doi.org /10.1021/acs.jnatprod.5b01055.

Nielsen J. 2001. Metabolic engineering. *App Microbiol Biotech* 55:263–283.

Nigam PK, Nigam A. 2010. Botulinum toxin. *Ind J Dermatol* 55(1):8.

Oren A, Gunde-Cimerman N. 2007. Mycosporines and mycosporinelike amino acids: UV protectants or multipurpose secondary metabolites? *FEMS Microbiol Lett* 269:1–10. http://dx.doi.org/10.1111/j.1574-6968.2007.00650.x.

Ozcengiz G, Demain AL. 2013. Recent advances in the biosynthesis of penicillins, cephalosporins and clavams and its regulation. *Biotechnol Adv* 31:287–311. http://dx.doi.org/10.1016/j.biotechadv.2012.12.001.

Peláez F. 2005. Biological activities of fungal metabolites. In An Z (ed.), *Handbook of Industrial Mycology.* New York, NY: Marcel Dekker, 49–92. ISBN 9781498783125

Petersen AB, Rønnest MH, Larsen TO, Clausen MH. 2014 The chemistry of griseofulvin. *Chem Rev* 114(24):12088–12107.

Pusztahelyi T, Holb IJ, Pócsi I. 2015. Secondary metabolites in fungusplant interactions. *Front Plant Sci* 6:573. doi:10.3389/fpls.2015.00573.

Rimando AM, Duke SO. 2006. Natural products for pest management. *ACS Symp Ser* 927:2–21.

Röhrich CR, Jaklitsch WM, Voglmayr H, Iversen A, Vilcinskas A, Nielsen KF, Thrane U, von Döhren H, Brückner H, Degenkolb T. 2014. Front line defenders of the ecological niche! Screening the structural diversity of peptaibiotics from saprotrophic and fungicolous Trichoderma/Hypocrea species. *Fungal Divers* 69:117–146. http://dx.doi.org/10.1007/s13225-013-0276-z.

Schmalenberger A, Duran AL, Bray AW, Bridge J, Bonneville S, Benning LG, Romero-Gonzalez ME, Leake JR, Banwart SA. 2015. Oxalate secretion by ectomycorrhizal Paxillus involutus is mineral specific and controls calcium weathering from minerals. *Sci Rep* 5:12187. http://dx.doi.org/10.1038/srep12187.

Schmidt-Dannert C. 2014. Biosynthesis of terpenoid natural products in fungi. *Adv Biochem Eng Biotechnol* 148:19–61. http://dx.doi.org/10.1007/10_2014_283.

Schreiber SL. 2005. Small molecules: The missing link in the central dogma. *Nat Chem Biol* 1:64–66. http://dx.doi.org/10.1038/nchembio0705.

Semeiks J, Borek D, Otwinowski Z, Grishin NV. 2014. Comparative genome sequencing reveals chemotype-specific gene clusters in the toxigenic black mold Stachybotrys. *BMC Genomics* 15:590. http://dx.doi.org/10.1186/1471-2164-15-590.

Shen B. 2003. Polyketide biosynthesis beyond the type I, II and III polyketide synthase paradigms. *Curr Opin Chem Biol* 7:285–295. http://dx.doi.org/10.1016/S1367-5931(03)00020-6.

Shigeura HT, Gordon CN. 1962. Hadacidin, a new inhibitor of purine biosynthesis. *J Biol Chem* 237:1932–1936.

Shimoyama A, Ogasawara R. 2002. Dipeptides and diketopiperazines in the Yamato-791198 and Murchison carbonaceous chondrites. *Orig Life Evol Biosph* 32:165–179. http://dx.doi.org/10.1023/A:1016015319112.

Singkaravanit S, Kinoshita H, Ihara F, Nihira T. 2010. Cloning and functional analysis of the second geranylgeranyl diphosphate synthase gene influencing helvolic acid biosynthesis in Metarhizium anisopliae. *Appl Microbiol Biotechnol* 87:1077–1088 http://dx.doi.org/10.1007 /s00253-010-2556-9.

Stadler M, Læssøe T, Fournier J, Decock C, Schmieschek B, Tichy HV, Peršoh D. 2014. A polyphasic taxonomy of *Daldinia* (*Xylariaceae*). *Stud Mycol* 77:1–143. http://dx.doi.org/10.3114/sim0016.

Strachan CR, Davies J. 2016. Antibiotics and evolution: food for thought. *J Ind Microbiol Biotechnol* 43:149–153. http://dx.doi.org/10.1007/s10295-015-1702-x.

Terabayashi Y, Sano M, Yamane N, Marui J, Tamano K, Sagara J, Dohmoto M, Oda K, Ohshima E, Tachibana K, Higa Y, Ohashi S, Koike H, Machida M. 2010. Identification and characterization of genes responsible for biosynthesis of kojic acid, an industrially important compound from Aspergillus oryzae. *Fungal Genet Biol* 47:953–961. http://dx.doi.org/10.1016/j.fgb.2010.08.014.

Tohge T, Watanabe M, Hoefgen R, Fernie AR. 2013. Shikimate and phenylalanine biosynthesis in the green lineage. *Front Plant Sci* 4:62. http://dx.doi.org/10.3389/fpls.2013.00062.

Unkles SE, Valiante V, Mattern DJ, Brakhage AA. 2014. Synthetic biology tools for bioprospecting of natural products in eukaryotes. *Chem Biol* 21:502–508. http://dx.doi.org/10.1016/j.chembiol.2014.02.010.

Vagstad AL, Hill EA, Labonte JW, Townsend CA. 2012. Characterization of a fungal thioesterase having *Claisen cyclase* and deacetylase activities in melanin biosynthesis. *Chemistry & Biology* 19:1525–1534. https://doi.org/10.1016/j.chembiol.2012.10.002.

Walter R, Ritzmann RF, Bhargava HN, Flexner LB. 1979. Prolylleucyl-glycinamide, cyclo (leucylglycine), and derivatives block development of physical dependence on morphine in mice. *Proc Natl Acad Sci USA* 76:518–520. http://dx.doi.org/10.1073/pnas.76.1.518.

Wang H, Sivonen K, Fewer DP. 2015. Genomic insights into the distribution, genetic diversity and evolution of polyketide synthases and non-ribosomal peptide synthetases. *Curr Opin Genet Dev* 35:79–85. http://dx.doi.org/10.1016/j.gde.2015.10.004.

Wawrzyn GT, Bloch SE, Schmidt-Dannert C. 2012. Discovery and characterization of terpenoid biosynthetic pathways of fungi. *Methods Enzymol* 515:83–105. http://dx.doi.org/10.1016/B978-0-12-394290-6.00005-7.

Wilkinson HH, Ramaswamy A, Sim SC, Keller NP. 2004. Increased conidiation associated with progression along the sterigmatocystin biosynthetic pathway. *Mycologia* 96:1190–1198. http://dx.doi.org/10.2307/3762134.

Yamada R, Yoshie T, Wakai S, et al. 2014. Aspergillus oryzae-based cell factory for direct kojic acid production from cellulose. *Microb Cell Fact* 13:71. https://doi.org/10.1186/1475-2859-13-71.

Yarbrough GG, Taylor DP, Rowlands RT, Crawford MS, Lasure LL. 1993. Screening microbial metabolites for new drugs: theoretical and practical issues. *J Antibiot (Tokyo)* 46:535–544. http://dx.doi.org/10.7164 /antibiotics.46.535.

Index

A

abiotic, 5, 29, 38, 56, 90–93, 97, 101, 102, 105, 172, 173, 203–205, 211, 214, 219, 220, 225, 232, 233, 243
ablation-based, 229
abscisic, 90, 104, 107, 109, 114, 221, 225
acetylation, 120
acetyl-CoA, 245, 247
acetyltransferase, 135
achlorophyllous, 192
acidulant, 248
acro-petal, 77
actin, 44, 49, 53, 144
acting, 127
actinomycetes, 209, 244, 253
acyl-homoserinelactonase, 204
acyltransferase, 248
adenylation, 245
adenylosuccinate, 243
aerobic, 29, 32
afforestation, 204
agriculture, 1, 27, 51, 55, 59, 113, 136, 145, 149, 169, 171, 172, 191, 194, 201, 202, 205, 208, 209, 211, 213, 224, 229, 230, 233, 237, 242
agrobacterium, 63, 81, 86, 132–134
amplification, 143
amylolytic, 248, 251
anaerobic, 29
anamorph, 47
anaphase, 128
anastomosis, 146
anatomy, 150, 192
angiogenesis, 101
anisogamy, 36
annealing, 141, 143
antagonism, 23, 107, 108, 121
antheridium, 36
anthracnose, 79, 86, 145
antibiotics, 29, 131, 212, 241, 242, 245, 250, 251, 254
antibodies, 68
antigen, 134
antimicrobial, 87–90, 95, 97, 100, 101, 105, 107, 108, 113, 116, 119, 181, 222
antisense, 64, 143
aplanospores, 35
apoplast, 13, 30, 49, 61, 72, 114
apoptosis, 45, 101, 106
appressoria, 41, 89, 90, 102, 116

(continued)

Arabidopsis, 4, 17, 19, 23, 25, 69, 92–95, 97–99, 104–110, 113, 119, 122, 176, 177, 179, 180, 187–189, 224, 229, 230
arbuscular, 146, 173, 176, 180–182, 185–187, 195, 201, 202, 205–212
arthro-spores, 34
ascogonia, 35
ascomycetes, 4, 32, 35, 37, 38, 41, 48, 132, 192
aseptate, 28, 35
asexual, 27, 28, 33–35, 37, 41, 46, 58, 151, 230
aspercryptin, 251
aspergillosis, 48
asymptomatic, 5, 79, 85, 140
ATPase, 44, 90, 109, 180
autophagy, 102
autoimmune, 241
auxin, 87, 98, 99, 105–109
auxotroph, 78
avirulence, 21, 23, 39, 60, 71, 73, 81, 86, 111, 115, 123, 177, 178, 189
axenic, 116

B

bacteriology, 20, 46, 121
bacteriophage, 9, 24
barcoding, 52, 53, 56, 67, 68, 71, 139, 144, 146, 150, 168, 169
basidiomycetes, 32, 37, 42, 72, 192, 251
biallelic, 141
bifunctional, 187
binucleate, 146
bioactive, 96, 98, 126, 135, 195, 202, 241, 242, 251
bioavailability, 201, 202
biochemistry, 1, 23, 33, 48, 49, 105, 106, 109, 111, 125, 130, 139, 185, 220, 224, 226, 232, 233, 239–241
biocommunication, 185, 188
biocompatibility, 38, 248
biocontrol, 29, 105, 180, 182, 187, 197, 199, 208, 219, 225, 228, 231, 232
biodiversity, 46, 52, 65, 73, 135, 184, 186, 210, 226
bioeconomy, 248
bioengineering, 135, 136, 230, 232
biofertilizers, 201, 206
biogenesis, 85
bioinformatics, 7, 9, 52, 56, 60, 62, 66, 140, 144, 147, 169, 229

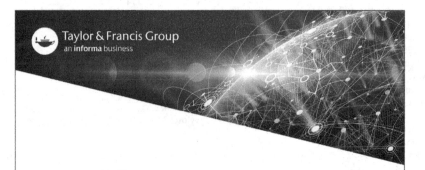

Printed in the United States
by Baker & Taylor Publisher Services